Linde

Also by Hans-Liudger Dienel

FLYING THE FLAG: EUROPEAN COMMERCIAL AIR TRANSPORT SINCE 1945
(*Editor with Peter Lyth*)

Linde

History of a Technology Corporation, 1879–2004

Hans-Liudger Dienel

Softcover reprint of the hardcover 1st edition 2004 978-1-4039-2033-1

First published 2004 by
PALGRAVE MACMILLAN
Houndmills, Basingstoke, Hampshire RG21 6XS and
175 Fifth Avenue, New York, N.Y. 10010
Companies and representatives throughout the world

PALGRAVE MACMILLAN is the global academic imprint of the Palgrave Macmillan division of St Martin's Press LLC and of Palgrave Macmillan Ltd.
Macmillan® is a registered trademark in the United States, United Kingdom and other countries. Palgrave is a registered trademark in the European Union and other countries.

ISBN 978-1-349-51457-1   ISBN 978-0-230-50953-5 (eBook)
DOI 10.1057/9780230509535

This book is printed on paper suitable for recycling and made from fully managed and sustained forest sources.

A catalogue record for this book is available from the British Library.

Library of Congress Cataloging-in-Publication Data
Dienel, Hans-Liudger, 1961–
    Linde : history of a technology corporation, 1879–2004 /
Hans-Liudger Dienel.
        p. cm.
    Includes bibliographical references and index.

        1. Linde Aktiengesellschaft—History. 2. Engineering firms—
Germany–History. I. Title.

TA217.L.55 2004
338.7'6657'0943—dc22
                                                            2004042838

10   9   8   7   6   5   4   3   2   1
13 12 11 10 09 08 07 06 05 04

Contents

Plates are between pages 194 and 195

Acknowledgements

The manuscripts have been sent to the publisher and my desk cleared. At the close of this project, which has lasted several years, amidst feelings of relief, happiness, and freedom, I would like to express my gratitude.

I would like to thank the Supervisory Board, Executive Board, and the employees at the Linde Corporation for the trust they placed in me when they asked me to write not only their company's history but also a piece of their own life histories. In particular, I would like to thank Gerhard Full and Dr Wolfgang Reitzle, the two board executives at Linde during the project, and Dr Hans Meinhardt, the long-standing chief executive of the Executive Board and Supervisory Board. Klaus Schönfeld, in Linde's central administration, supported the project from the very beginning, contacting several crucial figures in the refrigeration industry and ex-employees, as well as providing us with valuable advice on unavoidable historiographical problems such as how and where to draw the boundaries on our research.

In addition, I would like to express my gratitude to the members of the Linde family, which distinctively shaped the company's history through the second half of the 1970s and thereafter in second-tier management. I would like to extend special thanks to Professor Dr Hermann Linde, management board spokesperson until 1976, for his advice and help.

During the preparatory phase of the company history, the Linde Central Archive was reconceived and reorganized. It is likely to take a few more years before the entire archive with all of its fabulous but decentralized materials is in one location. Nonetheless, Dr Werner Jakobsmeier and his tireless efforts made it possible for the research to be completed.

In closing, I would like to express special thanks to the academic project members. Heather Cameron and Barbara Serfozo translated the book into wonderful English. Matthias Pühl, Marius Herzog, and Jutta Wietog (until September 2002) contributed a substantial part of the research and work. Their creativity and diligence have contributed greatly to the book.

Hans-Liudger Dienel
Berlin, January 2004

Preface

Company histories are enjoying growing interest. The field has expanded and become more professionalized.[1] In Germany during the last few years, several large firms have shifted from popular hagiographic commemorative volumes, so common 20 years ago, to historically researched studies from independent historians. In addition, an increasing number of histories have been published that have not been commissioned, for instance, on no longer existing firms whose archival documents have been stored and rediscovered in economic archives.[2]

In addition to the periodic jubilees, which invite a regular appraisal of a company's history, many firms' anxieties about lawsuits from foreign and forced labourers during the Second World War have encouraged further historical studies.[3] Company histories have also attracted growing attention from social, historical, and internal company groups. For companies, these histories are an attempt at self-confirmation, an appreciation of the company's traditions, culture, and identity. Jubilees usually provide the occasion to initiate such historical studies.[4] In this global era marked by stiff competition, where companies and their ubiquitous products seem interchangeable, company culture is something unique and deserving of its own history. The same is also true for businesses, which have completely shifted the focus of their economic activity, such as Preussag, which now operates as TUI, a market leader in tourism. Moreover, company histories have become historiographically relevant to economic, social, regional, cultural, and period histories.[5] In this sense, the relative retreat from the state and state institutions in the last two decades has left its traces on the historical guild. The growing influence of companies on our society has become more evident and necessitates a better historical understanding. Constant economic crises marked by mass unemployment, unpaid taxes, and changes in regional structure secure companies and their histories – the successful above all – great interest. Hardly any daily papers exist today which do not provide their readers with a column on 'successful regional companies'.

What is the core of a company history? Is it the company's economic development, its economic struggle to exist, as Peter Borscheid so strongly argues? Borscheid has correctly criticized recent company histories that examine a company from above or outside, either analysing it as a 'a sub-department of the national economy or one of its branches', thus, from a macro level perspective which does not properly consider the company's internal affairs and its economic motivations; or – even worse – examining 'the architecture of the firm's centres and the design of its products' without referencing the economic context of the firm's successes. Borscheid points to the necessity of a microeconomic approach for analysing companies and presents the clear advantages such an approach provides.[6] In addition, new institutional economics developed during the 1990s has directed attention to the idea of companies as one of three basic forms in economic processes alongside markets and states; and business culture – including contracts, norms, work ethic, attitude, and style – is also analysed for its influence on a business's successes.[7]

A possible danger of this focus on a business's economic success is, however, an inadequate evaluation of the business's driving forces, which are not exclusively economically motivated. What about companies driven by questions of technological possibility and attainability, whose founders were inventors and prioritized technical or scientific goals over primarily economic concerns, as for example in the case of the Linde Corporation? What about firms driven by ideological convictions, such as political, religious, or community-run businesses? Successful businesses often have some ideology and pursue in part or at some point a dream made up of notions of a good life and job, better products, or even of strengthening the nation's economy or military, or building international networks and culture.

In the present company history of the Linde Corporation, the economic struggle to survive stands at the centre of this portrayal, but this should not overshadow the company's other motives. Rather, this focus should integrate and make comprehensible these motives – not only in the sense that they explain the firm's economic successes but they also reveal the firm's character and uniqueness. At the beginning of the twentieth century, Wilhelm von Siemens wrote in a private note about his company, 'the goal of industry like the goal of life cannot be defined', and, at all events, it was definitely not 'to earn money and distribute the spoils'.[8] For the changing image of a company and its activities, the perspective of those involved in the firm is just as important as economic definitions. An understanding of the business's driving forces in a larger sense enables psychological, ethical, and aesthetic entry points in addition to microeconomic approaches.

Company histories are closely related to portrayals of business leaders as well as biographies of founders, owners, and central business figures. Company histories are often family histories, histories of how companies have been handed down from one generation to another and how families have risen and fallen in companies. The growing interest in company histories and in businessmen and women, thus, encourage each other. In his thick biography on Hugo Stinnes, Gerald Feldman writes that for a long time historians had very little interest in company histories dealing with individuals and families.[9] Company histories were either left to those outside the historical guild or written as popular portrayals.[10] This has changed substantially. Business leader (family) biographies have improved and developed to deal with business history issues, such as the important problem of ownership control in corporations.[11]

A long-standing problem that has affected company histories is the danger of too great an intimacy with the contractor financing the project. Luckily, the historical guild has reacted both sensitively and sensibly. Recent company histories have consistently been criticized for their uncritical depictions, for example in the portrayal of the Oerlikon machine tool factory[12] and the rebuilding of BMW (Bayerische Motorenwerke).[13] The firms contracting these histories could have been told that an independent, critical history of their firm would have provided a positive portrayal of the firm as open and self-critical, as in the instance of the Volkswagen history under National Socialism edited by Mommsen and Grieger.[14] Historians can learn something from art criticism, where contracted work still yields good results.

A good company history requires a comparison with other companies and how these companies defined and executed similar business endeavours. There exist, however, few such comparative works due, not in small part, to the typical way that company histories are financed.[15] The present study looks at Linde's competition and even dedicates an entire chapter to Linde's competitors in refrigeration technology in order to clearly portray the Linde Society's unique position as a science-oriented engineering firm.

This company history can also be accredited in part to a special Jubilee: 125 years ago, seven business partners, including engineering professor Carl Linde, founded the Society for Linde's Ice Machines, the present Linde Corporation. This history is nonetheless long overdue since the 100-year Jubilee did not provide the occasion for serious reflection on the company's history.

In the summer of 2000, the company's Executive Board provided me with the necessary resources to complete a multi-year independent project on the Linde Corporation's varied history. I was also asked for suggestions on how to expand the company's archive. During the course of the project, I had good access to archive materials. The only restrictions were on recent correspondence, in particular from the Executive Board and Supervisory Board. Therefore, certain aspects of the company's recent history could not be as intensively researched and developed as I had hoped.

In the past, the Linde Corporation had, in general, little interest in its own company history. During the 1960s and early 1970s, things changed quite dramatically when the Executive Board under Hermann Linde contracted a systematic analysis of certain aspects of the Linde Company history. Erik Jaeger in Höllriegelskreuth was the most important individual to work on the history during this period. His work provided much more than just the basis for this history. Due to his efforts, many important files and documents were also collected and preserved. After he left, the collection went into hibernation and was saved only through fortuitous circumstances. Hermann Linde himself saved over 6,000 letters from the company's correspondence in his grandfather's copybooks, a part of which he had typeset and included in a second edition of his grandfather's autobiography.[16] In addition, Hermann Linde published a private volume, which included the collected family letters of the second generation led by Richard Linde during the Nazi period and documents on the process that led to Richard Linde's resignation from the company.[17] Hermann Linde's resignation from the Executive Board and thus the family's withdrawal from the company impeded the cataloguing of archival materials and work on the company's history. Historical interest was too close to accounts of family traditions in the company. Recently, however, the company has become more aware of how important these historical documents are for creating a company identity. Thanks to Dr Werner Jakobsmeier the historical documents from several work groups and the Höllriegelskreuther factory photo collection were preserved and catalogued in the newly created central archive.

Published literature on the history of the Society for Linde's Ice Machines is manageable. Important for the insider perspective is Carl von Linde's autobiography, *Aus meinem Leben und von meiner Arbeit* published in 1916. Of all the commemorative volumes contracted by the company, the most noteworthy is that published in 1929 for its 50th Jubilee during the founder's lifetime. The volume is rich in details, self-critical, as well as historically researched. In contrast, the commemorative volumes for the 75th Jubilee (1954) and, even more so, that of the 100-year anniversary (1979) display the most minimal standards characteristic of commemorative volumes at the time. They were used as self-portrayals of the company and its products, offering very little concrete information. Other sources, most notably the numerous autobiographical writings of various Linde employees, provide detailed information. Among these works, Erik Jaeger's *Aus der Geschichte der Werksgruppe TVT* (1979) stands out. Interviews were also conducted with several important former employees, which have now been collected in the archive.[18]

Recent secondary sources from historians include the following: Mikael Hard (1994) has written a most useful study on the interaction between refrigeration and brewing technology with special attention to Linde. Hans-Liudger Dienel (1995) provides insight into Linde's influence on the international development of refrigeration and gas liquefaction and separation engineering until the Second World War. The master thesis from Marius Herzog (2002) examines the structural changes of the Linde Company between 1954 and 1984 from an organizational–sociological perspective.

Archival materials on the history of the company as well as its various undertakings during its 125-year existence vary in range and scope. An initial catalogue of the collection for the headquarters in Wiesbaden exists. Apart from a small core collection, the files were not systematically collected. In Höllriegelskreuth, a comprehensive collection on plant construction and the international gas splitting sector of the Society for Linde's Ice Machines exists, but it is also, for the most part, uncatalogued. The directors in Höllriegelskreuth, Richard and Friedrich Linde, Rudolf Wucherer, and Hermann Linde among others, left behind important business documents and correspondence.

For this study, several key figures were interviewed for an insight into the company's history. Interviews as documentary sources in company histories are still not usual; historians themselves are still not accustomed to using such sources. For the history of the Linde Corporation, the use of contemporary accounts was absolutely necessary due to the difficulty of accessing certain files and archival materials. Interviews, which were for the most part anonymously cited, allowed us to grasp and describe abstract long-term processes, which could hardly be found in files and printed materials. By using this source and the methods of qualitative social research, this study has become an interdisciplinary project – one of the project members is a sociologist. The border to 'oral history' was always clear to the author; several times contemporaries' accounts were unable to provide exact chronologies of events – in particular, when they happened decades ago – and naturally, memories are subjective and, not least of all, affected by the interview situation. Interviews conducted by Werner Jakobsmeier with retired employees from process engineering and plant construction were most useful. These interviews provided crucial information in various areas of the Linde history, for example in the use of foreign or forced labourers in plant construction.

The purpose of this book is to portray the history of the 'Society for Linde's Ice Machines' from its founding in 1879 to its corporate transition in 1965 and up to the present time. The Linde Corporation is one of the oldest companies in plant construction in Germany and has a fascinating, international, and pioneering history. Its founder, Carl von Linde, straddled the fence between science and practice. He shifted between university teacher and contracted businessman. The business that bore his name was from the very beginning something relatively new for its time, a public limited company. Even its first 15 employees worked predominantly overseas. The closeness with science, the familiar atmosphere of an engineer's office, technological market leadership, and licence production conducted with various manufacturers distinguished the Linde AG during the first 50 years. The business's rise starting in 1879 took place during a long-term economic crisis, which did not, however, adversely affect the young industry's special economic situation. Starting around 1895/1902, the Linde AG established a second front with gas splitting, whose sales exceeded those in refrigeration by 1920. During the first 50 years, the founder had a strong influence on his business. During the next 50 years, Linde's sons, sons-in-law, and grandchildren were at the steering wheel, and, like the founder, they were also science-oriented and progressive engineers. During the first 100 years the Linde AG, like many other Germany businesses, followed a path between owner-run and managed business since the Linde family was unable to use its large stock holdings to wield sole control of the company's future development.[19] While sales rose, the core business and climate remained relatively constant. Nonetheless, the business's character and organization has changed substantially since the 1920s due to increasing manufacture via additional purchases in peripheral areas, and as labour groups increased their independence. Since the middle of the 1960s, however, the Executive Board and Supervisory Board have discussed a fundamental reorganization of the business, which was for the most part executed during the period 1970–97 by Hans Meinhardt, who has been chairman of the Executive Board since 1980. Since the mid-1970s the business has been run by managers, who have strategically concentrated the business's activities and reorganized the area of forklifts (materials handling). But, the strengths of the business, its customer service and its concentration on exchange between science and practice remained crucial. The forklift business grew over time to become the most important area of sales, but the change was less radical compared to other businesses such as Preussag, Mannesmann, and Hoechst. In its 125-year history, and especially in the last decades, Linde has bought or (co-)founded several businesses. Its biggest and most important acquisition to date was the takeover of AGA in 1999. With this acquisition, the gas and engineering department, the core activity of the firm, was once again the company's first priority.

Organized in seven chronological chapters, this company history describes Linde's path from engineering office to multinational technology firm with over 45,000 employees. Each chapter is dedicated at the same time to a main theme and thereby extends the periodization in both chronological directions. The development of the business culture and its strategies are in the foreground – its employees' working methods, the interaction between science and technology in business, the handling of competition and political and economic conditions, and the confrontation with the business's foci and strategies. In general, company histories must choose whether to orient themselves around political changes or internal company developments. Both possibilities have advantages and disadvantages. We have oriented ourselves, for the most part, around political events that coincided with turning points in the company's development.

Chapter 1 examines Carl von Linde's path from scientist to businessman in refrigeration technology, and describes how he developed his business style. In order to further develop and market his technological innovations, Linde strategically chose partners to help found an engineering office as a limited company, then a relatively new format. Although the founding of the company in 1879 occurred during a long-term cyclical depression in the German Reich, the young refrigeration industry enjoyed a special economic position. The licence allocation method that Linde chose – allowing leading machine production companies in all the major industrialized countries to use his refrigeration technology – allowed the company to quickly rise as the world leader in refrigeration machine production. It was only in the 1920s that Linde decided to buy a machine factory and produce his own refrigeration systems. At the turn of the century, at the latest, the refrigeration industry lost its sense of being a research-driven industry, and Linde, a technology- and science-oriented engineering firm, suffered under increasingly stiff competition. This chapter portrays the various attempts, cartel agreements in particular, the company made to secure its market position.

Chapter 2 deals with the development of gas splitting and gas liquefaction as the Linde Society's second area of development. In 1890, Linde simultaneously moved from the company's Supervisory Board to its Executive Board and returned to research and development. In addition, he changed his location to Munich, where he shortly thereafter started work on air liquefaction. In 1895 he achieved the liquefaction of air at industrial levels, and in 1902, air separation via rectification. Linde returned anew to operations and established a division for gas liquefaction engineering in Munich, the second largest area of development in the firm and after 1920 its largest area of sales. Linde thereby hoped to secure his market position through constant innovation rather than price. Towards this end, air liquefaction and separation engineering was a better long-term prospect compared to refrigeration. At this point, Linde was large enough to conduct most of the production itself. This chapter portrays how Linde carved a space for itself in the worldwide gas market oligopoly. A further focus is the ascent of the second generation of the Linde family into management positions as they eventually replaced the founder.

Chapter 3 examines the company's history under National Socialism, looking mainly at the implications of the economic crisis, the politics of self-sufficiency, the rearmament-oriented four-year plan, the war economy, the racist persecution of Jewish employees, the use of foreign workers, and the destruction of the factories by air bombardment. The boom in domestic contracts starting in 1935, in particular within the field of synthetic petrol production, was thwarted by the political alienation of the world market and the decline in exports from approximately 50 per cent (1928) to 15 per cent (starting in 1935). Linde itself did not produce any weapons, but did provide oxygen installations for V2 missiles, propellants, and the manufacture of synthetic rubber; oxygen and acetylene; electrical welding devices; mobile workshops for the army and air force; and emergency generators and ship motors, all of which were crucial products during the war. The focus of this chapter is to portray the careful and tactical way that the firm's directors dealt with political and economic demands and with their position in the National Socialist system.

Chapter 4 deals with the period of rebuilding and the economic miracle after the Second World War. A slow start due to the delayed rebuilding of factories and the steady loss of overseas markets was followed in 1948 with rapid growth and a strong international position. This chapter describes the climate during reconstruction, the focus on the future and the silence surrounding the National Socialist past, the delayed transition to the third generation and the weak leadership of the 1950s and 1960s, the poorly planned diversification of products in which the company was not a market leader, and the brilliant rise of plant production. The firm's production in 1955 of hydraulic parts and hydrostatic machinery, necessary for the production of forklift trucks and the development of diverse industries, was also not strategically planned. In the 1960s, the firm experienced its first symptoms of crisis in an economic period still marked by growth. In the company, two groups emerged, the profit makers and the loss makers. The second group proposed increasingly radical suggestions for reform.

Chapter 5 is dedicated above all to the structural upheaval in the company from a technology-oriented engineering office to a multinational firm oriented around economic strategy. At the management level, this was on the one hand the setting-up of developments that had already been taking place for several decades in the company, but on the other, it was a true break that led to Hermann Linde's departure from the board of managers, marking the departure of the last member of the Linde family and a break with the company's traditional orientation around development, research, and technology. What followed was a new orientation around Hans Meinhardt's strategic management policies. This chapter lays out the difficult confrontations over strengthening the central administration and centralizing important business decisions. The business strategies developed in the 1960s and set into practice in the 1970s had the goal of reducing the number of different Linde products and improving total profitability. Thus, the number of traditional areas of production was reduced to four main ones by 1984: plant construction, industrial gases, forklifts, and hydraulics and refrigeration technology. The 1960s and 1970s also marked, however, an expansion into the high-risk field of ready-for-occupancy factory construction, which paid high returns, as well as an expansion into the field of forklift truck construction. In 1976, plant construction accounted for approximately 40 per cent of sales, but this figure fell after Hermann Linde left the Executive Board.

Chapter 6 handles the 1980s and 1990s, which marked an economically successful phase of 'normal operation' in the reorganized firm. Linde profited from the economic recovery after 1983 and was able to play out its new strengths in technological innovation in the areas of environmental protection as well as in the development of ready-for-occupancy chemical plants. Due to additional purchases forklift production rose to become a market leader and occassionally yielded huge profits. For the first time, cryotechnology products became profitable in the field of superconductivity in medical technology. As the politics of perestroika in the Soviet Union began to erode the Eastern Bloc in 1986 and enable the events that led to the fall of the Berlin Wall in 1989, new markets and production opportunities became possible in the East for Linde. Although unification created special economic conditions in the FRG, hoped-for opportunities receded relatively quickly and the planned expansion of activities in post-communist successor states did not yield the expected markets. Nonetheless, Linde successfully increased its activity in the East European countries neighbouring the European Union, where Linde was able to take over large previously state-run gas companies.

Chapter 7 looks at the company in the twenty-first century. It begins with the purchase of the internationally active Swedish gas company AGA in 1999 for 6.9 billion DM, the largest takeover in Linde's history. The purchase of AGA propelled Linde to being the fourth largest provider of industrial gases in the world and the second largest provider in Europe. The integration of AGA was successful in contrast to the fusion of other large companies in which the transition costs exceeded the returns of rationalization. The comparable histories of Linde and AGA facilitated the successful establishment of a joint identity. The acquisition of AGA re-established gas and engineering as the single strongest area of enterprise in the company and thereby set the path for the company's future development, marking the company's return to its traditional strengths for the first time in 30 years. On March 15, 2004, Linde sold its refrigeration branch to the American Carrier Corporation, a subsidiary of United Technologies Corporation in Hartford, Connecticut. With this sale, Linde ended an era in its history.

1
Linde as a Producer of Refrigeration Equipment

Boom years? Economic conditions for the rise of the Linde Company

The industrial situation in Germany

At first glance 21 June 1879 seemed a very bad time to start a business. The young German empire, called to life in the Hall of Mirrors in Versailles in January 1871 at the end of the Franco-Prussian war, had just got through the hardest part of an economic crisis that was to last in a less extreme form until 1894. Despite everything, this was the day that after a failed attempt and a year of planning the 37-year-old college teacher Carl Linde with five partners decided to bring the 'Linde Ice Machine Company' to life in Wiesbaden. Linde needed first of all to believe in his own talents. This belief was built on the fact that at that point 20 refrigeration machines – Linde's only product at the time – had already been built according to his patent. The young company believed it would be successful even without advertising. Linde demonstrated his confidence in his new undertaking by giving up his tenure as a college lecturer before the company was formally founded.[1] His example shows how the successful founding of businesses, or spin-offs as universities like now to call them, is determined more by inner qualities than external conditions.

Linde also needed a lot of confidence. People in Germany at this time did not believe the future promised positive development. The short boom years from 1871 to 1873 were followed by an economic crisis that initiated a long period of insecurity in the population. This widespread feeling of insecurity, and not the economic conditions until 1894 themselves, was the reason that this period was dubbed 'the Great Depression' under Bismarck's government.[2] Additionally the company partners showed their bravery by choosing to incorporate as a publicly traded company. True the company was listed first in 1910 on the stock exchange but the course of the market in the years just before the listing did everything but inspire confidence. The stock exchange showed the effects of the boom and bust cycle: between 1871 and 1873 there were more publicly traded companies founded than in the 70 years previous. However after the boom in new companies and huge rise in the stock market the years 1873 to 1877 showed a decrease in blue chip stocks of approximately 60 per cent.[3] Insecurity that

1

would continue into the future was not new to Germans. Nineteenth-century industrialization did nothing to change it. What had changed was the dynamic of the development.

Emerging as the dominant political power from the loose amalgamation of sovereign states and free cities that had been the German Confederation of 1815 (Deutscher Bund) and later the North German Confederation (1867), the Prussian empire of 1871 was in essence a federal state with a central government and a lower house (Reichstag), elected by direct manhood suffrage. In terms of developing an economic framework, the new state introduced step-by-step standardized weights and measurements, created a single currency (the Mark), homogenized credit systems and formed an expansive, unified trade zone. As the second largest trading power, Germany followed Great Britain in tying its currency to the gold standard. Between 1874 and 1877 the Reichstag passed legislation protecting commercial trade, including trademark and pattern protection laws. In 1877 both the national patent law and the national patent office were created, which effectively centralized the patent registration system and removed the need to register patents in multiple German states.

The long-term economic upturn from 1850 to 1873 is often referred to in economic histories as Germany's industrial 'take-off'.[4] During this period, Germany's national product grew 2.4 per cent annually, as opposed to the 0.5 per cent from 1830–50. Economic growth was also illustrated by the increase in foreign trade, which tripled between 1850 and 1873. By 1871, Germany – an industrial late-bloomer and second-runner to nations such as France and Belgium – had become the third largest economy and second largest industrial nation in Europe. Only Great Britain continued to maintain its lead.

The general euphoria brought about by the political unification of 1871 generated great optimism about the possibilities of future economic development and ushered in an economic boom, the likes of which Germany had never seen before. The new liberal shareholders' law furthered not only the birth of larger joint stock banks, but also brought into motion a wave of entrepreneurialism. Along with this wave came enormous speculation, particularly in the railway industry, but also in heavy industry and the real estate market. Share capital grew by nearly 250 per cent from 1869 to 1873 and share prices climbed by 50 per cent. With the demise of the US economy and the ensuant Viennese stock market crash in May of 1873, Germany's entrepreneurial boom came to an abrupt halt in the autumn of the same year. The speculation fever of the Gründerzeit was followed by a founders' crisis: banks went bankrupt, railway companies were nationalized and competitively priced iron and iron products from England flooded the German market. In 1877, the German coal and steel industry was forced to shut off over 50 per cent of its blast furnaces. Although the worst was over by 1879, deep psychological wounds remained.[5] Starting in 1878 the government took leave of its free trade policies and returned to protective tariffs, a course taken by other industrial nations as well until the First World War. Attempts to limit competition in the domestic economic market went hand in hand with such protectionist measures. The number of cartels increased significantly in Germany, where, at that time, such entities were desired and supported both politically and economically.[6]

Estimates show the average growth rate of the net domestic product hovered at 1.22 per cent between 1874 and 1883, rising to 3.02 per cent between 1883 and 1890 and then to 3.62 per cent during the following decade. After 1885, industry, crafts, mining, and transport together made up 37.4 per cent of the net domestic product, whereas agriculture comprised only 35.3 per cent. As of 1905, employee numbers of the former also exceeded those of the latter.[7] Germany did not become a true industrial nation until the second half of the 1890s, when growth was stimulated by the flourishing electronics and chemical industries. Machine construction also continued to be a leading economic sector, its growth exceeding the industry's average increase throughout the duration of the empire. Between 1871 and 1913, the value of this sector's exports increased nearly 16-fold. A glance at the German share of international machine exports in 1913 illustrates this international success: at 30 per cent, Germany stood well above Great Britain and even the USA. Despite all the crises from unification to the beginning of the First World War, the German economy witnessed successes so great that Hans-Ulrich Wehler spoke of the 'first German economic wonder' for the period from 1895 to 1913, in which the total industrial output increased by 150 per cent. By 1913 Germany had become, second to Great Britain, the largest foreign trading nation in the world; in terms of per capita gross national product, Germany took third place amongst the large industrial nations, behind the USA and Great Britain. As of 1900, just over a quarter of Europe's total production came out of German factories.[8]

These economic successes varied greatly in their distribution throughout the German regions. A clear East–West demarcation could be seen in the degree of industrialization. There were also great discrepancies within economic sectors themselves. From 1870 to 1913, the producer goods industry grew in all twice as fast as the consumer goods industry. Both the electronics industry and metal processing for machines, ships, and transport had the highest employment numbers, comprising nearly 17 per cent of the industry and trade sectors – absolute numbers more than tripling between 1875 and 1911/13.[9] Despite these major increases, the demand for industrial jobs particularly was not to be satisfied as the German population increased rapidly up to the First World War: by nearly 83 per cent from 1816 to 1871 and then by 58 per cent until 1910. This growth would have been even more significant had nearly 5.4 million Germans not emigrated over these 95 years, primarily (90 per cent) to the USA.[10]

The refrigeration industry as part of German machine construction

During the Wilhelminian empire, German machine construction became a leading industrial branch boasting high export numbers and recovering from the First World War more rapidly than its French and English competitors as well as other industrial branches within Germany. For many Germans, indeed for many people throughout the world, heavy industry, machine construction, the chemical and electronics industries had been, since 1900, 'the four cornerstones of German industry and technology, bound together through a network of struts and arched by science', according to Joachim Radkau.[11] This image distorted the actual industrial scale during the Wilhelminian period to the benefit of these branches, particularly their large-scale enterprises. The output value of mill production in 1900 surpassed that of mining and the chemical industry, which by 1914 made up only 2.3 per cent of German industrial production. Until 1926, machine exports remained higher than chemical exports, but lower than textile exports.[12]

Within the machine construction branch, motor construction (steam engines, steam turbines, gas motors) and its related compressor and refrigeration machine construction were internationally renowned as a German speciality. The success of exports as of 1870 in these three branches was indeed impressive. Oriented towards export, these factories stood politically in contrast to heavy industry and agriculture, which promoted protective tariff policies.[13] Over a half of the refrigerators produced in Germany before the First World War were exported, the most important market being England.[14] Up until 1913, nearly 80 per cent of the in-use refrigerators in Russia came from Germany. German refrigerators were exported even to the USA, where imports had grown since the turn of the century.[15]

Persistent in the technological and industrial historical literature are images of technologically advanced, but economically amateur, German engineering companies such as Brown Boveri Cie.[16] Such stereotypes, however, fall asunder upon closer case study analysis. First of all, particularly in industrial complexes where companies like BBC and many other refrigeration production plants were to be found, the organization was, economically speaking, entirely rational, aimed as it was towards technical solutions and maintaining a technical lead. In other cases, such as the steel industry, the importance of technological development for entrepreneurial success has been relativized or revised.[17]

A second common thesis of industrial history is the 'Main line of technology history ... increasingly complex processes developing and a trend towards making everything highly scientific.'[18] However, as of 1900, the opposite rang true for the refrigeration industry. Fundamental technological innovations were no longer numerous and remained manageable, refrigeration technology having increasingly become 'common property'. Indeed, amongst engineers, refrigeration technology was, by the 1920s, already considered to be sluggish in terms of innovation.[19] The most important technological innovations in refrigeration since the First World War came from the USA and included the move towards rapid, simple vertical compressors, high-performance chillers, and the development of assembly line production of small refrigeration appliances.[20] Hardly any German firms bought American licenses. However, in the Twenties nearly all manufacturers attempted to produce refrigerators similar to the American model, often suffering major losses as a consequence.

The number of refrigeration factories increased in phases. In 1870 there were two to three manufacturers in Germany, 20 in 1888,[21] then 35 in 1904,[22] and 55 in 1937. Over a half of these came about following the First World

War.[23] Parallel to this increase, the refrigeration industry shifted its emphasis from large to small refrigeration systems. By 1935 large refrigeration systems comprised only one-fifth of total turnover in the refrigeration industry, small refrigeration systems making up two-fifths and another two-fifths going to the refrigerator industry. The difference between refrigerator and small refrigeration system manufacturers was significant. Often, refrigerator manufacturers did not have the traditional German machine construction background, but rather immigrated after the First World War to Germany. The three areas of refrigeration technology – refrigerators, small and large refrigeration systems – were thus separated.[24] Assessing in 1929 the increase in manufacturers, Fritz Linde stated: 'There is no doubt that currently too many German machine factories are occupied with manufacturing refrigeration systems and are making life difficult with unhealthy competition.'[25] This was particularly true of small refrigeration systems. In contrast, the number of manufacturers of large refrigeration systems had already doubled between 1889 and 1894 and had since remained constant.[26] Certainly the golden 1880s of few manufacturers and large profits were days gone by.

In his thorough comparison of the development of leading English, American, and German industrial firms from 1880 to 1940, Alfred Chandler attributes the economic strength of German machine construction firms primarily to their organizational structure. According to Chandler, machine construction operations adopted American businesses' recipe for success, namely to invest equally in production capacity, management, and the marketing of their products.[27] To characterize German machine construction more exactly, Chandler refers on the one hand to the model of 'organized capitalism' and points to the 'shared belief in the benefits of industrial co-operation' as decisive for economic success.[28] He downplays factors such as a highly developed technical education and other state policies.[29] Chandler points to a second structural characteristic, namely a much stronger diversification than that found amongst American competitors.[30] He considers the establishment of or lateral entry via a market niche and resulting expansion of a product palette to be a typically German development. As the example of refrigeration systems illustrates, after the 1880s, a number of machine factories successfully expanded their product palette around refrigeration systems. However, along with this, a number of specialized businesses continued as before.

Thirdly, Chandler established that in comparison to American businesses, there existed a stronger, more familiar relationship in German management – particularly in machine construction operations – which did not hold the same negative consequences as that of England's patriarchal capitalism. Rather, according to Chandler, there existed already amongst German business families a much greater acceptance of what could be called an 'attitude of patriarchal-family self-servingness'.[31] The example of the Linde Company substantiates this thesis. The Linde Company was, however, an exception amongst German refrigeration manufacturers.

It was not only by far the most important business in the branch, but it was also the only major company that in the Twenties contracted out its production. Other engineering offices from the 1870s disappeared or drew up plans for and sold refrigeration systems to numerous manufacturers. The Linde Company's flexibility and potential co-operation capacity with all manufacturers was one of the underpinnings of its great economic success. As a market dominator, the Linde Company shaped the structure of the refrigeration industry well past the Second World War.

Markets for refrigeration

First, we will look at the refrigeration industry's most important early markets. Until 1900 breweries were the most important customers and accounted for almost 90 per cent of all refrigeration units purchased in Germany. The US refrigeration market was completely different. Not only was it much bigger, it also served different types of consumption, namely private ice consumption and cold chains for meat and foodstuffs. Because the American model was so important to refrigeration development in Germany, it will be referred to in the text to provide a comparison to the German model.

American natural ice consumption was greater and more varied than anywhere else in the world.[32] Its large refrigeration industry continued to grow mainly because of private consumption. In contrast, the German refrigeration industry developed primarily in response to industrial rather than private refrigeration needs. The demand for natural ice not only encouraged the refrigeration industry to expand, it also influenced the industry's technological development. Climate alone cannot explain the large demand for refrigeration that existed in the United States in the early nineteenth century. Rather, American culture, or the 'American way of life' provided a more compelling explanation for Europeans. Starting in 1850 European emigrants and visitors were astonished by the ever-present tinkling of ice cubes in restaurants in big cities and the everyday use of ice to cool foodstuffs. However until 1830, in the northern US, where large lakes and rivers ensured a good ice harvest, ice was used for cooling in a marginal way similar to Europe.

The first German ice business was founded around the middle of the nineteenth century. The largest ice consumers, the breweries, had until then procured their own ice. The first German ice trading company for American-style private consumption was the Berlin company Eduard Mudrack, which built an aboveground icehouse on Lake Schäfer in Berlin-Reinickendorf in 1856.[33] The 'North German Ice Works' then followed in the early 1860s on Lake Rummelsburger east of Berlin. Natural ice remained much more expensive in Germany than in the US, especially during the summer months. This high price prevented mass consumption. Private ice consumption grew only in Berlin, where ice prices matched the low prices found in the USA. The director of the Karlsruhe covered market, Hans Meidinger, argued that higher prices in Germany could not only be blamed on its small ice market, but above all on the old-fashioned and costly basements in which the ice – often exposed to ground water – quickly melted. In 1868, reasonably priced aboveground American-style icehouses could still only be found in Berlin.[34] American technology for ice harvesting and storage was adopted only slowly. It is remarkable how long the German technical manuals continued to favour underground storage and ignore the successes of American aboveground storage.[35] Moreover, ice prices in Germany fluctuated so greatly that ice imports from Norway during warm winters could also turn a profit. In this way, mild winters also encouraged the artificial ice industry. The lack of ice after the warm winter of 1882–83 was a major impetus for German refrigeration machine production. The dividends for Linde's ice machine company increased by 108 per cent during the boom year of 1883. Often, slaughterhouses, cold stores, and breweries ran an ice business on the side. In 1902, 93 municipal slaughterhouses sold ice alongside their regular business.[36]

From the very beginning, the history of German refrigeration technology was shaped by its tendency to look at America with both longing and expectation. In 1859 a German chemist compared the annual consumption of ice in Paris, 20 thousand tonnes, with that in Germany. In contrast, America, noted the author, exported 75,000 tonnes of ice per year to Central America alone. American consumption varied widely in German reports – an indication of the Germans' own fascination and willingness to suspend belief when it came to the American use of ice. The 'tremendous trade that North Americans do with ice, [is] proof of the importance of ice in civilized nations'.[37] The successes of the American model encouraged industrial investment in German refrigeration technology, but these investments remained unprofitable.[38]

Breweries

In Germany breweries using bottom-fermentation were the most important customers for natural ice and later for refrigeration machines.[39] They ranked high above both private consumption and the meat industry. Breweries used ice neither for refrigerated transport nor, at the beginning, for decentralized storage or pubs, but rather for the production of beer itself. From 1840 until 1870 many breweries switched from the 'English' top-fermented brown beer to the bottom-fermented lager.[40] It both kept longer and was increasingly popular. Lager-style fermentation required lower temperatures and the exact regulation of the cooling process. At the beginning, underground basements were adequate for the secondary fermentation but only if brewing took place during the winter. By 1870 the lager had as the 'Bavarian method ... nearly edged out all other types of beer at least here in Germany'.[41] Beer consumption subsequently increased in the territories of the German empire from 14.5 million hectolitres (1850) to 23.6 (1870), 49.3 (1890), and 66.6 million hectolitres in 1900.[42] Increased production in the large breweries was even more remarkable.[43] Expansion of lager breweries was limited by the capacity of their storage cellars, which were, if possible, stone cellars. Thus, the large breweries switched to building lager beer cellars with on-top refrigeration and natural circulation. The innovative brewer Gabriel Sedlmayr started the trend with his Munich Spaten Brewery in 1842.[44]

The Franziskaner Brewery followed in 1845. Sandstone cellars nevertheless persisted as the main form of storage. In 1884 the Bürgerliches Brauhaus in Pilsen continued to cool two-thirds of its then 2,100 fermentation vats, which each bottom-fermented 25 hectolitres, using the subterranean method.

Until the 1860s, the summer brewing of lager in Bavaria was illegal due to inadequate refrigeration. Bohemian natural ice cooling methods were then adopted for summertime brewing.[45] In 1878, however, ice was still too expensive to brew lager the entire summer. 'Obviously', according to Lintner's beer brewing manual, 'ale and draft were brewed but not lager'.[46] In the 1840s the demand for ice in Bavarian breweries shot up. In 1868, the Drehersche Brewery in Klein-Schwechat near Vienna used over 31,500 barrels of ice, approximately one kilogram per litre of finished beer.[47]

Apart from its inconsistent availability, natural ice also presented breweries with immense hygiene problems because it brought germs into the fermentation and storage cellars. Copper floats filled with ice were used to regulate the fermentation temperature. A leak or a complete 'drowning' of the float in the beer vat occurred regularly. The subsequent contamination altered the fermentation, often spoiling the beer.[48] The old beer cellars were cooled by placing ice directly against the barrels. This resulted in a humid environment threatened by mould. After the introduction of refrigeration machines, brewers replaced the ice floats in the hot brew and in the beer vats with a cold-water float that used hoses pumped with cooled water.[49] In 1895 natural ice was used in the Munich breweries only in beer transport, during which ice was not a hygiene problem.

Better hygiene and the more economic use of storage space (after the blocks of natural ice were removed) were the two main arguments for refrigeration machine makers against natural ice.[50] Until the 1890s, favourable price comparisons between artificial ice and natural ice could only be made after warm winters when natural ice prices were inflated. If the price argument was used, it mainly focused on natural ice's huge price fluctuations and the more consistent supply ice machines provided. Warm winter days led to a brewer's sleepless nights, since he would already have to start worrying about his summer lager.[51] 'Another four weeks of this weather', wrote the Linde engineer Robert Banfield to his colleague Rudolf Diesel in December 1880, 'and the brewers will panic and order machines'.[52] Two especially mild winters in Germany in 1883/84 and in the US in 1890/91 marked the refrigeration industry's breakthrough in its competition with natural ice for the breweries.[53] Breweries in Denmark, France, and England were also among the first users of refrigeration machines.[54] One reason why Germany rose to become Europe's leading refrigeration machine producer was its large beer brewing industry. Its independent and insulated operation helped introduce refrigeration equipment quickly, while the decentralized use of ice in the United States allowed natural ice to remain economically viable for a much longer period.

Starting in 1870, the transportation of frozen and refrigerated meat was one of the largest refrigeration uses in the US since slaughter for the entire country was done in a small number of cities (Chicago, Cincinnati) and then distributed by rail. In Germany, however, the slaughter and sale of meat remained in the hands of independent butchers until the last third of the twentieth century. Although butchers were forced to use a supervised communal slaughterhouse after 1868 for hygiene reasons, they remained independent.[55] Because they served regional consumption, refrigeration was unnecessary. The independent butchers' authority provided separate slaughtering spaces, slaughtering blocks, and storage areas. Although the technical facilities were modern, they were not developed for conveyor belt slaughtering techniques but rather to allow each butcher to slaughter his entire range of animals. Because local slaughter and sale remained regulated, over 700 communal slaughterhouses were founded by 1890.[56] In 1907 there were 68,153 butcher's shops in Germany that did their own slaughtering, of which four had over 100 employees.[57] During the imperial period, the communal slaughterhouse along with the gas and electricity company, the garbage disposal company, and the streetcar were the basic municipal amenities of a proper city or communal economy. This was called municipal socialism. In 1908, 97 per cent of German cities with over 50,000 inhabitants had their own slaughterhouse.[58]

In Prussia slaughter fees were limited to the production costs plus 1 per cent amortization to protect butchers under the new slaughter regulations. As a consequence, slaughterhouses remained modestly equipped. The communes were, however, suddenly prepared to purchase expensive refrigeration machines and cold stores once the new communal delivery law allowed cities to raise amortization fees to 8 per cent in 1893.[59] After 1893, slaughterhouses became the second largest purchasers of refrigeration equipment.[60] In 1897 30.9 per cent and in 1929 65.3 per cent of all German slaughterhouses already had a refrigeration system.[61] During the first international refrigeration conference in Paris in 1908, the main topic in the German papers was slaughterhouse refrigeration, a topic on which the Germans were the leading authorities.[62] Many cities hoped to increase revenues by building larger refrigerated rooms in slaughterhouses so that they could lease cold stores and sell ice.

These were not, however, the only reasons that private and public cold stores were less widespread in Germany than in the United States. Because of local meat production and sales, cold stores for meat imports and wholesale trade did not exist. 'Completely underdeveloped here', remarked a disappointed Carl von Linde in 1903, 'is the storage of vegetal foodstuffs like vegetables and fresh fruit, which fill massive cold stores in America'.[63] The agricultural crisis in Germany in the 1890s led to the founding of a powerful lobby that 'with basic force' pushed through taxes and other measures to block frozen meat imports.[64] Interestingly enough, Germany did become a major net importer of a range of foodstuffs, including, in descending importance, eggs, butter, milk, fish, poultry, and game.

Until 1900 the cold chain ended at the door of the cold store. After the turn of the century these refrigeration chains slowly expanded to take in the large department stores, hospitals, expensive hotels, and delicatessen shops, and then, in the 1920s, to bakeries and butchers' shops. The beginning of the First World War marked a substantial improvement in German cold store conditions because the state needed a 'state socialist economy for supplying and distributing' foodstuffs, meat in particular. The Central Purchasing Association (*Die Zentrale Einkaufsgenossenschaft, ZEG*) seized the largest portion of cold stores to store frozen meat,[65] had new cold stores built and renovated,[66] and funded research projects on the refrigerated storage of meat, fish, fruits, and vegetables.[67] In April 1916, they were responsible for purchasing and distributing meat throughout Germany.[68] In February 1917, 16 cold store directors started a lobbying association.[69] In 1908, the German Association of Refrigeration Technology (*Deutscher Kälte technisher Verein*) also took advantage of the increased interest in refrigeration technology.[70] A cold store boom for meat operations, however, did not happen, in part because local slaughterhouses increasingly avoided the ZEG's centralized system.[71] Nevertheless, the Reichstag decided in 1920 to maintain a 'national meat reserve'. This provided cold stores with a dependable client.[72] During the Twenties, however, a series of cold stores still went bankrupt and only three new large-scale cold stores were built during the Weimar Republic.[73] Cold store owners remained envious of the US and dreamt of the possibility of 'totally exploiting Asia' to supply big cities in Germany.[74]

Starting in 1933 autarky plans from the First World War were revisited, a national educational programme 'Fight Spoilage' was set up, and the manufacture of a 'people's refrigerator' (*Volkskühlschrank*) was suggested.[75] In 1936 the Imperial Institute for the Care of Fresh Foodstuffs was founded in Karlsruhe and later moved to Munich in 1941.[76] It quickly became the largest refrigeration technology research centre in Germany.[77] In 1938 the delegate for the Four-Year Plan for the Fishing, Tinplate, and Refrigeration Industry pushed through the first large-scale freezing operation for fish and fruit in Hamburg as well as accelerated cold chain production.[78]

In the attempt to explain Germany's weak frozen food industry, it became apparent that the traditional German way of eating was an important factor in refrigeration engineers' lack of interest in how refrigeration affected organic substances. The effect of cold on food was less important than the study of machine operation. Thus, Americans were at the forefront of using refrigeration for freezing foodstuffs into the 1930s.[79] Since there was so little interest in Germany in refrigeration's effect on organic substances during storage and transportation, there was a huge difference between foodstuffs' refrigeration and the many innovations refrigeration machine factories made for using refrigeration in the production of goods. The main focus of refrigeration was kept on the production process of foodstuffs until the end of the nineteenth century. The most important area was the already discussed beer industry. Refrigeration, however, also played an important role in sugar and chocolate factories before 1880. Then, in chronological order, came: dairies;[80] margarine factories;[81] mineral water, wine, and champagne production;[82] mushroom cultivation; bakeries;[83] pastry makers; and ice cream factories. Similarly, refrigeration machines served the needs of organic cooling in silk cultivation, mortuaries, flower cultivation (delaying blooming), and the storage of fur garments during the summer.

In Germany, the textile and chemical industries became major refrigeration users.[84] The cooling process achieved a status similar to that of the heating process; new uses for refrigeration machines were constantly being found, especially in the separation of gases and liquids, the production of azo dyes, the removal of wood fibre from frozen hemp and flax, the mercerization of cotton with cooled sodium hydroxide, and in the lard and oil industries,[85] in tanneries and in leather working, and in the production of tobacco, rubber, gelatin, perfume, soap, dynamite, photography paper, and matchsticks.

Another German development after 1883 was a freezing method for mining. The method was implemented in the construction of tunnels and underground transportation systems too.[86] The Americans, in contrast, pursued the use of refrigeration machines in the drying of blast furnace air (1894) and creating long-distance cooling networks (1900). Both of these pursuits, despite euphoric early reports, failed to fulfil expectations.[87] Artificial ice skating rinks, however, proved to be more successful, starting around the 1880s;[88] in 1908 there were 23 ice skating rinks around the world, and in 1915 the United States alone had 25.

An important reason for the strength of the American cold chain and its operations was the American agricultural department's major support of cold storage. Cold chain expansion allowed foodstuffs production to be centralized and also encouraged the foodstuffs industry to grow in America where, in contrast to Germany, it was already one of the country's largest businesses.[89] The food industry and agriculture played a larger role in the US compared to Germany. In 1917, the US had six meat companies among the top 200 companies (according to assets), and in 1930 five (Germany had none). Germany, meanwhile, had 11 breweries in 1913 and 15 in 1923 among its top 200 companies (the US had two in 1917 and none in 1930).

Although the natural ice industry had barely had any importance in Germany, it was one of America's biggest industries after 1830. It prospered from the enormous private demand for ice, followed by the needs of the meat transport industry and the breweries. America was a model for German refrigeration technology even prior to the Machine Age. German refrigeration engineers looked enviously at America, above all at its largeness and argued that whatever made money in the USA, must also be profitable in Germany. Starting in the 1860s, America began to develop a closed-circuit cold chain starting with meat and then perishable goods – butter, cheese, milk, vegetables, fruits, and so on – that went from producer to customer, from the cold storage at the point of production, to ice rail carriages, and then to the small cold stores of branch stores, and finally to the refrigerators of the consumer. In Germany, the first big markets for refrigeration were beer production and later the production of various chemical products. Refrigerated storage in Germany happened only slowly and for a limited number of products. Temperatures continued to be higher and the storage periods shorter.

Linde's Ice Machines Company, which started in this economically difficult environment, became one of Europe's biggest refrigeration machine builders in just a few years after 1879. What was it able to achieve? We will find out in the next chapter.

The founder: Carl von Linde

Carl von Linde, the company's founder, walked the line between industry and academics, science and technological practice. He was a successful businessman, scientist and inventor. He did not simply cream the benefits of university research, he actually contributed to this research as a member of the academic community. His success in science and business was a result of his ability to pursue a number of professional identities, to move to and between them and endure their inherent tensions. In so doing, Linde set the tone of behaviour for his top employees, towards which both his colleagues and successors oriented themselves, up until the 1970s. Let us look at three important perspectives:

Linde as entrepreneur

An entrepreneur the vast majority of his life, Linde devoted the last 8 of his 11 years as a professor primarily to self-employment. This was then followed by decades of work in the refrigeration industry, which he pursued almost exclusively. Indeed, his life is best understood as that of an 'employed entrepreneur'.

Linde was born in 1842, the third of nine children to a Lutheran minister's family in Berndorf in Upper Franconia.[90] His father was the son of a shoemaker, his mother Franziska Linde (née Linde) came from a merchant family active in the pietistic brotherhood of Neuwied on the Rhein. Before marrying, Franziska Linde worked for three years as a teacher at a girls' boarding school in a French canton of Switzerland, later securing a French governess for her children. The only son of four to complete *Gymnasium*, Linde decided against his father's wishes not to study theology and chose instead in 1861 to attend the famous Zurich Polytechnic (where he studied mechanical engineering which he preferred to the humanities). After having participated in a student protest against the director Bolley, Linde was expelled from the polytechnic without official documentation of his work completed. However, he did take with him two letters of recommendation from the famous professors Zeuner and Reuleaux, who helped him find employment with Borsig in Berlin. Two years later, the 24-year-old Linde took over as director of the construction office for George Krauss' newly founded locomotive factory in Munich, and he finally asked for the hand of Helene Grimm, who had been the object of his affection since their schooldays. A distant relative of his mother's and daughter of Berlin's chief public prosecutor, Grimm was a good match for Linde. In 1868 Linde became an adjunct professor at the newly founded Munich Polytechnic, which he left in 1879 when he became the chairman of Linde Ice Machines in Wiesbaden, which he had founded four years earlier.

Figure 1.1 Carl Linde's private home, which he built in Wiesbaden in 1880, was the first headquarters of the young company.

This 'employee-entrepreneur' was unusual in 1879 amongst German businessmen. With his education in technology, he was in this respect typical, however as a former professor, he was extremely rare amongst German managers.[91] In 1890, at the age of 48, Linde exchanged his position as chairman of the board of directors for that of chairman of the Supervisory Board, a position from which he did not step down until 1931 at the age of 89. Linde had four daughters and two sons. Both sons obtained their PhDs, Richard in machine engineering and Fritz in physics, and later joined the firm, as did Linde's engineer son-in-law, Rudolf Wucherer. In 1897 Linde, the individual – not his family, was granted the title of nobility, adding a 'von' to his name.

In his autobiography, Linde describes his entrepreneurial activities as a necessary and absolutely involuntary response to the external demands brought upon him, which his inventions then solved. For him, true satisfaction was to be found only in the work of a scientist. Similar statements were also made by Werner von Siemens in his memoirs. Linde's life, however, reveals this kind of self-assessment as coquetry, as his life-work demonstrated in actuality his interest in both business and science. He was, after all, a responsible, artistic, and creative entrepreneurial engineer, who founded and led for decades not only the Linde Company, but numerous other businesses, with resounding success.

As chairman of the board from 1879 to 1890, Linde, according to his own assessment, 'put business on the front burner'.[92] He focused primarily on acquiring contracts, but also project realization and the development of new applications, personnel issues, licenses, as well as product and patent processing. Via rapid international franchising, within just a few years Linde turned the company into Europe's leading provider of refrigeration units, although it did not manufacture its own products. In 1890, at the age of 48, he chose to give up his gruelling schedule and joined the Supervisory Board, moving in 1891 back to Munich. Linde attributes this remarkable change on the one hand to a 'consideration for my health, which had weakened due to chronic head and stomach aches'.[93] Above all however, it was 'the desire to return to scientific work' which ultimately drove him to leave his position as chairman of the board.[94] Linde had already built an experiment lab in Munich in 1889 that he handed over for a limited time to the Polytechnic Society of Bavaria for the purpose of refrigeration testing. It was to provide scientific proof of the superiority of Linde's refrigeration unit, which it did indeed do. He probably had the lab built with the intention of taking it over later and moving to Munich. In any case, once back in Munich, he demanded the lab be handed over to him. His attempts to turn the lab into the company's own research and development department, under his direction and for the purpose of developing and improving refrigeration components and processes, met with strong resistance from the Linde Company's Supervisory Board, which considered such a lab unnecessary and too expensive. It took Linde years to successfully convince the Supervisory Board of his goal.[95] In 1892 he took on an honorary professorship, which became a proper professor position without teaching requirements in 1900, thus allowing him to resume lecturing on refrigeration systems theory at the Technical University.

However, just like a modern manager unable to rest while on holiday, Linde's return to science was short-lived and half-hearted. For one, he steadily increased his association affiliations and duties.[96] In 1892 he became district chair of the Association of German Engineers (VDI) and chairman of the Bavarian steam boiler inspection association (TÜV).[97] In 1894 he joined the board of directors at VDI, becoming its chairman in 1904–05. In 1895 he was appointed both to the board of trustees of the Physikalisch-technische Reichsanstalt and to the science and technology commission of the VDI. In 1899 he was appointed to the Jubilee Foundation of German Industry. In 1898 he joined the Göttinger Vereinigung, the forerunner of the Kaiser-Wilhelm-Society (later the Max-Planck-Society).[98] In VDI and the Jubilee Foundation Linde actively promoted thermodynamics and later, as chairman of the special committee for aeronautical engineering, he acted as a patron of aerodynamics.[99] In 1896 he became a member of the Bavarian Academy of Sciences[100] and in 1903 he joined the Deutsche Museum's board of directors. His activities in other associations came about in part due to external pressure and were not always entirely voluntary.[101] Having powerfully demonstrated how to transform mechanical heat theory into stellar industrial success, when Linde returned to calmer waters, the boards of numerous technical associations were eager to have him on their committees. Those boards that could not persuade him to join made him an honorary member.[102]

However, while he acted as a board member for all of these associations, Linde took on various entrepreneurial activities in Munich and considered incorporating a variety of companies after 1890. Working with a rather shady businessman, Silbiger, he started in 1890 planning for an air-conditioned hotel in Calcutta,[103] which was to be under 'my overall control'.[104] Together with a professor from Vienna, who had invented a 'sailing-wheel-flying machine', Linde planned in 1893 the creation of a helicopter factory and was prepared to contribute 100,000 Marks 'from my circle of friends'.[105] He helped establish a milk powder factory in 1894 with the chemist Dr Knoch, who had married into the family.[106] Linde also acted as project engineer for the installation of individual large refrigeration systems for the Wiesbaden headquarters.[107] The proposed projects were enticing to him, as they appeared to fulfil his desire for an exciting entrepreneurial endeavour.

During this period, Linde also conducted scientific-technical experiments at the Munich laboratory. From 1893–94 he tested carbonic acid refrigeration equipment that was built according to his plans at the Augsburg machine factory. The goal was to illustrate the inferiority of carbonic acid machines, which were built by the competition Riedinger and Hall (England).[108] It was his work with the liquefaction of carbonic acid that brought him to develop technology for the liquefaction of air in 1894–95. Linde recognized the vast technical and economic possibilities of gas liquefaction rather quickly. It offered him a new, untapped entrepreneurial field of possibilities, or, put another way, it demanded of him an entrepreneurial undertaking that was to challenge him for the rest of his life. Instead of passing further work on refrigeration engineering technology on to Wiesbaden, he opened up, together with his sons, a large branch in Munich that, in terms of turnover, caught up with Wiesbaden by 1920. At the Munich branch he altered his business strategy significantly from that of the Wiesbaden engineering office. Production was not handed over to licensed factories; instead Linde built his own manufacturing plant. Foreign business was no longer delegated to others through patent and licensing sales, but controlled by the establishment of subsidiaries. The creation in 1902 of the Institute for Technical Physics drew Linde back into academic research for a good two years. Yet from 1903 to the early 1920s, he took on a more active business role than ever, establishing numerous subsidiaries both in Germany and abroad, working to expand the company's market and to secure its gas engineering cartel. Linde was 80 years old before he began to slowly retire from making business decisions, yet he remained on the Supervisory Board of his company until his death at the age of 92.

Linde's business letters are the best source on his manner of conducting business. There are 3,010 handwritten letters from Linde spanning 1876 to 1929, which are archived in 11 books of correspondence. Whereas Linde's autobiography, which he published in 1916 at the age of 72, provides an edited version of the events of his life, his daily business correspondence reveals the everyday life of his profession.[109] Indeed, letter writing was an important aspect of Linde's generation.[110] Because his company and customers were scattered around the world, he was dependent upon written correspondence. Starting in 1885, he spent over three months almost every summer in the Bavarian Alps, maintaining contact with his employees and business partners almost exclusively by letter. Penning himself the majority of the letters that hailed from his Wiesbaden and Munich offices, he also wrote a great many during business trips.[111] While the boundary between his personal and business life was often grey, he made a clear distinction between personal and business interests. His homes in Wiesbaden and Munich were used regularly as conference locations for the company. He purposefully used his summer residence in Berchtesgaden for negotiations. As the archived guestbooks show, many of Linde's most important co-workers and business partners were at least once guests at the Oberbaumgart.[112] Guests at the remote location had little choice but to stay the night and strategic decisions were discussed during hikes.

Like many engineers, Linde abstained from politics.[113] Political commentaries rarely appear in his letters. Although Linde was a member of the founding committee of the National German Fleet Association – a pressure group for naval armament – he did not remain so for very long. He subscribed to the (national-liberal) 'Kölnische Zeitung'[114] and as an internationally active engineer, anti-French and anti-English slogans fell deaf upon his ears. His resistance to the establishment of a French-dominated international refrigeration association in 1908 was motivated purely by economics. He opened, after all, his autobiography by praising his French upbringing – and this in the year 1916. Up until 1885 he wrote all of his letters sent abroad in French; he later acquired perfect English.

Even as a businessman, Linde remained interested in specific technical questions. As his correspondence shows, unsolved technical problems are found throughout the entire course thereof. Of course, business matters came up twice as often as technical matters (1876–1929), which tended to decrease over time. But his detailed inquiries and directions remained pedantic at times. With the help of vivid and graphic, but not perfect, sketches Linde tried to clarify his technical ideas and questions. At least in his letters, Linde expressed his aversion to public praise and avoided it at all costs, appearing modest.[115] Despite all of his scientific honours, Linde referred to himself amongst professors as a 'guest in scientific circles'.[116] He sought unambiguity and clarity in his environment. There was nothing he shunned more than 'a connection in which future complications and conflicts could be suspected'.[117] In business, this attitude paid off: the Linde Company was renowned for its reliability as a business partner.

Linde was a hard-working, small-framed man prone to periodic physical collapse. He suffered frequent migraines that made work impossible, sometimes for weeks on end, up until he was 60. After the age of 60, the symptoms of his physical strain disappeared as he then began to embark on extended bicycle tours through the Alps.[118] With his health stabilized, he was able to live well into old age. His work schedule was wearying, leading Linde to complain frequently of his pains in his correspondence. Yet at the same time, his letters reveal a strong asceticism. Linde felt obligated to his duties.

Is Linde's work ethic to be explained by his Protestant upbringing? Raised by a Lutheran minister, his mother came from a pietistic background. Indeed, many of his relatives were deeply religious and two of his daughters married Lutheran ministers. Describing his father, the minister Friedrich Linde, Linde writes: 'His life was characterized by an iron sense of duty and the relinquishment of his own pleasures, which, in his demands of us children, at times appeared severe.' [119] At the same time, Linde characterizes his father as a man 'who, as a humble Christian, exhibited patience and love towards others' mistakes and weaknesses'.[120] This could also describe Linde's own religious attitude. In an obituary commemorating what would have been Linde's 110th birthday, a friend of the family, Pastor Johannes Schattenmann, points out that Linde 'was a product of the Lutheran church and the power of faith'. Religious and above all charitable, Linde was significantly more tolerant and less pious than his relatives.[121] His response to his brother Sigmund's divorce illustrated his liberalism.[122] Linde also helped establish a Protestant parish in Berchtesgaden,[123] although he himself did not regularly attend church. While religion provided the roots of his work ethic,[124] his sense of duty to his business was an expression of duty to both God and State.[125] His personal economic interests or even those of his family were secondary to this sense of duty. This humility gave Linde a modern sense of self as business manager, which stood in contrast to the businessman in private possession of his company.[126] When in doubt, Linde made decisions for the benefit of the company at the cost of his own personal gain.[127] When faced with a decision, he struggled to achieve a supposed objective position and rarely argued using personal judgement or will.[128] He referred almost exclusively to factual necessities, to which he and others were to submit. In cases where his personal opinion did emerge, it emphasized that which was of no consequence. When it came to difficult decisions regarding employees, Linde expressed his rather mild personal view to ground the toughness required of him as CEO of the Linde Company. His Protestant upbringing also played a role in moral considerations of business decisions. Competition should be just; mercy, if tenable in business terms, was required. Linde sought honesty in business interactions and often complained of others' harsh and unjust acts.[129]

The letters also illustrate the vastness of Linde's travels. In hundreds of letters he is either under way or has just embarked upon or returned from another journey. During his travels, he took with him his copy-books.[130] One such file contains letters sent on a regular basis from the Wiesbaden headquarters to Linde as he travelled. Most of them are from Shipper, his deputy Krebs or the commercial director Reichenwallner and written in a short, telegram-like style. Schipper not only sought to gain from Linde's knowledge, but understood the importance of integrating Linde in business operations by asking specific questions. After 1895, Linde arranged for regular tours. His regular 'Autumn Tour' took him to Dresden (visiting Zeuner), Berlin (PTR, VDI-Executive Board, GfMK-Supervisory Board), Hamburg (Kühlhaus), Rotterdam (Feldmann), possibly Antwerpen (Frigorifères), London (Linde Br. Ref. Co.) and sometimes Dublin (Guinness), going through Cologne (Eiswerk) on his return to Munich. During the first few years, Linde's engineers also spent more time on the road than at Wiesbaden headquarters.

This combination of traditional and modern characteristics typified Linde's business style, just as he himself symbolized a union of the nineteenth and twentieth centuries. The company was tailor-made to suit him. Three generations of family members occupied the company's most important positions. Moreover, since its beginnings as a stock corporation, the company survived without major financial support from the Linde family.

Linde as scientist

As Associate Professor of mechanical engineering theory since 1868 at the newly founded Munich Polytechnic, Carl Linde made his debut in refrigeration engineering when he authored two scientific papers in 1870 and 1871 in which he classified refrigeration processes and the machines that were on the market at the time, employing Sadi Carnot's principle of thermodynamic cycles to evaluate refrigeration machines. A glance at these articles offers a closer look at the scientist himself: Linde specified three refrigeration processes and thus three ideal types of refrigerating machines. The first two processes, which had long been considered by many to be a single refrigeration process, entail evaporative cooling via (1) vapour compression or (2) vapour absorption. The third process of cooling entails expansion through exergy.[131]

1. Evaporative cooling by means of vapour compression is particularly conducive to establishing a (compression) cycle. Liquid evaporation requires a supply of energy that is essentially absorbed from an environment, which is then returned to the environment during condensation. Evaporative compression refrigeration machines make use of the fact that the evaporation temperature of a liquid is pressure-dependent. In other words, as pressure drops, so does the evaporation temperature. Refrigerants are condensed at high pressures and condensation energy is then diverted to the surrounding temperature, for example through the use of cooling water. Afterwards the surface tension of the refrigerant is reduced, which evaporates at a lower temperature due to the lower pressure. The refrigerant absorbs the energy required for this process from both its surrounding as well as itself, and cools down. A compressor sets up the difference in pressure between the two heat transfer

processes; a tension-release valve reduces the pressure. Linde pointed to the Siebe Machine to illustrate vapour compression process.

2. Evaporative cooling by means of vapour absorption is the nexus of a chemical and physical process, in which the evaporated refrigerant is bonded by an absorbent. This process also creates a cycle. Linde pointed to the Carré-machine as an example of how the process of absorption functions. Liquid ammonia, a highly water-soluble gas, absorbs heat from its environment when vaporized in refrigeration. Ammonia vapour flows then to an absorber, bonding to water. The resulting low pressure accelerates evaporation in the refrigerator; the water–ammonia mixture is then thermally divided in a boiler (expellant). With the help of cold water, the ammonia vapour is liquefied in a condenser, from which it then flows back to the absorber.

3. In expansion cooling via exergy, a drop in pressure occurs within a cylinder in which a gas causes a change in volume and absorbs the required energy through refrigeration. In such a process, which is used in cold-air machines, air is suctioned out of the atmosphere and compressed by a cylinder. Air is thus warmed (air-pump effect) and then cooled with water. The resulting release of pressure lowers the surrounding air temperature. (A spray can illustrates this sort of cooling.) Windhausen's cold-air machine served as Linde's illustrative example for his paper.

Of all the refrigerators using these three processes in 1870, absorption refrigeration machines were the most developed. Yet literature on the subject failed to make clear distinctions between these processes: in both theory and practice, the procedures were mixed. Linde's first achievement was to classify these processes. Classification is one of the most important tasks of science because it allows for differentiation and comparison, yet it can also limit the scope of thought and action and therefore prevent the mixing of procedures.

Linde's second achievement was the creation of a comparative criterion of assessment. He was certainly not the first to refer to a cyclic process in refrigeration.[132] New however, was the radical claim to have at his disposal, with thermodynamics, an instrument capable of evaluating the development capacity for a refrigeration process. The fact that Linde's contribution was focused not upon constructive questions, but rather the theoretically possible efficiency of a process, made its practical application difficult, as the units tested by Linde achieved in practice only 5–10 per cent of the theoretically possible performance.

Linde divided his experiment into three steps: starting with a calculation procedure for caloric efficiency, he sought to calculate this efficiency under different operating conditions for three of the most common refrigeration systems, and then to conclude with a suggestion for an improved refrigeration unit. To calculate caloric efficiency, Linde defined an ideal process – today referred to as the Carnot process – in which maximum efficiency has the value of 1. For refrigeration machines this spells miniscule differences between the two temperature levels in the process: the smaller the difference, the larger the

Figure 1.2   Carl Linde in 1868, the year he was appointed associate professor at age 26.

Carnot efficiency, or the closer the value to 1. Calculating theoretically possible refrigeration performances at differing surrounding and cold temperatures led Linde to the simple conclusion that refrigeration machines should be left to function at the lowest possible temperature difference.[133]

By establishing an ideal cycle process, Linde created a comparative value against which relevant refrigeration processes could be compared and then evaluated in terms of their development potential.[134] Yet only Siebe's ethyl ether compression machine provided a successful formulation of these 'real' cycle processes. Linde also installed a cyclic process for Windhausen's air machine, which was, however, flawed, as he assumed the release of heat in air occurs after compression at a constant volume (isochor); in actuality, the release of heat occurred through almost constant pressure (isobar). Lacking sufficient preparatory research on the laws of absorption, Linde did not consider himself to be in a position to formulate a comparative process for Ferdinand Carré's absorption machine. He thus left this beyond the scope of his tests.[135]

The calculation of the caloric efficiency of both cyclic processes at different temperatures revealed the vapour machine to be advantageous. The vapour machine process achieved 80 per cent of the Carnot efficiency ratio, whereas the air machine process achieved only 40 to 75 per cent. However, Linde's calculations were theoretical and lay far above each machine's actual performance. Linde suspected 'detrimental resistance' in the compressor to be the cause of the major losses that occurred in practical operations. To estimate friction losses, which influenced the efficiency of a refrigeration machine to a greater extent than the specific system employed, he added to his calculations a formula for the piston friction value, defined as a function of the compressor profile. Because of the poor thermal properties of air – it is inefficient in terms of storing or conducting heat – air machines needed larger compressors than other refrigeration machines, which also meant that loss due to friction was higher. This translated into another disadvantage of the air machine in comparison to the vapour compression machine. Even when absorption machines could keep up with numerous compression machines in terms of efficiency, Linde criticized the high temperature differences of absorption processes and the large quantities of water required that were to be heated and cooled during operation. Both aspects exhibited a negative impact on efficiency. It is astounding that Linde felt it was barely worth mentioning the advantages of the absorption machine, namely that it could function without a compressor or steam machine.

Finally, at the end of his study, Linde offered his suggestions for improvements to be made to the refrigeration machine: first, it should be a vapour compression machine, which, secondly, in order to achieve the highest efficiency ratio, should function within the smallest possible temperature range. For this, he accepted large surface contact in both the condenser and vaporizer. To increase the actual efficiency of the machine, he suggested in conclusion the use of methyl ether, a refrigerant with a high vaporization heat. The only change to machine design suggested by Linde in his articles was the use of mercury for stuffing boxes, a moving pump seal between the piston rod and cylinder casing at the junction between high inner and atmospheric pressure, which was to prevent the refrigerant from escaping. Refrigerant escape posed not only an economic problem because of its high cost, but a high safety risk as well because of its explosive capacity. The waiting time for a machine was therefore long. Linde's improved refrigeration machine was supposed to be capable of manufacturing 30 kg of ice per PS – approximately 20 kg of ice per kilogram of coal input – which would have made his machine twice as efficient as the maximum performance of any previous refrigeration machine.[136]

Although Linde, in his argument, overlooked the practical application of his machines by focusing on the theoretically obtainable performance of each refrigeration process without suggesting much in the way of design improvements, he nonetheless successfully drew attention to the vapour compression process, his own process, which was to become the dominant process in the industry.[137] Later industrial developments in refrigeration technology proved Linde to be right, when compression machines, as of the 1880s, began to dominate the submarkets. However, to this day, absorption and air machines continue to hold their own in specific applications. It is ultimately irrelevant that Linde's assessment was based not only on an exaggerated thermodynamics, but also on the neglect of how heat dissipation functioned in the machines. More important for the future of refrigeration was the fact that from the start, Linde's work in refrigeration technology linked together the building blocks of mechanical heat theory with the development of actual machines.

Rudolf Diesel followed Linde's line of thinking 23 years later by developing new technical processes using mechanical heat theory and thermodynamic cycles. Diesel's improvements to the motor were also the result of theoretical considerations on how to optimize a cycle process. Just as was the case with Diesel, the results of Linde's initial assertions are gloriously justified, even though the individual steps in between appear in retrospect less than plausible. Science was to provide a new perspective on technical reality, clearing the path to innovation. Certainly, as with any paradigm, there is always the risk that a given position dogmatically distorts that which is viewed. However, when it came to evaluating machines, Linde was capable of assuming various positions, switching from the thermodynamic specialist to the inventor.

A proponent of theoretical mechanical engineering, for Linde, technological applications were to be ruled by scientific principles. Although a successful inventor and technical designer, he was only marginally interested in mathematics – in contrast to many of his colleagues interested in theory at the university in the 1870s. From 1870 to 1900, while the spectrum of professors in theoretical mechanics turned increasingly towards an emphasis upon practical application, the entrepreneur Linde increasingly focused upon his scientific-theoretical interests. Thus, in 1875, Linde had more in common with his experiment-oriented colleagues interested in practical applications, whereas by 1895, as the battle over theory vs. practice raged on at all German technical universities, Linde stood amongst the theorists. Linde's life reads like an attempt to prove the interpenetration of theory and practice. He successfully walked the line between theory and practice, which for him meant walking the line between academics and the industry.

His body and soul devoted to science, the core of Linde's identity, however, was in being an inventor. Turning down a position as a professor of technical physics in Göttingen, he wrote in 1895 to his friend, the Göttingen mathematician Felix Klein:

The only thing I consider more satisfying than scientific work with ambitious and productive men is the production of something in one's own professional area … the desire to combine new processes and machines according to my own ideas is, of course, still there. Significant financing is however needed for this, and I can take full advantage of the company (Linde, d.V.) in this respect, as it will benefit from the results. This would not be the case in Göttingen. I would not be able to pursue my desire to invent there.[138]

Linde as independent inventor

According to Thomas Hughes, the era of independent inventors took place before industrial systems were established. Their most important characteristic was indeed their independence, as the fact 'that they were not limited by any organization allowed them to devote themselves to problems, which once solved, became the core of new technological systems'.[139] Hughes claims that it was, however, precisely this radical creation of new industrial systems which essentially left them high and dry as their relevance in the industry after the First World War was limited.[140] In addition to independence, Hughes describes the typical inventor's approach to work as pursuing an invention as a scientist seeks discovery, collecting scientific proposals and knowledge through outside consultation.[141] As a rule, inventors instructed mechanics 'to transform ideas and sketches into technical and electrical models'. Generally speaking, inventors dreamed of living like a monk with his own small staff of disciples in a well-equipped private laboratory where one could work in peace, to invent – not to conduct research. Famous examples of inventors include the American Thomas Alva Edison and the cartoon figure of Gyro Gearloose, who was never at a loss for a new technical solution to any number of problems confronted in the world of Donald Duck. Hughes describes the inventor's image as that of a pragmatic being, regularly evaluating the competition's patents and purposefully seeking weaknesses in the contemporary technology. In the days before public research funding, it was an inventor's spectacular imagination and slick self-representation which provided him with the needed support and attention to continue his work. According to Hughes, such independent inventors were especially typical of the USA,[142] although he does mention one German example: Rudolf Diesel.

A number of the characteristics named by Thomas Hughes as typical of the independent inventor can be seen in the case of Carl Linde. As a pioneer in machine technology, Linde was clearly set on experimentation, using theory as his guiding principle. He developed original methods of invention, surrounded himself with teams of mechanics and assistants and became a partner in companies that benefited from his patents.[143] The Munich laboratory served only occasionally as his primary workspace; more often it was at the client's site of assembly. His success as an inventor was a direct result of the close co-operation he cultivated with his clients – an important difference to the people-shy inventor portrayed by Thomas Hughes. Moreover, Linde's identity was manifold: not only an inventor, he was also a scientist and entrepreneur. His success can in large part be contributed to the fact that he could distinguish between his scientific, technical, and entrepreneurial tendencies, which allowed him to move back and forth, walking the lines between science, invention, and business management.

In the pages that follow, we will take a closer look at the inventor Carl Linde. Told in retrospect, the stories of inventions are often cut short, portrayed simply as success stories. However, a more detailed look at the difficulties in solving a problem, at the setbacks and errant paths taken will illustrate Linde's pragmatic approach, his tenacity and learning ability. We will look at early inventions, starting with the design and construction of the ammonia compressor from 1871 to 1877, then the development of a complete refrigeration system for breweries and other foodstuffs. Similar observations will be made on the invention of low-temperature technology.

The development of Linde's refrigeration

There are many accounts of the experimental phase from 1871 to 1877,[144] most of which are little more than embellishments of Linde's autobiography. The first test machine is for the most part described as having been a 'complete success' of theoretical predictions.[145] Yet this kind of hindsight is unconvincing as it fails to explore the numerous complete redesigns. Additional sources illustrate the difficult hurdles that were to be cleared,[146] pointing to numerous intermediate models that eventually led to the double-acting ammonia compressor model, which was based on the gas pump. During these six years, Linde's attention was focused primarily on the design of this compressor; other refrigerating components and thermal issues were of secondary importance. He concentrated on problems in machine design and building, working particularly on the duration of his machine's operational dependability.

How did Linde, following the publication of his papers, come to want to build his own refrigeration machines? The answer to this question lies in the network of brewers in which Linde found himself. The director of the largest Austrian brewery (Dreher), August Deiglmayr, visited Linde in the summer of 1871 together with a Viennese machine manufacturer and suggested he design a refrigeration system for the fermenting cellar at a branch brewery in Trieste that could be built in Vienna. Linde however, whose technical experience taught him caution, waved the suggestion aside: a refrigeration system with a sustainable operational reliability could be developed only through long-term preliminary testing, which he could only conduct in Munich. In response to this, Deiglmayr convinced his uncle, Gabriel Sedlmayr – a prominent Munich brewer – to cover the experiment costs and provide space in his Spaten brewery for testing. Linde was thus presented with the opportunity to conduct years of experiments, which led to a refrigeration system that was superior to all other competing products, especially in terms of operational durability and longevity. In return for providing the means to develop the system, Sedlmayr was granted joint proprietorship of the patent for which they both registered on January 17, 1873.[147] There are two drawings of this machine at the MAN Archives. One was drawn solely by Linde, the other together with Gabriel Sedlmayr – which indicates that Sedlmayr's contribution to the system's development was more than simply financial.[148] The Bavarian patent required the implementation of a system within the period of one year. Linde and Sedlmayr thus ordered the components from Maschinenfabrik Augsburg in January of 1873. The ice machine was completed on 7 October and sent to Munich following Linde's inspection.[149] At the brewery however, the system's assembly was delayed, which meant the required certified proof of the system being put into operation was

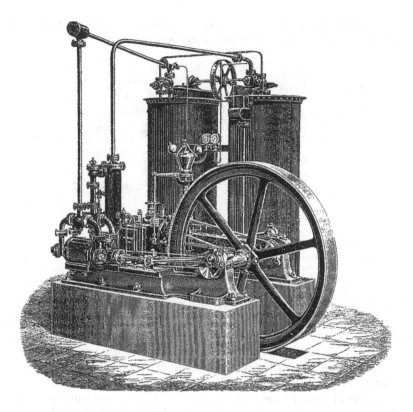

Figure 1.3 Standardization as a strategy for success: Linde cooling machine Type 7. Linde's type classification was adopted by the competition.

handed in to the patent office by Gabriel Sedlmayr at the last minute on 20 January 1874.[150] In response to leaks in the system that appeared on the first day of operation, the chronicler of the Spaten Brewery, Fritz Sedlmayr, exclaimed: 'Problem after problem.'[151]

Stuffing box development

Due to difficulties in sealing movable parts, the manufacturers of compression refrigeration units (similar to vapour machine manufacturing) for a long time insisted on low – as in barely above atmospheric pressure – steam pressure. The development of high-pressure machines such as Tellier's methyl and ammonia vapour compression systems and others failed due to sealing problems. Refrigerant loss was a cost factor,[152] as well as an operations[153] and safety problem.[154] The most important part of the compression refrigeration unit to be sealed was the stuffing box between the piston rod and the cylinder casing; together they formed the only movable seal between the higher internal and surrounding atmospheric pressure. As of 1870, ammonia in refrigeration machines was still volatile. Linde's new stuffing box design in 1875 was perhaps his most important contribution to improving the vapour compression process.[155]

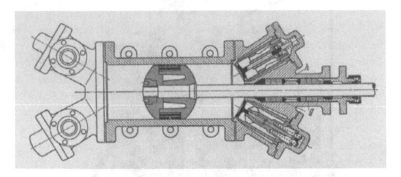

Figure 1.4 A model for decades: Linde's horizontal, double-action ammonia compressor with stuffing box (g) and distance piece (h)

Like his colleagues, Linde held much respect for high pressures. This was illustrated in the sheer size of his first seal constructions, which were larger than his entire refrigeration machines! His first idea was to attempt to 'avoid movable seals in their usual form', such as with pistons, stuffing boxes, and taps.[156] In his 1871 article, Linde had already addressed this idea and come up with a complex mercury sealing system. His patent specification of 1873 did not elaborate on his 1871 proposal by offering a structural design.[157] The mercury sealing system of 1871–73 was a monstrosity. The two compressor pistons were sealed by a mercury steep bath and surrounded the parts to be sealed with a water-cooling jacket. A lavish piston mechanism moved the two compression pistons in synchronization with a third displacement piston in a water bath. The resulting built-up water counterpressure was to prevent mercury from being pressed through the seal. This sealing construction weighed more than double that of the actual refrigeration compressors.[158] But it was not only the dimensions which presented a problem; a second problem was posed by the compressor, which could not exceed a maximum of 12 revolutions per minute so as not to disturb the metastable water/mercury/methyl ether equilibrium.[159] The system was so sensitive that Linde forbade common machinists to touch or work on it.[160] Despite all the tremendous technical effort, the machine was not gas-tight. Leaking methyl ether caused explosions in engine rooms and Gabriel Sedlmayr was forced to put an end to the tests in the spring of 1874 after an employee was seriously injured and the brewery's machinist refused to continue. A trainee employed by the Spaten Brewery at that time reported that methyl ether was compressed and expanded in rubber sacks under the bubble caps.[161] This seems hardly possible; more likely the trainee was merely describing one of the hapless attempts to make the chosen design gas-tight.

The first test machines provided mostly frustrating results. This is where Linde's tenacity played a role. In an attempt to keep Sedlmayr from withdrawing his support, Linde took on part of the further test costs himself.[162] Together with his assistant Friedrich Schipper, Linde designed a new compressor that was much more simple in design than the first. The water displacement pistons proved to be superfluous and disappeared. Linde retained, however, the communicating two-cylinder design. Glycerine replaced mercury and water as the sealing liquid.[163] The new design was above all much easier to assemble and service.[164] Linde ordered the new compressor in the spring of 1875 at Maschinenfabrik Augsburg. With the complex sealing system removed, the compressor weighed and cost about only half as much.[165] Linde registered the new machine as a Bavarian patent, which he received on 25 March 1876. It lasted for ten years.[166] On this day, Krauss, Gabriel Sedlmayr, and Carl Linde signed a licence agreement with Maschinenfabrik Augsburg to manufacture Linde's refrigeration systems.[167] In November of 1877, the five-year contract was extended to 1891[168] – when the official German patent (Deutsche Reichspatente) was due to run out, which Linde had obtained in August 1877.[169] In July of 1876, Maschinenfabrik Augsburg produced a brochure with Linde machines in seven different sizes. The first was sold in September 1876 to the brewer Anton Dreher for his subsidiary in Trieste.[170]

Figure 1.5 The first refrigerator sold, delivered to the Dreher brewery in Trieste in 1877 (today in the Technical Museum of Vienna)

Immediately following completion of the first machine for Dreher, Linde once again completely redesigned his system. The vertical chiller construction for Dreher had proved itself operationally reliable, but was limited by a low rpm. While he had shown that it was possible to overcome the sealing problem, the communicating glycerine cylinders were nonetheless a clumsy solution. Linde now wanted to and could focus his attention on achieving the greatest possible reduction of the machine's dimensions without sacrificing performance. He had since learned that all he needed was a tiny pressure chamber for the sealing liquid and he could avoid the problem of sealing the piston rings by using double-acting pistons like those used in gas-pump technology.[171] The size of refrigeration machines could thus be greatly reduced and their construction simplified. Linde built a small glycerine pressure chamber in the stuffing box between the ammonia and atmospheric pressures, a so-called lantern, which on the outside was subject to a higher pressure than the ammonia gas pressure. Leakage in the stuffing box led to glycerin entering the ammonia cycle. Similar stuffing boxes were already described in the 1860s.[172] Through experimentation, Linde developed a 'rubber packing with leak-proof cotton lining' and continually trickled glycerine onto the piston rods to lubricate and seal them. Linde wrote in 1877 that such a pack would hold 'amazingly well' for 14 days and had 'proven itself first-rate': 'no plucking, no replacing'.[173] Linde continued to labour over further improvements to the stuffing box throughout the 1880s and into the 1890s. After significant improvements were made in 1884, 'ammonia loss at even the highest occurring tensions in practice was genuinely minimal'; this meant ammonia loss remained below 10 kg per day, even for large machines.[174]

Further developments in refrigeration machinery

Let us go back to the year 1876, when Linde arranged his double-acting compressor horizontally. Thanks to this kind of space-saving design, it was possible to install the compressor in a low-ceilinged fermentation cellar, where large vertical machines like those common in America would not fit. Linde significantly increased the unit weight with a new stuffing box and compressor arrangement and thereby lowered the weight-dependent machine price.[175] As of 1877, Linde's vapour compression machines offered the best value for their performance on the market.[176] The third new model became the standard Linde machine until way beyond 1930.[177] However, Linde did not apply for a patent for this third model. Instead, he focused on converting the Bavarian patent from 24 March 1876 for his second design, the vertical single-acting cylinder, into a national patent (No. 1250). The reason for this was clear: Linde could not get a patent for a horizontal, double-acting compressor in 1876 because others, such as Pictét, had already done so.

Prompted by a question posed by Anton Deiglmayr, the thorny six-year path from Linde's initial work with refrigeration machine construction to the completion of a functional double-acting compressor was rife with setbacks. Signs of success did not appear until the fifth year of experiments, when in 1876 construction of the standing ammonia compressor was completed for the Spaten brewery. It was in the sixth year that Linde and his colleagues finally made their breakthrough with a successful design solution. During this period, thermodynamics played a minimal role.[178] Linde essentially ignored thermal problems, such as evaporator and condenser design, and chose a simple solution: the spiral pipe condenser. In thermodynamic terms, this condenser model was less favourable than the ones used by Tellier and Pictét. But mechanical problems were more compelling – Linde the former chief engineer of a locomotive factory and steam engine specialist was challenged more in those years than Linde the thermodynamics expert. Nonetheless, Linde's scientific background was relevant to his development of refrigeration machines. It is often forgotten that Linde's area of expertise – the theory of machine construction – was based only minimally on mechanical heat theory in 1870. As an instructor of kinematics and the theory of machine construction, Linde taught his students to use typology in creating various designs. The exorbitant kinematics of his first test machines may be seen as a negative consequence of this theoretical background. But the surprising range of chosen design solutions and his willingness to refrain from using minimally successful variants in design illustrates Linde's vantage point as a scientific examiner.[179] The scientific world also indirectly fostered the development of Linde's refrigeration machine. Linde's prominence as an academic engineer and professor inspired the science-oriented brewers Sedlmayr and Deiglmayr to support him. Let us look now once more at the year 1877, when, after having developed the stuffing box and double-acting compressor, Linde turned towards developing application-oriented apparatus for use in the artificial production of cold.

Ice manufacturers

One of the primary functions of refrigeration systems in breweries and ice factories was to produce artificial block ice, and later 'plate' ice. This ice was used instead of natural ice sold to other users. The two major technical problems posed by ice-making were how to produce crystal-clear ice and the mechanization of production.

The cloudy appearance of artificial ice was due to the formation of tiny air bubbles that materialize in water when cooled quickly. Brewers, and more importantly, their customers, disliked this milky quality. Clear ice, so-called crystal ice, was preferred, as dirt or any other pollutant was immediately visible. This kind of visual examination had its origin in natural ice-buying behaviour; such tests were hardly necessary for artificial ice. Another reservation concerning cloudy ice had to do with its larger rough surface, which made it melt somewhat faster than clear ice.[180] Calling his ice works 'Crystal Ice Factories' because it 'sounded very hygienic', Linde was well aware of the popularity of clear ice.[181] As of the turn of the century in Germany, one had the choice of cloudy ice, clear ice (up to one-third could be clouded) and totally clear crystal ice.[182] Ice factories tried for several decades to produce ice that was as clear as possible. Over the years, Linde tested three different ways to manufacture clear ice: slowing down the freezing process by enlarging the equipment, while producing the same amount of ice[183] purifying water, as air bubbles are difficult to form in distilled water[184] and introducing a shaking device that kept water in motion during freezing, which, however, often froze itself. Linde tested the first so-called finned device at his Barmer ice work. Finned or shaking devices were experimented with for decades,[185] however, the use of purified water was established in the mid-1880s, as the distillate from steam engines was readily available. The condensed exhaust steam thus contained traces of grease from the steam engine cylinder. This desired 'clean' crystal ice was in reality dirtier than cloudy ice.

The second technical problem to be overcome in ice-making was the mechanization of both the supply of water and ice excavation. In his autobiography, Linde refers to his first successful realization of an automatic can ice-make that remained essentially the same in the decades following.[186] Yet during the first five years, Linde danced to the tune of another principle of construction: the rotating ice generator. Linde was dissatisfied with the can ice-maker's stop-and-go operational nature – fill up, freeze, lift out – and wanted to develop a cycle for ice production like that used in refrigeration. He therefore designed a rotating semi-automatic ice-maker out of wood. For years, Linde persisted with this rotary ice generator for the production of clear ice.[187] In July 1877, Linde presented his first design to Maschinenfabrik Augsburg, asking them to construct a small test drum.[188] When its completion was delayed, Linde became increasingly euphoric about his rotary ice generator and cancelled his order, giving the contract to build a large rotary drum to the wagon building company of Rathgeber in Munich,[189] so that he could exhibit this together with a refrigeration machine from Sulzer at the Paris World Fair as the Linde System. In January 1878, Linde wrote to Sulzer: 'All operations are now completely automated, so that the operator himself barely has to come into contact with the ice. This new construction has exceeded my expectations.'[190] Inspired by Linde's enthusiasm, Sulzer insisted on building the rotary drum himself.[191] In October 1878, Linde started a series of experiments at the Spaten brewery.[192] Friedrich Schipper, who had in the meantime switched over to Maschinenfabrik Augsburg, led the experiments. Linde continued to express optimism towards his clients.[193] But Schipper could not overcome the problems with the labour-intensive rotary generator, which had existed since its inception. In August 1879 he complained of a residual air column in the ice plate;[194] in December he complained of the ice becoming cloudy,[195] because the air bubbles 'are not being blown or wiped away'.[196] Linde decided to equip his own first ice work in Barmen with the horizontal can ice-maker instead of the risky rotary generators.[197] Linde's business correspondence shows the intensity with which he pursued the creation of clear ice. Within one year, he wrote 16 letters to Maschinenfabrik Augsburg, 8 to Schipper and Sedlmayr and 16 to Diesel and the Sulzer brothers. Yet in the summer of 1880 Linde put an end to work on the rotary generator and admitted to his brother Sigmund in 1881: 'I have yet to come up with a successful process for clear ice, the engineers of the Paris company Diesel claim to already have found one.'[198] This wish was to remain unfulfilled.

Since the mid-1880s, Linde had been testing a 'finned apparatus', which was to provide an escape for air bubbles through the constant motion of a stick in water as it freezes. However, the vane itself tended to freeze, which posed a problem. In 1890, Linde experimented with a centrifugal device to produce clear ice and told his engineer Negele 'Currently, our most important task is to solve the clear ice problem ... it is more important than our acquisition business.'[199] Following his definitive retreat from the principle of the continuous production of clear ice, Linde focused on the mechanization of can ice generators already on the market so as to at least partially maintain continuous operations. Linde remained partially loyal to his belief in automation and continuous operations by optimizing the filling, insertion, lifting out, and emptying of ice cans. In the beginning, there were problems with the travelling cranes, but soon this semi-automated ice production was established everywhere.[200]

In 1890, Linde moved on to trying to produce clear ice with condensation water from steam engines. Linde's engineers experimented for years with condensation water filters (animal charcoal, sieve, wood shavings) to eliminate the traces of oil picked up by the condensation water from the steam engine cylinder.[201]

The development of comprehensive refrigeration systems for breweries

The previous section illuminated certain aspects of Linde's approach to developing things. His preference for continuous and automated processes – a consequence of his roots in theoretical machine construction – became clear again in his design of a rotating ice-maker. His success however, owed to his ability to reject failed ideas, even when they were theoretically convincing, and his readiness to work together with the clients who used his machines. Linde recognized the value of this kind of interaction from the start and cultivated his contact with the brewers. The fact that Linde himself did not own his own laboratory until the beginning of the 1890s was presumably quite conducive to a form of co-operation in which new devices and techniques were developed together with his clients, mostly individual major breweries. In this way, Linde could count on amicable teamwork, particularly with Sedlmayr and Jung, who both sat on his Supervisory Board. He maintained a close friendship with Feltmann in Rotterdam (Heinecken), and a good business friendship with both Jacobsen in Copenhagen (Carlsberg) and the brewmaster of the world's largest brewery, Samuel Geoghegan of Guinness in Dublin.[202] These were the breweries where Linde could experiment. While assembling equipment at the Baartz brewery in June 1879, Linde's assistant Banfield received a letter from him with the instructions: 'Show no doubt. No mention of experiments should be made in front of Baartz.'[203]

Experiments using direct cooling without ice to refrigerate fermentation and storage cellars had been conducted since the 1870s. In an attempt to break into the business, Linde forwent the offer of refrigerating services and guaranteed his first true customer, Dreher, the much riskier prospect of a constant temperature no higher than 5 degrees Celsius for his fermentation cellar in Trieste.[204] His first test using vertically composed 'pipe walls' was a failure.[205] As a result of this failure, Linde designed a cellar cooling system of forced convection using fans. It was through his co-operation with the breweries that after six years of experiments, in 1882, he decided to install the so-called stationary cooling system with condenser pipes beneath the ceiling.

At first glance, the fan-type cooling system appeared more efficient, easier to control and therefore more modern than stationary cooling with cold pipes. In use however, the fan-type cooling system proved to have a few glitches: the fan power brought heat into the air current, cold air was blown out of the cellar and warm outside air was constantly being sucked into the system through the forced convection. In an echo of his machine engineering roots, Linde engaged his colleagues and clients, even Zeuner, in a discussion over the specific advantages of various fans.[206]

Linde experimented for years on direct cooling using forced convection. He began in 1877 at the Dreher Brewery in Trieste, 1878 at the Spaten Brewery in Munich and 1879 at the Aktien Brewery in Mainz. In Trieste Linde produced cold air in which ice-cold water trickled down cascade-like into the air cooler and was then blown through ventilators into the fermentation cellar. Schipper was, at the end of 1877, dissatisfied with the fermentation cellar in Trieste and was 'depressed in every way'.[207] At the same time, the first machines built by Sulzer in a Basel brewery also failed to perform as expected. 'The need to dissassemble our refrigerating equipment there', wrote Linde to Sulzer, 'will be exploited mercilessly by our competitors'.[208] To make matters worse, at the start of 1879, Sedlmayr lost his patience and stopped all the fan experiments done in his fermentation cellars, despite Linde's promise that the first usable results had begun to appear. The salient problem with the ventilation system was that of obtaining the correct air direction. Sedlmayr's system 'sucked more [air] than it blew', a problem for which no satisfactory solution existed.[209] Linde harboured no illusions about the potential consequences of disappointing his most important guarantor amongst brewers.[210] Linde's personal relationship to Gabriel Sedlmayr remained unharmed; it was rather his relationship to Sedlmayr's head engineer Pfitzner which cooled.[211] At the end of March 1880, Linde removed his test machines from Sedlmayr's brewery.[212]

A u-turn in fermentation cellar cooling systems brought Linde together with Feltmann, the manager of the Rotterdam Heinecken Brewery, to whom Gabriel Sedlmayr had introduced Linde in October 1877.[213] To be on the safe side, Feltmann had ordered a refrigeration unit in November 1877 for ice production.[214] At the beginning of March 1880, Feltmann went to Munich to take a look at the fermentation cellar experimental laboratory at the Spaten Brewery.[215] While there, he noted the disadvantages of forced convection and insisted on natural circulation according to the model of top-icing for his operations. Linde tried to convince Feltmann otherwise, writing: 'The comparison is clumsy, as top-icing is much greater in surface area.'[216] But Feltmann insisted and thereby brought Linde to develop a stationary cooling system using condenser pipes laid in the lower part of the cellar ceiling. By April the two began discussing the introduction of a coil cooling system with brine.[217] In July 1880, Linde conferred with the Danish brewer Jacobsen, whom he had met through Feltmann,[218] over the design for quadrant pipes to be used in stationary circulation.[219] In

August of the same year he called for ventilation systems to be used only for fermentation cellars, whereas natural circulation he considered sufficient for storage cellars.[220] In the long run, forced convection was retained only for those rare cases in which the dry conditions of a fermentation cellar were especially important.[221] By August 1884 the change had been completed and Linde wrote to Zeuner (in reference to the Schlachhof project in Leipzig), that he had become 'so convinced of the use of a condenser pipe system, that he would be sorry to see its introduction thwarted by any unfounded objection'.[222] In a letter from 1892, Linde set the layout rules for stationary circulation and ventilators were no longer mentioned.[223] Ventilation was retained only in those areas that not only had to be refrigerated, but air conditioned as well. In many cases it was not only the dissipation of heat, but moisture drainage, which was at issue, such as in the particularly humid fermentation cellars of Southern Europe. Forced convection was also used in cold stores, where many products needed a dry atmosphere. This afforded Linde another application of his experience in storage cellar refrigeration.[224]

It was Linde's close relationship to the use-oriented brewers which permitted the development of a complete refrigeration system, first for breweries and then for numerous manufacturers of foodstuffs. Indeed, Linde attached great importance to his contact with the leading brewers. The satisfaction of these top clients and their suggestions were one of his top priorities. Without this kind of contact, Linde probably would not have found the simple yet strong solutions that made his equipment stand out. Both the ventilator issue and the problem of ice production were the sparks for the development of a highly sensitive system.

This kind of research and development strategy conducted in co-operation with clients demanded the on-site presence of engineers and a continuous written correspondence with headquarters. A highly developed correspondence culture was thus cultivated in the Linde company. Indeed, the practice amongst development engineers to write two or three long letters a day to the Wiesbaden headquarters about problems and possible solutions continued into the Thirties, well beyond the era of Linde's active participation in the company. Linde's experience provided suggestions which led to improvements made to the management of contemporary knowledge in a company that was geographically spread out.

Condensers

Issues of heat transfer provide a final example of Linde's entrepreneurial and engineering approach. Up until the 1890s, questions of heat transfer played virtually no role in his experiments. In early refrigeration machines – so-called immersion condensers – ammonia was passed through and condensed in pipe coils immersed in cooling water tanks. These machines were enormous and inefficient, but performed reliably. Competitors and developments in the USA provided the stimuli to improve the efficiency of ammonia condensers so as to enhance the economical use of cooling water. In 1892, the Linde Company installed the first atmospheric condenser in Germany. It remains unclear how Linde came to create this condenser, which partially vaporized the cooling water and put it to efficient use.[225] In the winter of 1890, Linde was offered a plan for an atmospheric condenser by the engineer Eduard Theissen, who had been working at the Maschinenbauanstalt Humboldt in Cologne.[226] Linde, however, was suspicious of this offer from the competition and turned it down. In January 1893, Theissen, who was no longer at Maschinenbauanstalt Humboldt, returned to Linde with the idea of a disk capacitor, only to be turned away again. According to Linde, technology had since moved away from disk capacitors and towards atmospheric condensers.[227] Whether or not Theissen's 1890 proposal held any significance for Linde remains unclear. Within the company, Linde often referred to Theissen's achieved level of efficiency.[228] Linde, concerned as he was about precise measures, was dissatisfied with the atmospheric condenser because it 'was not possible to get precise measurements with such a condenser'. Qualifying this concern, he stated: 'I do not of course expect that the facilities be built according to the needs of precise measuring.'[229] The first 'rain device', as Linde called his atmospheric condenser, was installed at the Löwenbrau Brewery in the summer of 1892, with two ventilators blasting air through cooling surfaces. The design for this condenser came from Linde himself, who also hired his own illustrator in Munich for the purpose.[230] The condenser was a success. Thereafter, all major new works installed by the Linde Company received atmospheric condensers. Other new condenser models came from the United States, first in 1912 to Germany, which was a leader in thermodynamic theory but not practice.

Linde's business correspondence shows how thermal issues were granted comparatively less importance than the design problems posed by the construction of condensers. Certainly Linde's business correspondence provides a rather thin indication of his everyday activity in technology. Nonetheless, it is worth noting that mechanical problems are discussed in nearly one-sixth (almost 500 letters) of the entire remaining business correspondence, whereas there are only 220 letters in which refrigeration processes and thermal issues are mentioned.[231]

The general decline of technical issues as a topic of Linde's correspondence indicated not only the changes in his duties as an engineer and entrepreneur, but also pointed to the fact that many design problems in refrigeration technology simply lost their relevance after 1880. Linde often complained in his letters about issues of price becoming increasingly the focus of competition in the industry.

In short, in Carl von Linde the characteristics of the technical scientist, the inventor, and industrial engineer were all combined. In many ways, his activities as an entrepreneur seemed to speak more to him than his activities as a professor. His Protestant work ethic, grasp of practical design, strong personal interaction skills with regards to his colleagues and clients and finally, his foresight and well-educated instinct for business and markets lay at the root of the company's success. His scientific background and self-styled scientific profile contributed to this success, but do not alone explain the success of Linde's Eismaschinen. As CEO, Linde made decisions as an entrepreneur, not as a professor or researcher. His strength, however, lay in the fact that he maintained a basic technical and scientific curiosity that he could satisfy in his professional life.

The founding and organization of the Linde Company

The founding of Linde's ice machine company

With demand for refrigeration particularly high in the brewing industry and the terrain cleared for a new company to establish itself in the refrigeration industry, Linde's Ice Machines was born in 1879. Windhausen, Pictét, Ferdinand Carré and others had shown that refrigeration systems, despite recurring technical problems, could compete with the natural ice industry. A serious competitor to both, Linde skilfully put his competitive advantages – which grew from his collaboration with breweries (such as his on-site work with Sedlmayr) and machine manufacturers – to work for him as he built the company up. From the beginning, he insisted on flexibility, particularly when it came to choosing his foreign licencees.[232] He thus retained strategic advantages and put them to good use. He kept a step ahead, moving on to the next goal once a given goal had been achieved. It was important to him 'to on the one hand render the natural ice industry obsolete and on the other hand develop entirely new uses for refrigeration'. Linde received a constant flow of requests from machine manufacturers who themselves lacked the qualified employees for such work.[233] It should thus come as no surprise that many of his business partners encouraged Linde to devote his time to a patent utilization company. This was, however, completely out of the question for Linde.

In January 1877, Heinrich Buz of Maschinenfabrik Augsburg (MA) joined the financial management group of Gabriel and Johann Sedlmayr and Georg Krauss, which had been taking in earnings from licence fees (15 per cent of sales). The first three members paid 4,000 Marks each in return for one-sixth of Linde's patent rights. Buz however, paid 12,000 Marks for one-third of the remaining rights held by Linde. Linde's total share was thus one-third and the Sedlmayrs merged their shares together to comprise another third. The search for other partners and financial backing began when costs grew (reaching 36,250 Marks by the end of 1876) due to experiments, travel, and patent registration associated with the methyl ether and ammonia machines.[234]

Thanks to a contract in the spring of 1878 for the construction of two refrigeration plants – one for the Aktienbier Brewery in Mainz – Linde met Carl Lang, who sat on the Supervisory Boards of many breweries in the Rheinland as a technical advisor. Lang had made a fortune as General Construction Director of the trans-Balkan railroad to Constantinople. It was, however, the banker Moritz von Hirsch[235] in Paris, who secured the permission to build the railroad from the Turkish government in 1869. When Linde regretfully informed Lang of his plans to abandon his work in refrigeration technology because he could no longer bear

Karl Lang

Heinrich von Buz

Gustav Jung

Georg Krauss

Carl Sedlmayr

Figure 1.6   The Linde Company's first Supervisory Board.

the weight of the double burden he carried, Lang offered Linde the prospect of funding for his own company. In addition to this, he suggested bringing Moritz von Hirsch on board as a partner. Initially hesitant, Linde eventually became enamoured of the idea and tried to convince Buz, Krauss, and the Sedlmayrs to sell their shares to Lang and Hirsch. While Buz and the Sedlmayrs agreed upon 100,000 Marks, Krauss demanded twice as much.[236] Mired as he was in the ethos of a civil servant that values security, Linde also shocked Lang in the negotiations that followed by demanding protection for individuals against commercial risk. When, in mid-1878 the fermentation cellars at the Aktien Breweries in Mainz also faced failure, talks between the two deteriorated rapidly. Linde nonetheless persisted and articulated his 'credo of the entrepreneurial Engineer': 'I cannot put a stop to my passion for my machines, although I'd like to believe they are merely objects and that I do not need to continue thinking about them or hear what happens to them.'[237] In his correspondence with Lang, he admits that 'he is the one exhibiting any true interest [in founding the company]'.[238] Linde backed down somewhat from his demands regarding the financial security of individuals, but not from those concerning the aim of the company. Linde was thus able to continue with his plans to set up a project-oriented engineering – instead of a patent utilization – company.

Various drafts of a contract were sent back and forth between Linde and Lang from September 1878. As has already been noted, in December of that year, before a final partnership agreement was completed – let alone signed – Linde handed in his resignation as a civil servant for the end of the winter term after his business plans had become public due to an indiscretion on Krauss's part.[239] Finally, in January 1879, Linde invited Buz to become a partner in the company that was to have a nominal capital of 400,000 Marks. Hirsch however, changed his mind at the last minute and insisted on a reduced nominal capital stock of 200,000 Marks. Buz, Sedlmayr, and Krauss were to become members of the company and would in exchange receive 100,000 Marks each for giving up his rights to Linde's patent. This change of heart probably had to do with the difficulties in getting a patent issued on time to the French company Sartre & Averly for a Linde machine to be built. Linde expressed his 'deepest personal objection' to this kind of company organization and tried at first to increase the capital by 50,000 Marks. However, he had already stuck his neck out too far to pull out of the project and eventually agreed to 'undertake the project even in its extremely reduced form'.[240] As Buz, Sedlmayr, and Krauss agreed to these conditions, the owner of the Aktien Breweries in Mainz, Gustav Jung, joined the circle of partners in May 1879. Because Wiesbaden was the site of the planned company headquarters, it became desirable to have Jung join, as another Supervisory Board member from the region was expected.[241]

The founding capital for Linde's Ice Machines was 200,000 Marks, making it one of the smallest of 45 German stock corporations founded in 1879. The total amount of founding capital in Germany that year reached 57 million with an average of 1.27 million Marks per corporation.[242] Linde Ice Machine's founding capital was also rather minimal in comparison to the amount made available for the second Pictét-Corporation in 1880 (3 million Francs). The Linde Company, like most others, did not have capital available as cash, but was rendered secure by the filing of its patents. The company suffered in the first months from cash flow problems, forcing Linde to provide funds through his own personal finances.[243]

Both the allocation of shares in the company's early years and Linde's relationship to his most important financial backer, Moritz von Hirsch, are anything but clear. In the spring of 1879 Hirsch intimated that he did not want it to become public knowledge that he was a joint owner of the company, which meant he was never publicly presented as such. Even in his business correspondence to third parties, Linde never mentioned Hirsch's name. Hirsch's silent partnership meant that although he provided the company's primary financial backing and together with Lang paid Linde 100,000 Marks,[244] he was the only original shareholder not on the Supervisory Board. His influence weakened over time as a result of this, particularly following the death of Carl Lang in 1884. While he did attempt in the year after Lang's death to place his own trusted representative on the Supervisory Board, he bowed to Linde's expressed desire to avoid bringing in any 'seed of discontent'.[245] In contrast to his quiet engagement in Wiesbaden, he publicly bought the French rights to the patent for 50,000 Marks and founded in 1879 in Paris an ice works company and also an engineering company, which hired a rather fresh-faced Rudolf Diesel in 1880. In England in 1885, Hirsch made his participation in the financing of the Linde British Refrigeration Co. public. However, in Wiesbaden, Hirsch was sliding further and further into the background and Linde would have liked nothing more than to buy all of his shares 'at 300 per cent'.[246] Not having been kept up to date with the company's activities since 1880, Hirsch sought revenge by trying to block necessary moves to increase the share capital, such as was the case for the takeover of the Berlin-based company Rudloff-Grübs.[247] By the end of the 1880s, Hirsch began to prune the branches of his wide-ranging network of businesses and started selling his shares. In 1891, the year in which he founded the Jewish Colonization Association to support agricultural settlements of oppressed Russian Jews in free countries, he and the Linde Company made an agreement stating that Carl von Lang-Puchhof, who had been promoted to take his deceased father's place on the Supervisory Board, should represent Hirsch's interests at the general shareholders' meetings.[248] What happened afterwards is difficult to establish, as the paper trail of business correspondence between Hirsch and Linde ends in 1892.

Linde's Ice Machines was founded with a total of 200 shares sold at 1,000 Marks each. Hirsch and Lang together held 25 per cent of the shares, while the others – Buz, Linde, Krauss, Sedlmayr, and Jung held 30 shares each.[249] Carl Lang was the first chairman of the Supervisory Board until his death in 1884, when Gustav Jung took over the position. Jung held the position until he died only two years later. Gabriel Sedlmayr kept a low profile, often sending Carl Sedlmayr to represent his interests. Following the deaths of Lang and Jung, their sons took over their positions and Adolf Jung became the chairman of the Supervisory Board. In the company's early days, Carl Lang was, after Linde, the most important figure at Linde Ice Machines. Lang's connections with various breweries proved to be most fruitful and even helped Linde get on the Supervisory Boards of other companies. It was Lang who determined the company's headquarters and structure. Although generally friendly, the relationship between Lang and Linde was not free of tension. Indeed, during an argument between the two in March 1881, Lang strongly suggested to Linde that he 'serve the company in pragmatic questions' and warned that Linde was 'not indispensable'. In other words, Linde was free to go, should he choose to do so. Linde apologized to Lang the next day, chalking up his behaviour to exhaustion and stating that he 'would prefer to leave at once', but that he felt 'a duty to the well-being of the company'.[250] Lang also felt obliged to the company and therefore prevented Linde from putting up private money to help found an English company. In fact, a subsidiary of the Wiesbaden company emerged.[251] Linde, unlike others such as Pictét, refused to be isolated from the business functions of his company and therefore became the leading figure of Linde Ice Machines following the deaths of Lang and Jung, having the final word on all questions concerning business in the company.

Developing the organization of the Linde Company

Linde's Ice Machines began originally as an engineering company, which was rather typical in the early days of refrigeration machine manufacturing when independent inventors sought to market their products. Others, such as Franz Windhausen, also chose to start with this rather modern, streamlined means of conducting business. By foregoing his own production facilities until 1897 when a department for gas liquefaction was established in Munich, Linde was able to increase flexibility and lower costs.[252] Building the refrigeration units was thus left to the machine manufacturers. One important difference between Linde's and Windhausen's mode of conducting business lay in the fact that the Linde Company cultivated contact with its clients instead of leaving this to the machine manufacturers. Using in-house mechanics to assemble and perform maintenance work on the machines increased client confidence and trust in the Linde Company.

For over 90 years, the Linde Company was characterized by strong bonds between a few families with interests in the business and its Management and Supervisory Boards. As the company's successes grew in the early 1880s, so too did the flow of outside capital into the company as the number of shareholders increased. Although banks were the largest shareholders of the Linde Company well into the turn of the century, a few families retained their long-term personal commitments to the company. In 1879 Carl Linde agreed to act as CEO for ten years without salary, but retained the right to a share of the net profit.[253] Up until 1888, Linde was the only member of the Executive Board. With the end of his contract in sight and driven by the desire to return to his scientific work, Linde began to prepare his long-term colleague Friedrich Schipper to step in as his deputy. In order to properly train Schipper, Linde held onto his position as CEO for one extra year and remained in Wiesbaden for another year after that, giving up his position before returning to Munich, thus providing Schipper with the necessary initial support and guidance.[254] In 1890, Schipper took Linde's place on the Executive Board and Linde became chairman of the Supervisory Board. Schipper retained this position for 39 years. His deputy, August Krebs, retired in 1904 due to ill health. Schipper stood alone on the Executive Board for four years until Linde's son Friedrich[255] stepped in as deputy in 1908. The Executive Board did not expand until after the First World War, when Ernst Volland became a member. Volland was the director of the machine manufacturer Sürth, which had been acquired in 1920 by Linde Ice Machines. Eventually in 1928 the number of authorized officers increased from three to seven, with the addition of Hugo Ombeck, Otto Hippenmeyer, Linde's son Richard[256] and his son-in-law Rudolf Wucherer.[257] The Technical University in Munich served as an 'elite academy' of sorts, where all members of the board (excepting Friedrich Linde, who had a PhD in physics and Volland, whose training was as a locksmith) before the Second World War completed degrees in engineering. In 1924, Friedrich Linde was designated CEO. This was little more than a formality, as Schipper had already been transferring much of his leadership and responsibilities to Linde beforehand. Friedrich Linde held this position until after the Second World War. Indeed, the Linde family was for decades the determining force behind the company.

The Supervisory Board was also characterized for decades by the continuous presence of a few families with strong interests in the company. Although common in family-owned businesses, such continuity is rather rare for a company with such an inflow of outside capital. The seats held by the brewers Sedlmayr and Jung remained in their respective families until after the Second World War. Both of Linde's sons as well as his sons-in-law Carl Ranke and Rudolf Wucherer took over positions working for him. The seat held by Buz, the director of the Maschinenfabrik Augsburg, was also kept in his family.[258] Following his death in 1906, Georg Krauss's seat however, was given in 1907 to Professor Frommel, in 1915 to Wilhelm von Oechelhäuser,[259] and in 1930 to Johannes Hess.[260] In 1917 another seat on the Supervisory Board was created and occupied by Theodor Plieninger, the general manager of the chemical factory Griesheim Elektron AG and later a member of the board of IG Farben. This seat on the Supervisory Board was a concession to the waxing importance of the chemical industry and the waning importance of breweries for the Linde Company, which had, since 1897, expanded into the field of gas liquefaction and developed a strong relationship with the chemical industry. After Plieninger left office in 1930, his seat continued to be filled by representatives from major chemical companies.[261]

The Supervisory Board held over 198 meetings between 1879 and 1930, on average almost four times a year, meeting more frequently in the early years. In time, the number of shares held by a few influential families sank, as did the number held by the original members of the Supervisory Board. Large banks became the new controlling shareholders after 1900, as the other shareholders were in no position to provide the finances for the necessary increase in share capital. The banks, however, refrained from exercising influence on the board, allowing for the continued promotion of retired members of the Executive Board to the Supervisory Board, which helped cultivate the familiar atmosphere under which the company flourished. Members of both the supervisory and management boards were absolutely prepared to make decisions that were in the company's best interest rather than their own. The influence exercised by the Linde family on the company since its beginning was therefore not related to the number of shares it held, which decreased in percentage after the turn of the century, but was rather a direct result of the Linde family's technical, scientific, and commercial competencies and the Supervisory Board's desire to maintain strong personal bonds.[262]

Table 1.1 Linde Company shares held by influential families. Linde AG Archives, Wiesbaden, 1942

Family	Share value (1 000 Mark)	Share (per cent)
Linde (including Ranke, Seiler, Michaelis, Wucherer)	1 785	7.1
Other families on the Supervisory Board (Sedlmayr, Jung, Meyer, Hess, Proebst)	1 392	6.3
(former) directors of Refrigeration Technology at Linde (Otto Wagner, Hugo Ombeck, Friedrich Schipper, Otto Hippenmeyer, W. Brückner)	155	0.9

Source: Dienel, *Ingenieure zwischen Hochschule und Industrie*, p. 157.

Both the structure and number of employees for Linde's Ice Machines were shaped by the fact that until the department of Gas Liquefaction was established in Munich in 1897,[263] the company did not manufacture its own products. Originally, the company was essentially a planning and engineering company, leaving the manufacturing of refrigeration units to licensed machine manufacturers. The company was therefore staffed primarily by engineers, draughtsmen, mechanical engineers and a few businessmen, the total staff remaining relatively limited in numbers until the First World War (see Table 1.2). After the First World War, the Linde Company made moves into the machine manufacturing and small refrigeration unit industries, acquiring or building large production facilities such as the machine factories in Sürth near Cologne (1920), Walb in Mainz-Kostheim (1926) and Güldner in Aschaffenburg (1929).

Starting with only one employee and a single-roomed office in Wiesbaden in 1879, by the end of the first year Linde had three employees, and by 1914 there were 620. The Linde Company continued to grow, becoming a major business with more than 1,000 employees after the First World War. Records

Table 1.2   Number of employees 1879–1914

Year	Employees	Year	Employees
1879	3	1895	147
1882	18	1896	138
1883	34	1897	173
1884	53	1898	198
1885	50	1899	217
1886	48	1900	251
1887	57	1901	231
1888	56	1902	212
1889	77	1903	181
1890	116	1904	202
1891	131	1905	206
1892	130	1906	226
1893	138	1910	>500
1894	145	1914	620

Sources: For 1882–1906: Calculations based on labour costs documented at the Linde Zentralarchiv Wiesbaden, IV: Protokolle der Generalversammlung, in which the average annual salary was at 1,500 Marks throughout 1895, from 1896–1906 at 2,000 Marks; for 1879 and 1910: Archiv der Linde AG, Wiesbaden 879 1979. Wiesbaden 1979, p. 85; for 1914: Ernst Barth, *Entwicklungslinien der deutschen Maschinenbauindustrie von 1870–1914*. Berlin 1973, p. 111.

show that by 1940, there were over 5,000 people employed in Germany by the Linde group.[264] Of the total 620 employees in 1914, 340 were labourers and 280 were salaried employees, so-called 'company civil servants'.[265] Compared to the average machine manufacturer, where the percentage of salaried employees usually lay between 7 per cent and 12 per cent, the Linde Company's percentage of salaried employees was extremely high at 45 per cent. This high figure points to the emphasis placed on planning and engineering in the company. This percentage dropped sharply after the gas liquefaction manufacturing facility in

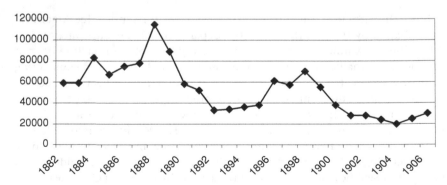

Figure 1.7   Sales per employee 1882–1906 (in Marks)[266]

Sources: Linde Zentralarchiv Wiesbaden, IV: Protokolle der Generalversammlung. Employee calculations based on documented labour costs in which the average annual salary was at 1,500 Marks through 1895 and from 1896–1906 at 2,000.

Munich was built and again after the First World War when the Linde Company acquired numerous machine manufacturing companies. Stabilising at 25 per cent through the 1930s, the percentage of salaried employees continued thereafter to remain comparatively high.

The increase in staff was hardly continuous; yet despite all the emergent crises in both the national economy and the refrigeration industry before 1914, there were no long-term reductions in personnel at the Linde Company. Setbacks were short-lived – even when, as a consequence of the crisis in 1900–01, nearly 20 per cent fewer people than the previous year were employed in 1903. By 1904, the staff numbers bounced back, increasing by 10 per cent. Turnover was relatively high for the low numbers of employees (see Figure 1.7), fluctuating between 20,200 (in 1904) and 113,107 (in 1888) Marks per employee from 1882 to 1906. Assuming that the ratio of added value to employee functions as an indicator of productivity can be positively correlated to per-employee turnover, then the Linde Company witnessed its highest productivity rate during the 1880s. Following this period, the trend spiralled downwards until the First World War. This drop in productivity was by no means an indication of poorer work quality; it was caused by a change in price structure. During the 1880s, the Linde Company enjoyed the benefits of a large market in the brewery industry for its refrigeration machines and was, as the market leader, able to dictate prices. Soon, however, the refrigeration market in the brewery industry was saturated. Furthermore, the durability of the Linde machines meant replacements were not needed for a long time. A glance at the annual net profit during the 1880s shows success. However, this was surpassed by the period from 1896 to 1901 in which turnover reached 13.78 million Marks in 1898 and the net profit grew to over 1 million Marks. The annual net profit from 1908 through the First World War was also higher than that of the 1880s.[267]

At Linde's Ice Machines, it was the engineers who carried most responsibility. Their duties included the sale, sizing, installation and maintenance of a given refrigeration system. When problems arose, the first person to be called upon was the engineer responsible for that system. This kind of all-encompassing approach meant the engineers were often on the road, leaving a small core team of chief engineers at the Wiesbaden office. As Carl Linde himself was often on the road, this core team made many decisions in his absence and had their own project engineers and draughtsmen to assist them.

It was only through a high degree of independence and the ability to endure stress that such far-reaching duties could be met. Engineers with these capabilities were often also quite inventive, a quality of which Linde, as a businessman and their superior, tended to be critical. Linde thus often found it difficult to handle the otherwise necessary independence of his engineers when they were creative with technology. There were tensions with even his most prominent colleague, Rudolf Diesel, who worked for the Linde Company from 1880 to 1893.[268] After receiving his degree from the Technical University of Munich in 1879,[269] Diesel followed Linde's recommendation and completed a one-year internship at the refrigeration department of the Sulzer Brothers in Winterthur that was financed by Moritz von Hirsch. Two years later, we see Diesel's unique sense of independence in his appeal to mechanical engineering students at the Technical University of Munich: 'Don't go into the civil service! Don't restrict yourselves to engineering!' With these 'passionate, fiery words', Rudolf Diesel, the 'Iceman', advised in 1881 against 'engineering the specialties of machine manufacturers' and called for the leadership in factories to 'demand more independence and efficiency'. In 1883, Diesel took on the job of representing the Linde Company in France – a temporary position that was based on commission.[270] Diesel insisted upon his right to work for other companies at the same time and made use of this privilege, much to the dismay of his employers in both Berlin and Paris.[271] In addition to his regular work, Diesel continuously designed new inventions in refrigeration technology. As discussed earlier, Linde and the Sulzer Brothers in Winterthur had corresponded with each other intensively on the problem of clear ice production from 1880 to 1882. Other problems arising in the process of refrigeration Diesel tried to solve included the extraction of glycerine (1881–83), the extraction of lactose (1881–83) and paraffin (1881–82) from crude petroleum, as well as the extraction of fats and carbon dioxide.[272] However, he remained unsatisfied by these innovations and was in constant search of a more fundamental invention. He thus turned to fields beyond his studies in thermodynamics and refrigeration technology, working from 1880 to 1888 on an ammonia-powered steam engine, the 'ammonia engine'.[273] Due to its capacity for a greater difference in temperature, Diesel expected his ammonia engine to be twice as efficient as the ordinary steam engine. However, the operational dangers of working with the highly poisonous ammonia proved too great. In 1886 he wrote of his 'idea for a solar-powered machine'.[274] Diesel established his own refrigeration laboratory in Paris before Linde built his lab in Munich.[275] In contrast to his instructor and employer Linde, Diesel maintained his interests in theoretical issues while working as a refrigeration engineer. He was also better versed in mathematics and loved to play with the fundamentals of thermodynamics. A close follower of the literature on heat theory and refrigeration technology, he extracted from them constantly and corresponded regularly with German and French professors on thermodynamics.[276] It was Diesel who provided Linde with books on refrigeration technology. Whereas Linde concentrated on questions of construction as a refrigeration engineer, Diesel consistently explored the possibilities of application for the mechanical theory of heat to refrigeration technology. The following statement was typical of his approach: 'find a law governing the relationship between pressure and temperature in mixtures of ammonia and glycerine to make it easier to control cycles in ice machines'.[277] In his attempt to come up with an original invention, Diesel went from general notions to the concrete, focusing specifically on unsolved problems. In 1881 he wrote: 'My studies are directed at transforming, as much as possible, the chemical industry into physics – to use the processes of evaporation and crystallization (in other words, simple differences in temperatures) instead of chemicals to generate clarification and purification.'[278] Linde however, had difficulties with his employee's inventiveness and openly rejected Diesel's more radical idea of the ammonia and later diesel engine. Bitterly disappointed by Linde's reactions,[279] Diesel quit after Linde refused to collaborate on an oil engine. In a personal note in his files, Diesel gave expression to his bitterness:

> Although ripe for the picking, Linde refused to discuss my idea with anyone else and didn't lift a finger to help find others who might be interested; he wanted instead to keep the issue quiet and strongly advised me not to give my paper 'Theory and Construction, and So On' to the Privy Councillor Zeuner in Dresden. He repeatedly informed me, both verbally and in writing, that my invention was not compatible with my work at the Linde Company until I finally left because of these hints constantly directed at me and found myself for a moment in the most precarious of situations.[280]

This was certainly an overreaction written in the heat of the moment when quitting his job.[281] In fact, Diesel attracted supporters with Linde's help. Diesel's risky diesel project was probably taken on by the Maschinenfabrik Augsburg precisely because of Linde's support.[282] Upon his resignation, Diesel received a letter from Linde stating 'it goes without saying that he sends his best and warmest wishes'.[283] Linde's lack of trust and 'wait and see' attitude towards this exceptional employee whose inventions lay beyond the bounds of the company's business is understandable. Diesel's behaviour may have raised eyebrows, but was not out of the ordinary: following the establishment in 1898 of his Allgemeinen Gesellschaft für Dieselmotoren, he based his relationship with his employees on Linde's example.

A great degree of independence was in general desired within the Linde Company. The idea of a salaried engineer with far-reaching duties and responsibilities originated from around 1870 and lasted longer at the Linde Company than it did elsewhere in the industry. This professional mode was a result of the fact that the Linde Company was essentially an engineering company without production facilities of its own. It was a mode that survived longer at Linde than in other industrial branches such as electrical or mechanical engineering, where defined working hours, hierarchies and lower wages as well as specialization and the resultant delineation of competencies had already become established.

Above-average wage increases were another means by which the Linde Company sought to establish loyalty amongst its engineers. Starting wages in the 1880s for erecting engineers and chief machinists at Linde were rather low, beginning between 1,200 and 2,200 Marks.[284] On a par with the wages of university assistants in technology,[285] like the company's chief machinists and assembly mechanics,[286] they were of course far above average wages in Germany at the time.[287] The average engineer's income in the Berlin metropolitan area was not much higher,[288] whereas the Maschinenbauanstalt Humboldt (Cologne) hired 11 engineers in 1908 at 3,300 to 5,400 Marks per person.[289]

It was, however, the increase in wages that was decisive. As in other companies, financial rewards were higher for increases in sales than technical achievement. In 1882, Linde was looking to hire a production engineer for the ice works, offering an annual salary of 6,000 Marks. With a salary of 2,400 Marks, the ice works managers earned more than some engineers, the office manager Reichenwallner receiving 3,000 Marks in 1882.[290] Chief engineers earned considerably more: Schipper received 4,800 Marks in 1882, following his job at the Technical University, August Krebs started at the Linde Company with an annual salary of 3,600 Marks, whereas the associate professor Moritz Schröter had to be content in 1885 with 3,180 Marks.[291] Crucial to a chief engineer's earnings was his percentage of corporate profits. Schipper, for example, started in 1885 to receive 2 per cent of net profit for the entire company, which in the first year amounted to an additional 15,200 Marks on top of his base salary.[292] Including their profit-dependent share, August Krebs (Schipper's deputy in Wiesbaden) and Rudolf Diesel (as manager of the Linde Company's Berlin branch) each received 16,800 Marks in 1890. Authorized officers earned approximately half as much; in 1891 Robert Banfield received 10,000 Marks as his annual salary, which included 1 per cent of the profits – similar to that received by other authorized officers such as Scharnberger and Reichenwallner.[293] The percentage of profits for Carl Linde and his authorized officers cost the company more in the first ten years than the total sum of fixed salaries for all employees. The highest salaries went to branch managers abroad. The manager of the English branch, T.B. Lightfoot, earned 40,000 Marks in 1898 plus 20 per cent of the branch net profit, which can be attributed to his business achievements and higher wages in England. In Paris, Rudolf Diesel started in 1883 with 6,000 Francs (4,800 Marks) and a 2.5 per cent profit-sharing bonus; in 1889 he earned 30,000 Francs (24,00 Marks) – considerably more than a well-paying professorship. In 1911, the manager of the Internationalen Sauerstoffgesellschaft (International Oxygen Company), a subsidiary of the Munich branch, received 6,000 Marks in salary and a profit percentage of 10 per cent.[294]

Linde expected his engineers to focus all their energy on the success of their projects. Each project was led by a chief engineer, a principle that effectively structured the business and rendered further internal organization unnecessary. Linde set an example for his engineers by investing an inordinate amount of time and energy in his work despite his weak physical condition. Attempts to emulate him, however, took their toll on others. Unable to cope with the high levels of stress, some engineers were dismissed due to physical or psychological illness, or, in some cases, they died.[295] Nonetheless, the majority remained tied to the Linde Company for years, continuing with their work well into old age.

For Linde, hiring people he met at the Technical University in Munich such as Friedrich Schipper or Rudolf Diesel or those recommended to him by Moritz Schröter or his former instructor Zeuner was an important means of protecting the company's intellectual capital. Other such hirings included the sons of valued employees as well as Linde's own relatives, including his two sons, one of his brother's sons-in-law, two nephews and two sons-in-law. The

Munich graduates appear to have been given preferential treatment, as up until 1930, only two engineers without a degree from the Technical University in Munich were granted commercial power of attorney. Those without an academic background had it even harder than those with degrees from universities other than the Munich Technical University. Only Albert Hoefle, an engineer who started with Linde in 1896 with a degree from an industry school in Augsburg, made it to a top-level position, becoming in 1928 (35 years after his start) the manager of the Wiesbaden engineering company.

Employees who wished to do so were encouraged to publish articles in various commercial, technical, and science journals, which not only facilitated work satisfaction but also helped to promote the company as a leader in both science and technology. Company secrets, however, were not to be revealed in such articles. Linde was interested primarily in those articles which not only presented the latest technology, but provided scientific backing to Linde's ideas. A number of such articles, in which the pros and cons of carbon dioxide vs. ammonia machines were debated or in which a criticised idea was defended, were authored by Linde but published under a different name. He would decide which of his employees was best suited to the article's contents. Breweries' journals were fed articles like this and even employees outside Germany were expected to comply and publish Linde-friendly articles. Linde, however, saw this activity as having a positive effect on his employees: 'an overview of the current theoretical and practical state of affairs in our field is gained, which is of direct use for us. I have learned a lot in the last weeks'.[296]

All articles were carefully edited to prevent new ideas or construction details in drawings from being published. In the interest of confidentiality, engineering diagrams and drawings were passed on to licensors for personal use only. The same was true for their own formularies with diagrams and tables of refrigeration technology, which were considered confidential within the company itself. This kind of secrecy was common for engineering companies that sold knowledge and experience rather than finished products. The willingness to prevent company knowledge from being passed on to outsiders was an essential aspect of the company ethos, more important than perfecting thermodynamics. To prevent any company knowledge whatsoever from landing in the hands of the competition, a clause was added to Linde's employment contracts which prohibited an employee from working elsewhere in the refrigeration industry for the rest of his life upon leaving the Linde Company. Essentially debarring employees from practising their profession elsewhere, this stipulation was later modified, limiting the debarment to three years. Certainly the most effective way to avoid the betrayal of company secrets was to cultivate a quasi-lifelong relationship with the company itself, which was for the most part successful. However, within a few decades this led to a situation in which the company leadership was rather advanced in age. Serving 50 years on either of the boards was not uncommon for the first generation of Linde management. Both Linde and Wucherer continued to fulfil their duties on the Supervisory Board until they were 90 years old. Other examples include Friedrich Schipper in Wiesbaden as well as Richard Linde and Rudolf Wucherer in Munich-Höllriegelskreuth, who all kept their positions in management until well after Reuther, who was sent to Vienna to act as manager of the newly founded Linde-Riedinger after their eightieth birthdays.[297] New duties were also given to men over 70, for example in the Herman Corporation.

Whereas Linde maintained informal contact with his chief engineers and invited them all to his summer residence in Berchtesgaden, with his project engineers he cultivated a collegial atmosphere in which exchanges over their work could flourish. Even young engineers were brought into the detail-oriented correspondence. Linde took an interest in the private lives of his engineers and on occasion even conferred with their wives when difficulties arose. In contrast to a patriarch like Alfred Krupp, he refrained from intervening in their personal lives to help them find happiness. Although Linde liked to present himself in public as an engineer, within the company he preferred to be addressed as 'Honourable Professor', an awkward salutation at best, especially for the self-confident Rudolf Diesel in Paris, who slowly grew accustomed to using it. Among the brewers, Linde was known as 'the Professor'. Linde himself addressed only his closest colleagues with the personal salutation 'Dear Mr. ...'. Usually, he avoided any salutation whatsoever in his business correspondence with the Wiesbaden office (after Department B – Gas Liquefaction – had been built in Munich), giving directions telegram-like. This style of writing, which was essentially an imitation of that used in civil administration, was meant to cultivate a bureaucratic structure in which duties were to be carried out irrespective of personal relations.[298]

From the beginning, it was the engineers who held the best positions at Linde's Ice Machines. They were responsible not only for the technical aspects of the company, but also for numerous other business matters. It should therefore come as no surprise that the company had little need for financial managers in the early days of its business. In 1882 at the Wiesbaden office, there were ten engineers (including Carl Linde) and three draughtsmen, but not one financial manager employed. For a short while, Linde's brother Gottfried took care of the financial side of the business, but soon left to establish his own ice works and cold store.[299] Up until the 1950s, Gottfried Linde's successors as financial manager were all businessmen with a commercial apprenticeship under their belt, but no university background. The first to succeed Gottfried Linde was Johann Reichenwallner, who acted as financial manager until 1915, followed by Gaston Emmerling who held the position until 1928 when he was replaced by Otto Flössel. In 1910, nearly one decade after it had been built, Department B established its own financial department. Peter Eggendorfer was the first financial manager for Department B, a position he held until 1957. (His nephew later became a member of the Executive Board.) Another financial department was established with the takeover of the machine manufacturer Sürth in 1920, which was integrated into the Linde Company as Department C and was managed by Karl Brunke, whose background in finance made him an ideal financial manager. The fact that before the Second World War none of

Linde's business or financial managers was appointed to the Executive Board is illustrative of the higher value Linde placed on his engineers.

As a consequence of this policy in which engineers were expected to fulfil administration duties, even the best engineers were limited in the amount of time they could devote to the technological needs of their projects. Linde complained to Schipper in July 1893 that the company was beginning to lose flexibility: 'We are losing our ability to confront and quickly put an end to strikes because our most important technicians are now caught up in administrative work and a younger generation of the same quality is nowhere in sight.'[300] The approach of providing good, highly competent engineers who cultivated close client contact was on the one hand a great competitive advantage for the Linde Company. On the other hand, the policy was counterproductive insofar as administrative tasks ended up taking too long to be completed. The preference given to technology on all levels also led to the creation of a particular image of the technician in the company. Even those managers lowest on the totem pole with a business background had to be concerned with technical questions and were included by Linde in discussions over design and development.

This primacy of technology over business, which characterized the Linde Company well into the second half of the twentieth century, was not limited to the Wiesbaden headquarters and the Munich department of Gas Liquefaction. At other branches of the Linde Company (such as engineering companies, independent cold stores and oxygen works) that were similar in organization to the parent company, businessmen were seldom hired as managers. This was, however, not the case at Linde's subsidiaries and acquired subsidiary plants, which essentially received the latest technology from Wiesbaden instead of developing their own independently. Consequently, employees with business skills were more highly valued at the subsidiaries. Most of the managers hired for the Gesellschaft für Markt- und Kühlhallen, which was founded in 1890, were businessmen, as was Karl Brunke, the manager of the new Department C, or Maschinenfabrik Sürth. The former owner of the small refrigeration unit manufacturer Walb, Friedrich Walb, was kept and made CEO in 1926, when the company was taken over by the Linde Company.

Things were different too at the Linde British Refrigeration Co. Ltd, which was founded in 1885 as the first Linde-associated company abroad and boasted the self-taught English refrigeration pioneer T.B. Lightfoot as managing director. He pursued a strict regimen to make the British company turn a profit, which was felt by the technical director from Germany, Robert Banfield. Seeking independence for his technical department, Banfield complained to Linde of the unusual authority given to Lightfoot. Linde responded with criticism, explaining to Banfield that Lightfoot was hired for the express purpose of providing the British company with direction. He warned Banfield of the possibility of losing his position in England, should he insist on challenging Lightfoot:

> I am surprised by your failure to see that after years of heavy losses in our company, it has become clear that we must hire a man with all the incipient

qualities of a *businessman*, that is, a man possessing great energy, strong initiative, a man with thorough knowledge of business needs and who is ruthless in his pursuit of the company's best interest. We believe Mr Lightfoot is this man.

Banfield was well-regarded by Linde, who hired him following his exams in 1877 and then sent him abroad as a sign of preferential treatment, but then brought him back to Germany.[301]

Along with the engineers and the business managers, there was also a number of travelling assembly mechanics sent to set up and then put the refrigeration systems into operation. Linde made an effort to have his engineers treat the self-confident assembly mechanics with respect and care. Writing in 1882 about a conflict with assembly mechanics, Linde stated:

> I have repeatedly indicated to our engineers and reminded them that they are to establish a friendly consensual and discussion-oriented approach of tact and respect with you assembly mechanics, whose practical experience and viewpoints are greater than theirs. I hope this will help resolve the problem of having young engineers work with the more experienced.[302]

Linde had to warn even his clients against treating his confident assembly mechanics poorly. Machinists with that kind of experience and knowledge of his machines were not easy to find on the market, nor did Linde want to risk leaking company knowledge by losing his assembly mechanics. In August 1883, Linde wrote to the brewmaster Fenzl in Wrexham (England) in support of his head assembly mechanic Schlaschter, warning that he: 'can get annoyed if he senses that he is being instructed on technical issues while carrying out his job'. Even if Fenzl should, based on his own expertise, have reason to expose a problem with Schlaschter's work, Linde asked that 'the greatest consideration be taken to handle this delicate situation and to avoid stepping on the assembly mechanic's toes'.[303] Similar problems arose later at the Munich Department B branch. Looking back in 1954–55 on his travels between 1907 and 1909, when he brought machines into service, Richard Linde remembers noticing

> that the intelligent assembly mechanics won the trust of the erecting engineers as soon as they noticed that the mechanic had the larger picture in mind. This helped significantly when the erecting engineer's instructions ran up against the assembly mechanic's opinion, but was proven to be right. It was of course often the case when something unexpected occurred that the erecting engineer's explanation thereof impressed the assembly mechanic.[304]

Before moving on, let us look at research in the Linde Company. Until 1890, research and development was conducted entirely on-site at clients' plants or at co-operating machine manufacturers. Linde's experimental labs for new developments were chosen breweries,[305] the test stations at Maschinenfabrik

Augsburg, or the Sulzer Brothers. The company's own ice works and cold stores were often used for these purposes later on.[306] A major change came in 1887 with the establishment of the Munich test laboratory. With the help of Linde personnel and financing, Germany's first research laboratory for refrigeration units was established. It is quite possible that Linde's suggestion to the Polytechnical Association to establish this station was given with the intent of turning it into a test station for the Linde Company and eventually taking it over. In any case, he impatiently drove the Association out after his move to Munich in 1891 and planned to take over the station with his son Fritz, who was working on his physics dissertation in Berlin. When, in 1892, Fritz complained to his father of not having access to the appropriate equipment at the Berlin laboratory, his father placated him by reminding him of their future in Munich: 'we will be more comfortable at the test station'.[307] While at first resisting the notion of taking the station over due to a sense that it was neither affordable nor necessary, the Linde Company's Supervisory Board eventually gave in to Linde's insistence.[308] Along with his work in cryogenic engineering, Linde published articles on theoretical thermodynamics in the 1890s. His successes in gas liquefaction and separation engineering, however, soon put an end to this phase. He handed over the laboratory to the universities in Munich, where research in thermodynamics could be continued, and built a new development-oriented test station to improve the process of gas liquefaction and separation. Publicly, he presented this development laboratory as an institute for fundamental scientific research.[309] However, soon the gas separators were so large, that Linde had to move his work back to the breweries and brought his machines with him to be tested on-site. The calculations department grew in importance because the decision to build new types of machines was based on their performance estimates. American engineers visiting the Linde Company in Höllriegelskreuth after the Second World War were quite surprised to find no research institute for pilot plants in this innovative company. Until the acquisition of the machine factories Sürth and Walb, the Wiesbaden division of the Linde Company had very few of its own test stands. Not until the beginning of the Fifties were all of the research and development facilities in Mainz, Cologne, and Wiesbaden brought together under one roof in Sürth to form one central research institute. For the first time, laboratories for physics, chemistry, measurements, and material sciences were combined with test stands to research and develop the machines. In addition, there was also a library, archive, and patent office.[310]

Collaborating with licensees and subsidiaries

Linde himself and later the Linde Company left the assembly and execution of refrigeration equipment to major machine manufacturers both within Germany and elsewhere.[311] The licencing system allowed for a rapid increase in sales numbers for Linde's refrigeration units without much of his own capital being invested. Generally, contracts stipulated a 15 per cent commission for the Linde Company as well as Linde himself retaining the right of sale, which was more important to Linde than the commission rate.[312] In return, the Linde Company agreed to grant the licence as well as to deliver complete, detailed illustrations of its machines and models for the metal castings. In addition, he guaranteed regular information regarding improvements to the machines and practical application experience. For Linde, the value of patent protection was relatively less important than that for which there was no patent: practical experience and tips.[313] At the beginning of his collaboration with Maschinenfabrik Augsburg (MA) and the Sulzer Brothers in Winterthur, he agreed that MA would build his machines for Germany and the Sulzer Brothers for the rest of the world. Indeed, Linde turned down licencing requests from other German machine manufacturers.[314]

Much to the dismay of the Sulzer Brothers, Linde did, however, continuously seek further licensees abroad to build his machines. He was forced to do so by national patent laws, which would grant a patent on the condition that the patented machine be built within one year inside the given national borders. Linde also recognized, however, that decisive market success was dependent upon him having strong partners in each country. Linde's international success was the result of a consistent international orientation despite major setbacks and financial losses. During 1879, Linde spent 3,294.50 Marks in fees on his patents in Germany, Austria, England, France, Belgium, Italy, and the USA[315] and had won over a number of important business partners before the company was officially founded in 1879. His international orientation is reflected in his business correspondence: over 40 per cent of all his letters were sent abroad. Among the letters sent to his employees, the percentage sent abroad was over 60 per cent. In the early days of founding his business, Linde consistently sought international contracts and preferred to hire employees with international experience and who were at least bilingual, including the french-speaking Rudolf Diesel and the Englishman Robert Banfield. Of the letters 258 went to England,[316] 144 to Switzerland,[317] 138 to Austria (with its 1914 borders), 132 to France, 85 to Belgium, 77 to the Netherlands, 52 to the USA, 47 to Italy and almost 100 to other countries. Let us turn now to individual licence holders in foreign countries.

In France, Linde sought co-operation with the French manufacturers of the Carré machine, Mignon & Rouart[318] and Armengaud & Fils.[319] However, upon reviewing the designs, both companies turned down Linde's offer. Through the Sulzer brothers, Linde established contact with the Lyon firm Satre & Averly,[320] which had a licence to produce Sulzer ventilated steam machines. Although co-operation between the two ran smoothly, shortly after the Linde Company was founded Moritz von Hirsch took over the Linde patent for France and founded the Société Pour La Production de Glace et D'air Froid D'après le Systeme Linde.[321] Employees of this company included Diesel, Sauvan, and Robert Banfield (albeit for a short time).[322] Business was never particularly good, which Linde blamed on the French office's lack of technical orientation. In 1890, Linde bought his principal shareholder's licence rights and began working with the French company CAIL, which apparently enjoyed excellent relations with government and bureaucratic circles which was conducive to increasing sales of Linde machines. Linde's representative in Paris, Desvignes (Diesel's successor as of 1889), moved the subsidiary office before 1896 to CAIL property, which rapidly eased relations with the Linde Company.[323]

Work with the brewery facilities manufacturer Robert Morton began in the brewer metropolis of Burton.[324] However, Morton switched to a competitor after a short while and Linde signed a licencing contract with Thomas Middleton & Co. in London.[325] But even Middleton forewent manufacturing the machines, choosing instead to distribute equipment sent from Germany.[326] The Linde Company sold its English patent in 1881 to the Austro-Bavarian Lagerbeer Brewery,[327] which provided customers and collected a high 15 per cent from the Linde Company for distribution in England. Having gained much in the way of self-confidence in the meantime, Linde planned together with his Dutch friend Feltmann to found an English company without the involvement of the Linde Company. However, the Supervisory Board of the Linde Company demanded to see Linde's plans in March 1883, shortly before the English company was officially to come into being, and insisted that the Linde Company maintain overall control.[328] Pulling back, Linde began negotiations with Atlas Engine Works in Birmingham, which eventually led to a communal commercial enterprise:[329] the founding of the Linde British Refrigeration Co. in London with members of the Austro Bavarian Lagerbeer Brewery, Atlas Engine Works, and the Linde Company.[330] The new company immediately built two large cold stores at the London harbour, but it was the English refrigeration pioneer T.B. Lightfoot who brought the company out of the red after being appointed head in 1889.[331] Whereas Linde hindered Lightfoot's attempts to increase the intake of English capital, he was able to successfully increase Lightfoot's capital and profit share, despite resistance to this in Germany.[332] Starting in 1892, Lightfoot thus oversaw the construction and assembly of Linde machines in England, albeit with only average success.[333] During the First World War, Lightfoot eventually took over the subsidiary, renaming it Lightfoot Refrigeration Co., and never resuming its business relationship with Linde after the war.[334]

In Belgium, it was Carels Frères in Ghent that managed the realization of Linde machine patents. Linde probably came into contact with Carels via the Sulzer Brothers, whose steam engine Carels had the licence to manufacture.[335] Although their relationship got off to a harmonious start, problems arose once they began working on concrete projects. Perhaps the Sulzer Brothers' ill temper played a role in the demise of Linde's business relationship with Carels. In 1886 the Linde Company founded, together with a few Dutch and Belgian business friends, a cold store business in Antwerp, the Société Anonyme des Frigorifères d'Anvers, and used this in the future as their only 'base for the important distribution business to Belgium and Holland'.[336]

Licenced production stabilized in the Austro-Hungarian empire after 1881. In 1876, Linde issued a right of licence to the Prague machine manufacturer Noback & Fritze.[337] The company was interested in licence rights to persuade Linde to agree upon licensing with the Viennese machine factory G. Sigl for Austria-Hungary.[338] However, Sigl failed to manufacture a machine before the one-year patent limit was up, forcing the Linde Company to smuggle one to Austria.[339] In 1881 Carl Heimpel went to Vienna as an independent representative of the Linde Company. After 1890, a machine factory in Smichow near Prague and the united machine factories near Skoda in Pilsen began to produce Linde Machines.[340] This was followed by the Brünn-Königsfeld machine factory in Brünn and the Wiener Maschinenfabrik AG, formerly Tanner, Laetsch & Co.[341] After the First World War, the machine manufacturers Nicholson and Schlick also began to build Linde machines. Competition amongst Austrian manufacturers placing similarly constructed machines on the market was softened by the Linde Company's near cartel-like division of the market since 1913.[342]

By 1878 Linde was already seeking contacts in the USA, trying without success to win over the Pictét engineer Bürgin.[343] Then in 1879 Linde was approached by the German-speaking American brewery equipment manufacturer Fred Wolf from Chicago, who wanted licensing for America. The co-operation between the two was from the beginning a stellar success. At first, Wolf imported refrigeration equipment and steam engines from the Sulzer Brothers, starting later in the middle of the 1880s to manufacture his own refrigeration equipment.

Despite numerous failed collaborations and countless problems, the Linde Company was nonetheless able to secure strong partners in all the important foreign markets by eagerly searching out licence holders throughout the 1880s. Linde's most important partners in refrigeration machine manufacturing remained however, the Maschinenfabrik Augsburg and the Sulzer Brothers in Winterthur.[344] Both companies had already branched out into other areas, competing against each other, particularly in steam engine manufacturing. By working with both, Linde was able to obtain the manufacturing engineering know-how of both companies, playing each against the other so as to spur higher performance levels.[345] Moving back and forth between modes of co-operation, interlacing and control, the Linde Company's interaction with the two companies created what came to be known from the outside as the 'Linde Block', as all three were focused on seeing the Linde machines succeed. According to all sources, it was the quality and reliability of the Linde machine's mechanical details which led to their market success.[346] Certainly, the Wiesbaden engineering office played a role, but it was primarily the two manufacturers' factories which contributed to this success.

Linde had his very first refrigeration unit built in 1873 at Maschinenfabrik Augsburg, which was the leading machine manufacturer in Southern Germany. The business friendship between the director, Heinrich von Buz, and the expert Linde (whom Buz already knew beforehand) grew to last over 50 years. Buz sat on the Supervisory Board for 39 years and Linde often relied upon him in problematic situations.[347] The friendship between the two cultivated a friendly relationship between the respective businesses that survives today. Both Buz and the chairman of the Maschinenfabrik Augsburg Supervisory Board Max Schwarz joined in 1891 the Supervisory Board of the Gesellschaft für Markt- und Kühlhallen. Heinrich Buz's successor, his son Richard, and later Otto Meyer, both spent decades on the Supervisory Board at Linde.

Buz already recognized by 1873 the huge potential market for refrigeration machines and the opportunity to push factories to their fullest capacity. Enthusiastic about refrigeration, he travelled with Linde to England in 1876 to promote his machines at breweries there and became a shareholder of Linde's Ice Machines (15 per cent) in 1879. Buz saw to it that the best engineers from Maschinenfabrik Augsburg were put in charge of the refrigeration machines: as of 1873 this was engineer A. Krumper,[348] from 1878 to 1880, Linde's former assistant Friedrich Schipper and as of 1880 Buz's son-in-law Lucian Vogel.[349] Krumper and Vogel later became directors of MA, Schipper the CEO of Linde's Ice Machines. In 1885 and for a few years following, the refrigeration machine business comprised over 50 per cent of total turnover, making it MA's most important production branch. This decreased again after 1890; so too did the relative importance of the ice business.[350]

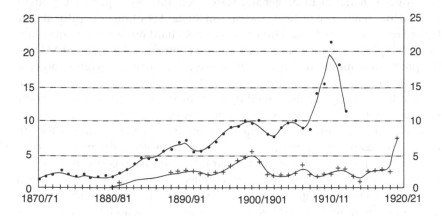

Figure 1.8 Sales of the Maschinen-fabrik Augsburg (MA) and its ice department (after 1899: MAN's Augsburg plant)

In 1899, Maschinenfabrik Augsburg merged with a former machine manufacturer Cramer-Klett in Nuremberg to become MAN. The Nuremberg factory had been building some refrigeration machines since 1882, but dissolved its ice department in 1903. Within its new group, the factory quickly became the leading business and grew faster than the Augsburg factory. Consequently, the relative importance of refrigeration machines at the Augsburg factory decreased. In 1912, MAN had 100 million Marks in turnover, of which 11 million was made in Augsburg, 2 million coming from refrigeration machines. From Anton Rieppels' point of view in Nuremberg, this last amount was negligible; for Heinrich and Richard Buz in Augsburg, it comprised a valuable portion of turnover.

The Linde Company insisted firmly upon its right to represent itself – instead of sending the licensee – to the clients. Abandoning this limit put upon the manufacturer would have spelled financial ruin for the Linde Company.

Exceptions were made in the case of regular, long-term customers or in regional cases, but only when discussed and agreed upon beforehand. After a patent had expired, the Linde Company also began to ease up on the fixed premium it received from the licensee, which was determined by turnover.[351] The licence agreement of 1876 was renewed in 1895, 1900, and 1910.[352] In 1915 the Linde Company's fixed premium dropped from 15 to 7 per cent and in 1920 this contract was renewed for another ten years, with the last fixed premiums falling in 1927.

Up until 1900, MA built 40 per cent of all the Linde refrigeration machines in the world, even though this percentage was dropping regularly. In 1890 they had to cede 20 per cent of the German market to the Sächsische Maschinenfabrik, by the end of the 1890s a few percentages had been handed over to Nuremberg and in 1912 the Atlas factories in Bremen. In 1924 and 1925 respectively, both the Broncewarenfabrik Riedinger and Maschinenfabrik Esslingen were brought in.[353] Heinrich Buz died in 1918, and his son Richard continued to cultivate the trusted relationship with the Linde Company. Richard Buz wanted to maintain the developing co-operation between the two companies, but felt that MA was not prepared to move into the growing markets for small refrigeration units or domestic refrigerators, and limited his work to that of traditional refrigeration. The Linde Company was itself building up its own production capacity for the household refrigerator industry. Relations between the two companies began to deteriorate when the Linde Company bought the Sürth machine factory. In the end it was the group policies of Gutehoffnungshütte, the new holder of MAN, which put the ultimate strain upon the relationship between the Linde Company and MAN. Director Paul Reusch wanted to consolidate the refrigeration manufacturing of three group factories – MAN, the Esslingen, and Riedinger factories – and move it to Esslingen, a goal towards which Richard Buz slowly had to acquiesce.[354] L.A. Riedinger and MAN's refrigeration machine businesses were combined in 1924, leaving Riedinger as the exclusive manufacturer of large Linde refrigeration systems after 1925. In April 1927, presumably against the will of the Linde Company, a merged MAN-Riedinger sold their production know-how to the Soviet Union.[355] A short few months later they stopped working with refrigeration machines, moved their manufacturing thereof to Esslingen and put an end to their long history with the Linde Company. Linde, for his part, had purposefully strengthened his basis for negotiations with Gutehoffnungshütte (GHH) by building up his own production capacity.[356] Having struck a new co-operation contract with the Esslingen factory, Linde was able to retain some of the more important aspects of his work with MAN, although the new contract did not stipulate a premium. The Esslingen factory, for its part, wanted no engineering support whatsoever. It was agreed that the Linde Company would buy all of its compressors and two-thirds of its equipment from ME, while ME agreed to produce exclusively for Linde.[357] This allowed the Linde Company to continue its absolute control over client relations and maintain authority over particular projects.

Exactly what was the nature of the technical co-operation with MA? Up until 1886, ice machines at MA fell under 'general machine construction', while some engineers in the engineering department had of course specialized in refrigeration machines since the end of the 1870s. Since 1886 there had been a separate department for ice; they had their own engineering department and as of the late 1890s their own test stand. They also were granted their own staff of assembly mechanics for work on the road[358] and were given an erecting hall. Part of the manufacturing of components was cordoned off in a small section of the erecting hall[359] but remained primarily in the department of general machine construction. From 1894–97 the ice department was given its own pipe production facilities to be used in the construction of atmospheric condensers. After the turn of the century, modernization of the ice department was discontinued as it contributed relatively minimally to MAN's turnover. MAN engineer F. Möller remembers 'an impractical, high building' to be used for pre-assembly that was already obsolete before the First World War.[360] The department of ice had a staff of 26 for its engineering department and test stand in 1920. There was also a calculations department of 16 and an administrative office of 5 people. The plans for new refrigeration development continued to lie solely in the hands of the Linde Company; the engineering department was to take these designs and 'update them for the production of workshops and storage buildings'.[361] After the First World War, Möller felt 'that the Linde Company essentially dictated the manufacture of refrigeration machines in Augsburg. A trusted Linde employee was constantly at the refrigeration department.' According to Möller, the ice department supervisors moved not to Esslingen, but to Wiesbaden with Linde after the Riedinger company and Maschinenfabrik Esslingen merger of 1924–27.[362]

Möller, a self-confident MAN engineer, viewed the dependence of engineers at Linde upon the company as unusual. For the Linde Company however, this kind of obedience was par for the course and therefore extended to their licensed factories. It was precisely this culture of satellization which led Carl Linde to send his nephew Wilhelm Heyder to the Sulzer Brothers in Winterthur. Friedrich Schipper was sent to work at Maschinenfabrik Augsburg from 1878–80,[363] and a former Linde engineer (who later sat on the Linde Board of Management) Otto Hippenmeyer directed the ice department from 1913–26.[364] Linde's former personal assistant Robert Banfield was sent in 1885 to the newly founded Linde British Refrigeration Co. to act as technical director. He was assisted by the engineer Weinberger and the manager Busch as his correspondence clerk. The engineer Götzendorfer was sent by Linde in the same year to an independent branch operation in Vienna owned by Karl Heimpel.[365] Linde sent a number of engineers to the USA to help Fred Wolf's production in Chicago. This concept of satellization, which had brought Linde so much success, was applied as well to Department B, Gas Liquefaction. The former Linde engineer A. Scholler sent confidential reports regularly from Piesteritz, the Bavarian nitrogen plant, to Fritz Linde on the Management Board. Similarly, Philipp Borchardt kept the

Linde Company up to date on in-house developments on ammonia production at the BASF facilities in Leuna and Oppau.[366]

Certainly, this kind of internal personnel communication reflected a desire to regulate and control while preventing important know-how from falling into the wrong hands. The pursuit of close working relationships with other engineering businesses was to be avoided, or at least made known as early as possible.

The Sulzer Brothers in Winterthur were also leaders in the manufacture of steam engines.[367] Furthermore, they built heating facilities – another important component of refrigeration machines. Linde initiated their co-operation in 1876 and by 1877 the first refrigeration machines were delivered, with a total of 2,500 by 1934.[368] The Sulzer Brothers was an export-oriented business with strong international contacts, from which Linde profited. Sulzer provided local licensed manufacturers Carels Frères in Belgium as well as Satry & Averly in Lyon, who were also important for patent registration in both countries.[369] The Sulzer Brothers were also active in establishing new contacts abroad through the export of refrigeration machines, seeking to break into new international markets such as the South American market for steam and diesel engines.[370] Refrigeration machine manufacturing reached its greatest relative production share around 1890.[371] Their relationship with the Linde Company was strained by the increase in foreign licensing, which had tightened the market for Sulzer. There were several occasions after 1893 where Linde himself was forced to clear the air by speaking one-on-one with Sulzer's CEO Sulzer-Steiner.[372] In 1912, Sulzer Brothers reduced their collaboration with Linde to an agreement on delivery quotas without commission and ultimately terminated their contract in 1917.[373] Whereas the co-operation between Linde and Sulzer Brothers ended earlier than that with Maschinenfabrik Augsburg, the innovative power of the Sulzer Brothers made for a considerably more intense working relationship with Linde in technology, at least up until the turn of the century. Also in contrast to MAN, the Sulzer Brothers continued in the 1920s with the refrigeration industry, increasing their capacities in both process cooling and household refrigerators.

Similar to the organization of MA, refrigeration engineering at Sulzer was for a long time a subdivision of the general mechanical engineering department. Directed by the innovative engineer Konrad Hirzel-Gysi until 1897, the engineering office of the refrigeration department was made independent before the turn of the century, while production remained in this department.[374]

Linde discussed technical issues concerning things such as compressor construction, refrigeration in fermentation cellars and frozen meat facilities for South America more frequently with Sulzer than with MA.[375] Linde fast became the most important supplier of refrigeration units for South America[376] and in 1886 Linde and the Sulzer Brothers began to tackle the building of marine refrigeration units.[377] After their contract with Linde ran out, Sulzer moved with success into the then-blossoming air conditioning industry, which complemented their historical experience in heating.[378] The developing technology of air liquefaction was another high-tech field in which Linde consulted the innovative Sulzer Brothers instead of MAN in 1894–95.[379]

Despite the fact that Linde cultivated a closer relationship with Sulzer Brothers than MA, the Linde Company did more business with MA than Sulzer Brothers. As exciting as the developments hailing from Winterthur might have been, they failed to materialize because the rather slow and ponderous Otto Henkel replaced the lively and quick-thinking Hirzel-Gysi as chief engineer. Sulzer Brothers cut their ties to Linde in 1912, let Otto Henkel go and hired new leadership.[380]

The personnel change proved to be a failure and Refrigeration Technology, in comparison to other production departments, remained a weak spot in the company's development. In terms of production, Sulzer responded only half-heartedly to the new department director Junod's requests to have his own production facilities for refrigeration engineering. Refrigeration thus remained a subdivision of general machine construction.[381] The financial crisis of 1929 put an end to the heavy financial investment made in producing household refrigerators. In this year, 12 years after the co-operation contract with Linde had run out, the new department head Rudolfo Landoldt complained to CEO Hans Sulzer of the years of technological dependence upon Linde. According to Landoldt, as a consequence, he was confronted with 'un personel sans initiatives'.[382] He described the chief of acquisitions Junod as 'sans orientation', the technical designer Hoppeler he considered useful only for obsolete horizontal compressors, and referred to Schlechtlin as 'est mediocre'.[383] Landoldt succeeded nonetheless in leading the department to new technical and economic triumphs.[384]

Considering the resounding success of this strategy of licensing, Linde was rather hesitant to begin founding subsidiaries. With no contracts for refrigeration machines in sight after the company had been founded in 1879, Linde funded the building of a number of ice works. He was, however, not interested in throwing himself fully into the ice and cold storage business, particularly at his own financial risk. The Linde Company opened a total of four ice works by 1882.[385] In the Linde Company's anniversary publication of 1929, Friedrich Linde wrote 'the company built the ice works primarily to demonstrate the company's potential, not to run them forever as businesses per se'.[386] Nonetheless, Linde's brothers Gottfried (1882) and Siegmund (1883) took over and successfully ran ice works in Cologne, Toulouse, and Barcelona.[387] When Gottfried sought Linde Company investment to increase his ice works in Cologne, Carl responded with 'Our company can invest only where and when we can open up a new area, whether this be in qualitative or geographical terms.'[388] By 1890, the Linde Company had sold off all of its ice works.[389] It was not until after 1896 that Linde decided – prompted perhaps by the drop in machine sales or by the desire to establish reserves – to establish massive cold stores and ice works in Nuremberg (1896), Leipzig (1901), Dresden (1910), Königsberg (1914), and Magdeburg (1937).[390]

In contrast to this hands-off treatment of ice works in Germany, the Linde Company was rather active in ice works abroad, seeing this as a means of breaking into the local market. Generally speaking however, the Linde Company often sold its shares once refrigeration machines had been sold to a given ice works. Linde considered these ice works, such as the Ice House Corporation in Antwerp,[391] to be starting points and an effective means of introducing their machines to a regional market, as can be seen in the plans for ice works in places like Bombay,[392] Bordeaux,[393] Pesth,[394] Florence,[395] Livorno, Venice,[396] and Saloniki.[397] Trying to keep a lead on Pictét, the Linde Company invested in a major ice works in Milan[398] and then in a crystal ice factory, H. Ritter & Co. in Trieste in 1894.[399] From 1900 to 1910, Linde worked with Erste Österreichische AG für öffentliche Lagerhäuser (the First Austrian Corporation for Public Warehouses),[400] while in England, Linde's subsidiary ploughed steadily ahead with its own ice works and cold stores.[401] During the First World War, the Linde Company sold off a series of investments at a loss, while others were expropriated.[402]

At the suggestion of the chief architect Georg Osthoff, in 1890 Linde founded the Gesellschaft für Markt- und Kühlhallen. Starting with a share capital of 1 million Marks, this increased to 7.5 million by the First World War and to 8.5 million by 1930.[403] Linde's trusted friend Max Schwarz was on the Supervisory Boards of both the Gesellschaft für Markt und Kühlhallen and MA.[404] Other Supervisory Board members for the Gesellschaft für Markt und Kühlhallen included: Heinrich von Buz (general director for MA), G. Proebst (Löwenbräu) and J. Hallbauer (Eisenwerk Lauchhammer), with Linde himself acting as CEO. However, the great euphoria over developing English–American cold store contacts was short-lived.[405] Georg Osthoff proved to be unreliable and was dismissed in 1890. His successors, the engineers Diesel and Sorge followed by the businessmen Krüger and Klint, lacked strategic and sales-oriented thinking. The same held true for the Supervisory Board. Furthermore, Carl Linde himself did not want to invest the money, energy, and time needed to turn his cold stores into a success story. He therefore started in 1895 to gradually reduce the Linde Company's shares in the Gesellschaft für Markt- und Kühlhallen until they were down to 6.6 per cent by 1930. Consequently, they limited themselves to model factories for large facilities and were no longer engaged in increasing the number of cold stores.

Linde and the Linde Company founded new companies primarily as a means of creating new markets for refrigeration machines. This goal lay at the heart of the motivation in developing numerous applications for refrigeration that originated with Linde's engineers: refrigeration of fermentation cellars (Dreher, 1877), beet sugar extraction using the strontium process (Waghäusel, 1879), milk refrigeration at the Schaik dairy (Delft, 1880), chocolate production at Stollwerck (Köln, 1880), aniline production at Meister, Lucius & Co. (Höchst, 1880), a skating rink (Frankfurt 1882), stearin refrigeration (Brüssel 1883), slaughterhouse refrigeration (Wiesbaden 1884), alkaline crystallization (Aussig, 1884), benzene extraction (Sheffield, 1884), paraffin extraction (Pechelbronn, 1885), freezing meat (Argentinien, 1886), marine refrigeration (White Star Line, 1888), carbon dioxide production at the Guinness brewery (Dublin, 1894), liquefaction of chlorine (1895) and a process to freeze asparagus (1903).[406] The Linde Company very rarely went beyond delivering the machines, essentially disregarding the rapidly growing markets. It saw its role in responding to the concrete needs of various industrial branches by providing tailor-made machines for specific purposes. Each new application was documented immediately in a new brochure. Its high-profile, more flashy brochures that were published by the bibliographical publishing house Eckstein in Berlin offered illustrations of similar-looking machines for use in a wide range of applications. In so doing, the company graphically instructed potential clients in the diversity of the machines' applications. Clearly adept at the art of technical applications, the Linde Company was, however, less adept at creating long-term strategies for stimulating demand. It wasn't until the oxygen industry began to develop at the start of the century that Linde's Ice Machines decided to use its machines in the creation of nitrogen, oxygen, and acetyline factories as a means of increasing demand.[407] When it came to marketing his refrigeration machines and developing new markets for them, the entrepreneur-engineer Carl Linde proved to be creative and tenacious. Nonetheless, he failed to gain footing in the American market. This was in large part due to the fact that Linde restricted his commercial activity to technological improvements rather than convincing consumers to use frozen goods in the USA.

These considerations provide insight into the pivotal change in Linde's business strategy that occurred with the acquisition of machine manufacturers such as Sürth, Walb, and Güldner. It was these acquisitions which turned what was essentially an engineering office into a corporate group of companies producing machines. The company's corporate identity was so powerful that further acquisitions were viewed as extensions of the engineering headquarters. This corporate identity also played a role in the success of foreign licensing and Linde's co-operation with his licensees. This identity was maintained primarily by the Munich plant construction engineering division and it was not until the company's structural reform of the 1960s that this began to change.

Founded in 1871, the machine manufacturer Sürth[408] built facilities for the production and liquefaction of carbon dioxide for the Linde Company from the 1890s and was also their competitor in the chiller industry.[409] Bankrupt by 1908, Sürth was bought by its earlier management and continued production, albeit at a much lower rate.[410] In 1916, the company was bought by a strong Linde competitor in the gas separation market, Deutsche Oxydric AG, which was eventually acquired in 1920 by the Linde Company.[411] The machine manufacturer Sürth thus came under the Linde Company umbrella in a rather casual and unintended manner. By purchasing DOAG, the Linde Company essentially aimed to buy out its major competitor in the gas separation branch rather than increase production capacity. When bought, Sürth had a staff of 530, only 10 of whom were engineers – certainly a radically different profile from that of the Linde Company. Sürth produced gas separators for DOAG, carbon dioxide facilities, (marine) refrigeration units, and compressors.[412] At first, Linde considered Sürth to be of little use as he had slated Augsburg and Esslingen for the manufacture of compressors. In order to keep its head above water, Sürth worked laboriously in repairing locomotives.[413] Finally, in 1924, the Linde Company commissioned Sürth to build high-pressure compressors for Linde's air liquefactors.

Major societal changes brought about the decision at the Wiesbaden headquarters to strengthen their presence in the market for small refrigeration units and domestic refrigerators. Because small refrigerator projects 'could be realized only modestly within the current engineers' work load' they decided to create a new and separate refrigerator department.[414] Within the same year, the Sürth manufacturer introduced the 'Rheinland' refrigerator as a standard affordable product to the market. However, sales failed for years to reach the heights hoped for. Reporting in 1926 to Höllriegelskreuth, Schipper complains: 'Sürth is a Danaide that swallows everything in sight.'[415] Wiesbaden headquarters tried to reform the inefficient structure at Sürth – which was not even capable of series production – by sending engineers to help. In 1925 Sürth began to produce compressors for use in refrigerators. Having little experience in the field of domestic refrigeration equipment, the Linde Company sought to expand its know-how in this field and in 1926 acquired a modest refrigerator factory that had been manufacturing mechanical refrigerators since 1910, the G.H. Walb & Co. in Mainz-Kostheim.[416] The owners of this company founded in 1876 were craftsmen: founder Georg Hermann Walb was a coppersmith by trade; his son Friedrich Walb never completed an official training or education. The company manufactured spray coolers, wine cellar appliances, cognac distillation equipment and, as of 1881, they manufactured refrigerators for butchers and hotels.[417] Starting in 1910, Walb registered a series of patents for refrigerator insulation and had become the representative in Southern Germany for Sürth's compressors to be used in small refrigeration units (Gotha model).[418] Following the First World War, he began to buy compressors from other manufacturers.[419] This distribution network for their refrigerators throughout all of Germany was one of the company's assets. After having acquired the Walb company, Linde sold off its antiquated factory and bought an empty factory in Mainz that was to be brought into line with modern production methods. Despite their purposeful attempts at breaking into the refrigerator industry, the Linde Company failed to make much headway in this field until the mid-1930s. Although the Walb company was competent at building solid refrigerators, it lacked the proper know-how to mass produce its products. Not until the 1960s did the Linde Company begin to see success in household refrigeration products, which historically was not one of its competencies.

Between 1925 and 1929 the Linde Company bought the Güldner engine factory in Aschaffenburg. Already prepared to help his friend Hugo Güldner, Linde was essentially compelled, as chair of the Supervisory Board, to do so once the company ran the risk of going under. The motivation to do so was deeply connected with the Linde Company's goal of having their own engine production facilities as a bargaining tool in their negotiations with the Gutehoffnungshütte (Riedinger, MAN, Esslingen).[420] In the long run, this buyout proved to be one of the best decisions made by the Linde Company. Beginning with diesel engines, the engineers at Aschaffenburg used this technology to develop tractor engines and forklifts. Since the 1970s they have not only seen the largest yields within the group, but have made up its largest division.[421]

One of the secrets to the Linde Company's success has been its co-operation with its business partners. These business relationships enabled Linde to rise quickly as a market leader in refrigeration units in Europe without much in the way of equity capital. Modern from the start, the Linde Company was organized as a corporation and therefore able to increase its capital resources according to demand. Its structure as an engineering company without its own production allowed the Linde Company to expand quickly, continuing to do so during the recession of the 1880s. Continuity in its personnel also contributed to maintaining solidarity within the company.

The Linde Company's well-paid engineers worked in a manner similar to their boss: they exhibited independent thinking in their design, development, and project planning and maintained their interest in theory, publishing frequently in numerous journals. Research was conducted not in a separate department, but on-site and in co-operation with the clients and company leadership, which continued to be dominated by trained engineers rather than businessmen. Consequently, technically less interesting businesses such as ice manufacturers and cold stores failed to catch the eye of the company leadership, which chose instead to focus on innovative applications for refrigeration in the chemical and foodstuffs industries. Once the engineering problems had been solved in refrigeration technology, Linde lost his interest and focused his research interests in the new field of gas liquefaction and separation engineering.

2
Linde Enters into Liquefaction and Separation of Gases

The founding of the gas liquefaction department

Laden as he was with the move to Munich in 1890 and with his duties as chairman of the board of directors in Wiesbaden, Linde shifted his engineering interests from refrigeration to cryogenic engineering. A new field for him, cryogenic engineering was technologically less discovered than refrigeration engineering at that time. Originally, the company was to finance small-scale research for the next five or so years. However, after 1895, when Linde began to make pioneering discoveries in cryogenic engineering, it became clear not only that the company would delve further into this field, but also that technology development would become a core competence. Indeed, competition for Linde was to evolve less around price than technological competence. This department of gas liquefaction and separation in Munich grew increasingly following the turn of the century. By 1920, it had become in many ways more important than the older Department A for refrigeration machines in Wiesbaden. This chapter recounts the founding of this area of operations, which, up until the end of the 1970s, constituted the core and identity of the company. Organizationally, this department, as an engineering department for technological development, was similar in some ways to Department A of the 1880s in Wiesbaden. The following describes the foundation, expansion and inner structure of the department up to 1930.

In 1890 Carl Linde moved from Wiesbaden to his summerhouse in Berchtesgaden. This was followed in 1892 by a move to the laboratory for refrigeration experiments in Munich, which became the institutional research centre of the company. The successful development of gas liquefaction techniques in 1895 led to the establishment of an independent department of the Linde Company, the so-called Section B.[1] Linde handed the old experimental station in Nymphenburger Street over to the Technical Univeristy from 1902–12 to establish a laboratory for technical physics (the Southern German representative of Wiesbaden's Department A was later housed there) and transferred the relatively new Department B to new facilities in the forest south of Munich on the Isartal highway, towards Höllriegelskreuth. This location provided great opportunities for expansion, which remained unexploited until recently. The Munich department was financially dependent upon central headquarters in Wiesbaden until around 1905, when it began to contribute increasingly to the group's profit. Since the most important employees of the Linde Company, particularly Linde's sons, worked in Munich, the leadership of the entire company group progressively moved to Munich. In 1924, Friedrich Linde took over as chairman of the board of directors of the company. The entry into the field of gas separation engineering and the founding of Department B were responses to increasingly tough economic competition in refrigeration technology and Linde's own shrinking attraction to the perspective of technology development. Linde arrived in Munich in 1891–92 not as a retiree, but as a man with scientific and developing entrepreneurial ambitions. The experimental station itself and the aforementioned negotiations on the building of an aircraft factory point to Linde's desire to come up with another technologically novel product. That he ended up with gas liquefaction and fared so successfully in doing so is in many ways due to lucky coincidences – the product could have been different. Gas liquefaction was naturally related to earlier endeavours and was therefore an ideal area to be developed, which even today allows the company to live on as a development-oriented engineering company. Linde's son Fritz finished his doctoral thesis in 1895 on the electric resistance of liquefied gases – a topic he certainly agreed upon with his father – before he went to the experimental laboratory.[2]

Linde's experimental work with carbon dioxide refrigeration machines put him on the path to gas liquefaction.[3] In 1892 the Guinness brewery in Dublin ordered a carbon dioxide liquefaction installation. Linde accepted the contract, although the company had yet to construct such installations, and this unleashed his interest in the area of industrial gas liquefaction. In the following year he held a lecture on the subject, in which he discussed the technical liquefaction of oxygen.[4] Linde had ordered a high pressure compressor from the Machinenfabrik Augsburg for the carbon dioxide machines.[5] In the summer of 1894 he considered constructing an air liquefaction installation, which he called an 'oxygen machine' in his letters, indicating the industrial use he had in mind.

Linde's two most important technical ideas were continuous cooling in a countercurrent and the use of internal work for cooling.[6] In contrast to the usual compression refrigeration process, Linde's idea was to have gas cooled through expansion pass through a countercurrent generator. The resulting circular process stabilized different temperature levels on two sides of the countercurrent generator: a warm and cold side. On the warm side (surrounding temperature) cooling takes place from the outside and therefore a continuous drop in temperature on the cold side becomes possible.[7] Linde wanted to achieve a drop in air temperature through the reduction of pressure under external work. His idea was to use a ram cylinder[8] similar to those used in air refrigeration machines.

In order to cut costs, Linde planned to reconstruct a marine refrigeration machine for his 'oxygen machine'. He wanted to replace the steam cylinder with a ram cylinder so 'the marine refrigeration machine remains usable'.[9] Yet before the work cylinder ordered from Sulzer arrived in Munich,[10] Linde discovered 'that alongside the equivalent of the external work, significant internal work can be applied in the refrigeration process, as soon as the temperature at the lower level of the countercurrent generator approaches the lower critical temperature'.[11]

Linde's discovery here was essentially a development of an effect already described by the English physicists Joule and Thomson in 1857 in which the reduction of pressure in real gases (as opposed to the ideal gases used in thermodynamics), could result in a cooling process even in the absence of external work below a particular temperature (the inversion temperature). The Joule-Thomson effect had become a standard in physics lectures,[12] but no one had yet considered its practical exploitation.[13] According to Carrol Pursell, this postponement by decades of establishing practical applications for scientific discoveries that had already found their way into textbooks was typical of the transfer of knowledge between academia and industry. Several decades often passed before the results of fundamental research found commercial applications.[14] Linde possibly learned about the Joule-Thomson effect through Hans Lorenz. In any case, Lorenz hints somewhat indignantly at this alleged theft of ideas in his autobiography. In 1894 he wrote – according to the autobiography – 'a series of thermodynamic papers about the variability of specific heats and Joule-Thomson effect ... It is certainly remarkable that at the same time Professor Linde based his ingenious process of air liquefaction on the Joule-Thomson effect.'[15]

The lag in the practical use of the Joule-Thomson effect in gas liquefaction was presumably due to the fact that the effect was very small. In the case of surrounding temperatures the temperature of the air barely dropped as much as one-quarter of a degree centigrade per one atmospheric pressure loss. Yet as temperature dropped, the effect grew. The ingenuity of the procedure was its simplicity. The complicated ram cylinder was abandoned and Linde used a 100-metre-long winding double steel pipe for a countercurrent generator.[16] The construction of the pilot installation was managed by Linde, his engineers Negele and Mönch as well as Linde's son, Fritz. In the spring of 1895 Linde was frequently ill and as a result, the first test run was conducted in May, and was to everyone's surprise a great success: after several hours of cooling in the countercurrent the installation produced 'litres of liquid air'.

By 1894, Linde had already recognized the great economic potential of gas liquefaction. After having successfully liquefied air, he moved quickly and purposefully towards its industrial applications. There was, however, an unexpected technical difficulty: the oxygen separated from the air with great difficulty. There was no demand for liquid air, with the exception of a few installations for physics laboratories. Developing liquefaction had taken only a few months, yet separating liquid air into its two main components was to preoccupy Linde and his colleagues for seven frustrating years.[17] Until the turn of the century, Linde worked with fractioned vaporization, until he had to acknowledge that it was impossible 'to reach an oxygen concentration level above 50 per cent in economically viable exploits'.[18] Since the evaporation temperatures of oxygen and nitrogen are very close to one another, no pure oxygen could be obtained from multiple vaporization. It was therefore not until 1899 that Linde introduced his oxygen-rich air – so-called Linde air – for the first time to the chemical industry and as an explosive.[19] But sales were sluggish. It was the rectification method, a sophisticated form of vaporization, that produced the desired purity level after the turn of the century.[20] Originating in the distillery and already in use since the 1860s for absorption refrigeration machines, the engineering of rectification columns was decades-old knowledge.[21] But Linde was unfamiliar with absorption machines and two-element compounds.[22] Several years of unsuccessful experimentation with fractioned vaporization had to follow[23] before Linde accepted the idea of rectification supported by his son and the chemistry professor Hempel. The first rectification column contained glass beads[24] which were replaced by sieve plates in 1903. In 1905 Linde built the first rectification column for nitrogen and in 1910 the first two-column apparatus with two exits for pure oxygen and pure nitrogen. The development of the two-column apparatus is primarily Fritz Linde's achievement.

At around the same time that this use of the rectification method was being developed, a huge market for liquid oxygen opened up: Ernst Wyss and others invented autogenous torch cutting and welding. This new joining technology fundamentally changed the entire fields of metal processing and construction, ensuring the creation of a new major industry: air separation for oxygen manufacturing. Linde immediately acknowledged the huge potential market for technical gases and decided to establish a monopoly on oxygen (and later nitrogen as well as hydrogen) production. In 1904 he founded the Vereinigten Sauerstoffwerk (VSW) in Berlin with two of his potential competitors, who produced oxygen according to no longer competitive chemical processes (namely the Baryt- and calcium-plumbite methods). He subsequently bought their shares and incorporated them in 1910 within the Sauerstoffwerke GmbH to further market his findings.[25]

Until 1930, oxygen remained insignificant in steel manufacturing. The steel industrialist Hugo Stinnes asked Linde in 1906 if he would consider experimenting with oxygen injection in steel production. In the autumn of 1907, Fritz Linde began to perform such experiments at a steel factory in Differdingen.[26] After long co-operation with Linde, Stinnes, Neubert, Thyssen, and Beninghaus bought the Pictét oxygen factory in Berlin-Wilmersdorf in 1909 and founded the Sauerstoffindustrie AG, Berlin, becoming then a direct competitor of Linde's.[27] Yet after experiencing great difficulties with the oxygen factory, the group sold its entire assets to the Linde Company.[28] Due to its high costs, the method of using oxygen for blast furnaces was discontinued in 1908. In 1910 however, Linde was able to make an attractive offer of oxygen at lower prices.[29] At the Belgian Thomaswerk Ougree-Marihave, oxygen installations from Air Liquide were up and running by 1913. Engineering journals regularly reported on the application of oxygen in metallurgy.[30] But it was only through a radical reduction in price for producing oxygen with the high-tech introduction of the Linde-Fränkl procedure that the method became economical in 1930.[31]

On the advice of Professor Frank at the Technical University Berlin, Linde began experimenting in 1905 with extracting pure oxygen from water gas.[32] In collaboration with Frank and his students as well as the director of the Bavarian nitrogen factory, Caro, Linde was able to separate water gas[33] into hydrogen, carbon monoxide, nitrogen, oxygen, and carbon dioxide. The three colleagues patented their method as the Linde-Frank-Caro process for water gas separation. The first major clients for the hydrogen installations were margarine factories using this in the solidification of fat. By 1912, argon, which constitutes 1 per cent of the air, was extracted for use in the production of incandescent lamps.[34]

The chemical industry grew along with cutting and welding technology to become one of the most important consumers of technical gases. The task here was either to eliminate undesired or to extract desired products from mixed gases.[35] Cryogenic separation competed mostly with chemical methods, as issues of price and purity were at stake.[36] The demand for exit materials for use in the synthesis of nitrogen compounds, namely calciumcyanamide (a fertilizer of the same value as sodium nitrate) and ammonia (as fertilizer and exit material for use in the manufacture of explosives), rose after 1905 to make them the most important on the sales market. For abstracting nitrogen, the top methods were the arc method,[37] the calcium cyanimide method,[38] which was also developed by Caro and Frank, and the Haber-Bosch synthesis invented by Haber and Bosch in 1911. In Germany the arc method did not move beyond the pilot stage because of its high energy costs. The other two methods required technical gases, which could be produced through gas separation,[39] and the Linde Company co-operated closely with the leading development companies, the Bavarian Nitrogen Factory (calcium cyanimide method)[40] and BASF (Haber-Bosch method). With the start of the war and the rise in demand for ammonia to be used in the production of explosives, the synthesis factories of both methods became the most important clients of the Linde Company. In 1914, general opinion held that the means of abstracting calcium cyanimide would lead more quickly to the goal of higher ammonia production.[41] However, large installations for ammonia synthesis using Haber-Bosch methods went into operation in Leuna and Oppau in 1916.[42] Gottfried Plumpe has recently argued that the war was an economic disadvantage for BASF's ammonia production.[43] It is difficult to find proof of this – for the Linde Company the war was in any case a period of strong gains and profit increase.[44] The production of poison gases during the First World War was in contrast exclusively done through chemical methods, while oxygen found great use in military hospitals and as a breathing aid in poisoned environments.[45]

The experiments with water gas separation led to the separation of coke gas.[46] Both gases were highly impure waste gases, from which valuable components were extracted through low-temperature separation. The coke gas consisted of hydrogen, nitrogen, methane, and ethylene and its separation put a much higher demand on the rectification columns than air separation. This became one of the specialities of the Linde Company, which controlled this area of the world market and built more than 50 installations during the 1920s.[47]

The complicated and highly successful separation of coke gas brought about 'technical momentum' in the development department of the Linde Company towards further work on the problem of obtaining ethylene from natural gas and oil. Thus the ethylene installations, which were in some respects similar to coke gas separators, became a selling hit after the Second World War. They produced the precursor to polyethylene, the raw material for the upcoming plastic age.[48] Gas separation proved to be a key technology with a remarkably broad range of applications in the constant production of new materials. With that in mind, Linde moved his company towards what he wanted to be, a technological leader in a potentially unlimited field of technological development.

Organization

The internal development of Department B at the turn of the century resembled in many ways the founding years of the headquarters in Wiesbaden. As in Wiesbaden, only a few large systems were planned and calculated in Höllriegelskreuth.[49] However, Höllriegelskreuth was clearly more research-oriented. For the first time, chemists and physicists were hired. In the years to come, research activities in Wiesbaden waned, while they increased in Höllriegelskreuth: by 1922 they had their own thermodynamic calculations department.

Carl von Linde managed Department B until 1908, when he handed it over to his son Friedrich. In the 1920s, as Friedrich became CEO of the entire company, his brother Richard, who had previously been in charge of installations, operations and calculations, took over as manager of Department B in Munich. Commercial issues were entirely secondary at Department B, both in terms of personnel and their effect on decision making. Under the direction of Peter Eggendorfer, Department B had 4 employees in 1911, increasing to 12 by 1924. The department did not establish its own personnel department until 1952, after Linde already had 1,000 employees at Höllriegelskreuth.[50] Up until then, the payroll office had been subsumed under the office of operations.

In contrast to Department A in Wiesbaden, Department B had from the beginning its own manufacturing plants.[51] The department employed up until 1930 nearly twice as many people as all other departments together and was led by a chief engineer.[52] The department of operations was under his command,[53] which included the payroll department, the departments of pre- and post-calculations[54] and the directors of individual sub-departments. The core of production was in coppersmithing. In addition to this there was a turning shop, iron-smithing, a locksmithery, a joiner's workshop, a stockroom, a boilerhouse, a storeroom with delivery facilities, and an oxygen works with filling station. Specialized as they were in heat exchangers, the production facilities in Höllriegelskreuth were not at all suitable for the manufacture of compressors. Linde brought in high-pressure compressors for gas-separator systems from various external manufacturers which could be played against each other.[55] In 1924 the coppersmithery received a large new hall and a training shop was established in the same year, predominantly for coppersmiths.[56] Instead of manufacturing in a series, the company produced a few commercial-sized systems simultaneously. Every morning at the start of the workday, Richard and Friedrich walked together through the manufacturing facilities to check on production progress. Richard Linde remembers that large water-gas separators in 1913 could be 'designed, built, assembled and put into operations in three weeks' thanks to the direct means of communication between development and manufacturing.[57] The drawings were 'of course simple rough pencil drawings', and their assembly was done primarily according to directions given verbally. The new director, Dr Alfred Hess, introduced a limited rationalization which came up against resistance from Richard Linde, who saw the introduction of new formsheets as superfluous and even went so far as to forbid the use of specific forms.[58] Hess was a man of exceptional organizational skills and possessed thorough knowledge of the then under-appreciated science of labour. In addition to his often controversial attempts to rationalize his own department, Dr Hess attempted as well to tighten the workflow of each assignment throughout the entire company. His attempt to institute an accordian system that incorporated the uniqueness of Höllriegelskreuther operations, which was essentially characterized by piece production and lots of technical handcraftsmanship, was supported by Jakob Vonbun (who joined in 1913). J. Vonbun also helped develop a concept for regulating the flow of work between the technical departments and workshops that ran so well that it was kept until the start of the 1960s. With the help of Ernst Dietrich (operations assistant 1923–26), Dr Hess introduced a series of new manufacturing methods.[59] Despite possessing its own manufacturing plant, Department B did not define itself as a factory, rather as an engineering company with an associated manufacturer. Problems in the technical-design departments were of primary concern, rather than those found in manufacturing and production technology. The largest technical department in Höllriegelskreuth was the engineering department, or rather a draughting studio, under the direction of Dr Emil Vogel.[60] This is where most of the work was done in what outsiders consider to be engineering: planning and construction. By 1930, there were 40 engineers working in this department.[61] However, despite its size, the engineering department was in effect merely a feeder service for the crucial departments of apparatus layout and erection supervision. The latter department was led by the engineering leadership of the company, which was also geographically removed from the engineering department and began slowly, after 1910, to differentiate itself into three small but powerful departments led by top-notch men. The increase in engineers from 1925 to 1929 is notable, as their process engineering duties could be passed on to others, allowing them to eventually take positions of leadership. In 1925 Walther Ruckdeschel began to work for the Linde Company, working first at the draughting table, as was par for the course at that time – he later was a member of the Executive Board from 1950–70. Later he took over important assembly and operational duties, particularly those relating to French, Belgian, and Italian clients. Carl Peter

Hochgesand started in 1926, and like Ulrich Hailer (began 1928) and Franz Hammerschmidt (began 1929), he too was part of the assembly department for many years.

Over the years, numerous departments were formally removed from the engineering department. The first to be removed before 1908 was the small bidding office under Ernst Mönch.[62] Until 1929, Mönch had only one employee for bids, replacement parts, shipping, and purchasing. Such a meagre staff was possible here only because the managing engineers processed and negotiated all of the major contracts themselves.[63] The department of gas works was separated in 1914. Led by Linde's son-in-law Rudolf Wucherer, this department built and managed the ever-increasing number of oxygen, nitrogen, and acetylene factories, which were often built near Department A's cold stores. By 1914 there were 20 such works built in Germany, by 1925, 42. Boasting a reliable turnover, these works have constituted the economic backbone of the company since 1920.[64]

Let us look now at the three small but influential development departments. From its inception independent of the engineering department, the tiny operations and erecting department was led by Richard Linde, later by Philipp Borchardt and Herbert Lindenberg. Fritz Linde and Rudolf Wucherer contributed to work on engineering questions in this department, as did Carl von Linde for the first few years.[65] The assembly mechanics were held directly accountable to the erecting department. They worked rather independently on-site and were the highest-paid labourers in the company.[66] Because of the erecting supervision department's direct link to on-site research with clients, this department became one of the most important reserves of practical engineering within the company. The Wiesbaden headquarters had already exhibited a similar structure in the 1880s.

The important department of plant calculations and sizing was at first under the direction of Fritz Linde, then his brother Richard, and was therefore linked to the erecting department. In 1922 plant calculations was separated out, given the title department of calculations and placed under the leadership of Dr Helmuth Hausen, who was assisted by Dr Rudolf Becker. This is where the difference between the 1920s Höllriegelskreuther and the 1880s Wiesbaden department becomes especially clear: central within the company, in the calculations department, theory-oriented employees were able to exercise great influence and created a link to university scientists. The calculations department itself made up the company's future leadership. The calculations for the commercial-sized gas separators lie at the core of the company's most valuable industry secrets. Only the most important managers had access to the calculations processes. Calculations was not affiliated with the engineering department due to security reasons. Sizing values were passed on (not given directly) to technical designers and they drafted out the plans; they did not conceptualize anything independently and were excluded from the company's inner circle of knowledge.

Department B had been running a small research department since its inception in 1897, the physical-chemical laboratory. This later became the department of chemistry, process engineering and patents, from which the department of research and development grew after the Second World War. This department was under the direction of chemists, including Dr Sieder until 1914 and Dr Franz Pollitzer until 1938.[67] A rather small department, they occupied a room on the ground floor of the administration building and had a small laboratory in the cellar. This is where Helmut Hausen worked, while the board member Rudolf Wucherer shared his office with an employee.

The fact that this small company had a number of development and project-oriented departments furthered the independence of the chief engineers, which in turn cultivated communication on technical issues at the leadership level. Discussions were plentiful because of the abundance of differing starting points. Moving up the company ladder occurred mostly via these three small development departments, as this was where direct contact with company leadership was possible.

The chemistry department did not produce any directors or board members. This was, however, due to the fact that many employees in this department with great potential were forced to leave Germany in the 1930s because of their 'non-aryan' origins.[68] The function of leadership selection was clearly shifted from the erecting department to the processes calculations department. This illustrates the fact that calculations became increasingly important in the 1920s for sizing the ever more complex gas separators. More closely tied to one another in the beginning, the departments began increasingly to differentiate their secrets: the sizing and erecting of complex plants including the realization of large projects, calculations and laboratory research. The central position of all three small development departments illustrates the self-image of the entire company.

In this chapter, we will look at these developments until 1930, during which time improvements to processes of gas separation were of the highest prestige in the company. Departments such as sizing and erection supervision were therefore of direct concern to the company leadership. Richard Linde's autobiography focuses nearly exclusively upon new commercial-sized systems and rarely upon theoretical issues in the company. Several improvements were introduced via the fortune of mishap. For example, in 1915 Richard Linde converted shut-down water-gas separator plants for the production of pure nitrogen, but could not bring the Swiss Burckhardt compressors to function 'at more than 60 instead of 120 ATM, as required'. However, 'it was shown that 60 ATM was enough for cooling. I later built a number of systems with 60 ATM.'[69] In 1929, the company festschrift looked back and boasted of the success of a theory for new 'calculations for countercurrent [condensors] requiring an ultimate pressure of 60 ATM'.[70] Richard Linde himself, however, later wondered if more theoretical work should have been done. With regard to important questions on the improvement of heat transmission in countercurrent [condensers], he later reflected that even after 1920, the Linde Company wanted to leave gas 'time for heat transmission' instead of increasing the velocity flow.[71] He and his assistant Philipp Borchardt would have failed in their attempt 'to develop a complex theory of these things', instead of measuring temperature differences in the countercurrent [condensers] 'through which the relationships would have been made immediately clear'.[72] On the other hand, Linde regretted the fact that the company began first in 1922 with Hausen, that is, 'much too late' to work with 'the thereotical aspects of rectification'.[73]

As was the case with Wiesbaden in the 1880s, the months-long work of on-site erection by the most important employees demanded communication with headquarters on developmental questions.[74] Carl von Linde had already required the erecting engineers in Wiesbaden to maintain daily letter contact.

Figure 2.1   At his desk: Carl von Linde aged 83 (1925)

The erecting engineers at Department B were also in regular contact with the engineering management in Höllriegelskreuth. This daily written correspondence constantly provided the engineers with new ideas and helped them to avoid making the wrong decisions, contributing greatly to the succes of the department. These open discussions over questions in gas separation were, however, limited to a small circle of personnel. As an engineering company, Department B was not interested in allowing the spread of knowledge regarding gas separation.[75] Linde was so concerned about information security that he went so far as to not even register certain important processes and refused to reveal technical documents for patent processing.[76]

Engineers and their independence

Linde's chief engineers published frequently. However, they were in actuality quite limited in terms of what they could publish about their research results. For Linde family members and those uninterested in a university career, this limitation was of little interest. Yet it must have pained even Carl von Linde to have been forced to wait until after the expiration of his patent in 1911 to respond to the snide writings of Pictét, Mewes and Gieses referring to his 'untenable' claims regarding the Joule-Thomson effect.[77] This policy posed much more of a problem for a number of employees such as Hausen, Pollitzer, and Karwat, who saw themselves as part of the scientific community. The director of the calculations department, Helmuth Hausen, suffered most under these limitations. After having received degrees in both electical engineering and technical physics at the Technical University Munich, Hausen wrote a dissertation on the Joule-Thomson effect[78] with the support of the Linde Company.[79] In 1928, already director of process calculations at Linde, Hausen was promoted to professor at the TU with his work on thermodynamics[80] and began reading up on cryogenic engineering, the thermodynamics of composite and thermal radiation. In 1934 he received the civil servant status of an associate professor. By 1935 Hausen had published 30 and by 1940 over 50 articles and tables, yet in a display of company loyalty, he refrained from revealing important approximation procedures for the calculation of polyphase columns.[81] Evidence that the sizing equations were documented accurately only in Hausen's calculation records is seen in the fact that important equations in the blotters differed at different dates.[82] This was true for his calculation procedures for two and three component mixes (oxygen/nitrogen or oxygen/nitrogen/argon). Hausen's equation was easier to solve than the more elaborate but published procedure from Hausbrand. Even more important were Hausen's iterative equations for gas mixtures with any number of components.[83] Hausen's condensation equations for polynary mixtures remained unpublished for 12 years. It was not until 1940 when the physicist V. Fischer published an article on the condensation of vapour mixtures of any number of components in the Annals of Physics that Hausen's calculation procedure was available to the public. He published his equations in 1943.[84] Within the company, his calculation procedures remained secret as well.[85]

Hausen maintained an effective balance by looking both to his company and the university community equally. After 1925 he participated in the sizing of nearly all of the commercial-sized gas separators and his employees became the leadership of the Linde Company in the post-war period. At the same time, Hausen participated in university research, particularly in the new field of process engineering,[86] absorbing the articles and dissertation work on this subject.[87]

As a private lecturer he held lectures and seminars and also oversaw doctoral dissertations. In 1950, at the age of 55, he became the chair of thermodynamics in Hannover. Hausen could cultivate his ambitious work in thermodynamics at Linde because he was simply one component of the company. As the head of his company, Carl Linde had had to rearrange his way of working since 1879. Hausen, however, had no commercial power of attorney – simply development engineers as his employers who could appreciate his calculations and use them in their work.

The security measures of the engineering-oriented Linde Company in which upper-level employees defined themselves as engineers were limited to central issues regarding knowledge advantages over the competition. Whereas Carl, Richard, and Fritz Linde were concerned on the one hand about details in Hausen's publications on gas separation,[88] publishing employees were on the other treated with great respect in the company. Hausen and other chief engineers therefore achieved an impressive rate of publication. It was, of course, within the Linde Company's interest to have the majority of German-language publications on low-temperature process engineering come from Linde employees. Fritz Linde's own rate of publication dropped drastically after 1920 when he began to focus on managing the entire company. His brother Richard, however, remained deeply involved with research despite his duties in business management in Munich and continued to publish many articles well into the 1950s. Even after having left the Linde Company, Hausen maintained friendly ties by acting as a consultant to the company.[89] Richard Linde's son Hermann published together with the 90-year-old Hausen the standard work in German on cryogenic engineering in 1985.[90]

Quite often, the conflict between an employee's research interests and the company's limitations led to a rather loose bond on the part of the researcher to the company – making him more of a chief consultant. This was the kind of status held in later years by Steinmetz at General Electrics or Walther Schottky at Siemens.[91] This was the kind of status held by Mathias Fränkl and Paulus Heylandt at Department B for gas separation.

Mathias Fränkl (1877–1947) is a succesful example of a practitioner of low-temperature engineering business.[92] Fränkl registered a patent in 1925 for an alternating converter in heat exchangers (regenerators), which would potentially reduce the costs of air liquefaction tenfold. Fränkl's idea was to first cool the heat exchangers within the converter directly as refrigerating reservoirs and then to release this cold. Fränkl got this idea through oberving cowper equipment in blast furnaces. This is where blasts from blast furnaces were preheated non-continuously. One major advantage of the alternator was that clinging frozen byproducts would be expelled, while the heat exchanger, because of its continuous operation, would slowly become clogged, demanding a shut-down for cleaning. Fränkl the Bavarian with no formal education[93] rose through the ranks as a self-taught man to become the director of a pipe works in Bochum. After the war, he retreated for a needed respite for two years as a farmer, eventually founding several small machine factories, such as the refrigeration machine company MAPAG in Augsburg, the AGA in Vienna,[94] and in 1924 an oxygen plant in Bamberg. Fränkl registered several patents, the regenerator patent being his most successful.

In 1926 Fränkl received a contract for his regenerators in the construction of an oxygen installation for the Maximilianshütte in Sulzbach-Rosenberg. However, the installation performed so poorly that the management of the smelting works brought the Linde Company on board. There were at first major reservations about the regenerators on the part of the Linde Company – Richard Linde had declared the principle impossible. The company was especially unimpressed by Fränkl's arrogance towards the edifice of thermodynamics. Fränkl was perhaps not entirely lacking in theoretical expertise, yet seemed to enjoy making fun of theory-laden engineers. Once, when presented with the second principle of thermodynamics, he stated: 'What? You still believe in the old Carnot? I've long since proven that he's no longer right.'[95] Ultimately, it was the chief theoretician at Linde, Helmuth Hausen, who, after examining Fränkl's work, pushed for co-operation.[96] In 1929 Fränkl built two more installations in Austria and Linde built its first major oxygen plant in Hannover in 1930. Fränkl continued to provide constructive ideas. From 1925 to 1934 he spearheaded 18 patents and the Linde Company 15 patents that shaped the process.[97] Starting in 1932, the Linde Company began to build large oxygen installations using regenerators from Fränkl's MAPAG in Augsburg.[98] The drastic reduction in the price of producing oxygen brought about by the Linde-Fränkl process was a precursor to the introduction of the LD process in steel production, which became the most important industrial use of oxygen after the Second World War. Since the 1950s, this process has pushed first the Thomas and then the Siemens-Martin processes from the market.

The example of Fränkl illustrates that the self-taught inventor engineer had the opportunity to succeed well into the twentieth century. The Linde theoretician Hausen admired his antithesis Fränkl 'above all because of the rapid results of his inventions', under which he himself suffered. Other inventions such as the first pumps for liquid oxygen also came from Fränkl. He worked on gas refrigerating machines using the Stirling process. He was, for the Linde Company, invaluable as an independent idea provider. His most important contribution remained, however, the regenerator. According to Richard Linde, there had been, since the introduction of double rectification, 'no more important progress made in air separation, ... than the use of regenerators'.[99]

A second self-taught man who benefited greatly from the Linde Company was the inventor and manufacturer Paulus Heylandt (1884–1947).[100] Raised in Erfurt, he heard of Linde's liquefaction of air as a 14-year-old student and began experimenting.

By the age of 18, Heylandt was already registering patents for liquid air transport containers and one year later built his first air liquefier with expansion machines. The highly compressed air was expanded in one step to atmospheric pressure, which produced external work.[101] In 1908 he received a patent for this system. Heylandt sold these air liquefiers primarily in England. In the

Figure 2.2 Paulus Heylandt in his Berlin laboratory, *ca.* 1935. Front right: cutaway model of the transport container he developed for liquefied gas

years following his first patent, he founded a number of businesses that were to make use of his increasing number of patents.[102] Heylandt's most important inventions were his transport containers, particularly the carburetter tank of 1917 which allowed for the transportation of liquid boiling oxygen. Transported by this means the volume carried was reduced by one-third compared with that transported via bottles.[103] In 1923 the Linde Company agreed upon a co-operation with Heylandt's company to invest in liquid oxygen containers.[104] Beginning in the 1920s, Heylandt experimented with Max Valier (who died in an accident in 1930) on rocket-driven cars. Starting in 1942, Heylandt worked intensively on the construction of the A4 liquid oxygen rocket. He had been a National Socialist enthusiast and party member before 1933. Heylandt was taken in 1945 to the Soviet Union as an expert, where he worked for two years as a development engineer before dying there. As of the 1920s, the Linde Company used and sold Heylandt's containers and air liquefier, the Heylandt System. Heylandt's independence and his co-operation with Linde's competitors was a constant bone of contention for the Linde Company.[105] Yet overall it benefited – as was the case with Fränkl – from Heylandt's independent inventiveness. Heylandt and Fränkl came to low-temperature engineering without an academic education under their belts. Yet neither was untheoretical in his work nor did they avoid contact with the academic world. The 'physicist' Heylandt was awarded an honorary doctorate in engineering in 1925 from the Technical University in Berlin, and Mathias Fränkl employed top engineers who later became professors at technical universities.[106]

New Linde Company competitors

As in the refrigeration industry, low-temperature engineering offered a number of opportunities for specialists. The Linde Company's greatest competitors came from the specialized industries of separated gases.

Welding technology provided an easy mode of entrance into the gas business.[107] Oxygen-acetylene welding originated with the French inventor Fouché in 1901. One of the reasons for the continued strong French position in low-temperature engineering can be traced back to their headstart in welding technology.[108] Linde was able to secure co-operation with the owners of the Linde Fouchés patent by joining the International Oxygen Association in an attempt to devlop the market together. However, welding technology in Germany developed independently of that in France, ultimately forcing a dividing of the market with the Linde Company.

The chief engineer at the Bitterfeld works for the chemicals manufacturer Griesheim Elektron,[109] Ernst Wyss, built in 1903 a device for welding with hydrogen and oxygen. Both gases were easily accumulated as byproducts of the plant's chemical processes.[110] In 1906 Wyss strongly recommended the Griesheim Executive Board to take on the oxygen industry, which would provide them with 'dominance in acetylene welding'.[111] The board agreed and in 1906 acquired the rights to an older patent for torch cutting. The Cologne-Müsener Mining Equity Association had been blowing 'frozen' blast furnace taps with a hydrogen-oxygen lance since 1900,[112] but had no sense for the unlimited application of this basic patent for torch cutting: the cutting of metal with an oxygen beam.[113] Thus, in 1906, Griesheim had in its hands the important cutting and welding patents and sold related devices and gases to end customers. Griesheim's first welding torches were finished by a company in Lübeck called Dräger, but as of 1910, Griesheim began producing its own.[114] Soon, the production of oxygen through chemical processes at Bitterfeld was not able to keep up with its exponentially growing demand and Griesheim made a deal with the Linde Company over supplying oxygen. Yet in 1907, Griesheim signed a blanket contract with Air Liquide in France over the supply of oxygen.[115] Bending to make concessions, Linde divided up the German market for oxygen supply in Germany with Griesheim.[116] Linde chose to co-operate with Griesheim on this matter out of fear of the power of the chemicals industry.[117]

Together they moved to block smaller competitors and began to swallow up all remaining independent plants.[118] After Griesheim and Linde together had bought up Deutsche Oxydric AG (German Oxydric Corporation, or DOAG), Griesheim's manager Theodor Plieninger became a member of the Linde Executive Board.[119] The oligopoly functioned well into the early postwar years, when competition strengthened. Forty independent works had been established by 1925 and organized themselves as the Deutsche Industriegas-Schutzvereinigung (German Industrial Gas Protection Association), only to be bought up in 1926 by Linde and Griesheim.[120]

The Nürnberglicht Gesellschaft (Nuremberg Light Association) was one major competitor in the early days. Having bought in 1905 a Linde air separator installation, the company began producing oxygen and by 1908 was producing virtual replicas of the separators themselves.[121] With assistance from the Diskonto Association, the Industriegasgesellschaft für Sauerstoff und Stickstoffanwendung (Industrial Gas Association for the Applied Use of Oxygen and Nitrogen, also known as Indusgas) was founded. It merged in 1910 with the Süddeutschen Industriegasgesellschaft mbH (Southern German Industrial Gas Association) to form the Deutsche Industriegasgesellschaft (DIAG) (German Industrial Gas Association).[122] DIAG merged in 1916 with Deutschen Oxydric GmbH, a manufacturer of oxygen using the outdated electrolyte process to become Deutschen Oxydric AG (DOAG).[123] From 1908 to 1914, and starting again in 1916, Linde sued this company for patent infringement, eventually taking DOAG over in 1920 together with Griesheim by means of share trading.

The company owned by the Lübeck watchmaker Heinrich Dräger (1847–1917) had been producing oxygen valves for Dr Elkan's oxygen plant in Berlin (who began co-operating with Linde in 1904) and for Griesheim since 1895 and 1903 respectively.[124] Dräger became the most important fitter in gas engineering. He produced oxygen recovery devices for medical and security purposes.[125] His brochures portrayed his workshop as scientific 'laboratories for physics, physiology and chemical engineering'.[126] Dräger limited himself to manufacturing specialized devices and avoided conflicts with the producers of oxygen installations.

At nearly the opposite end of the spectrum there was the inventor engineer Adolf Messer, who ventured into gas liquefaction through welding engineering, thereby putting himself directly in the line of fire from the big manufacturers, and persevered in spite of them.[127] Of course, he suffered many a beating at the hands of Griesheim's and Linde's repeated patent infringement cases. Although he lost the majority of these suits, his delay and manoeuvre tactics allowed his company to grow nonetheless.[128] He supplied installations on the one hand to independent oxygen works that produced outside the Griesheim–Linde block and on the other hand he supplied his industrial clients, such as shipyards, with liquid gas for welding devices. They received from Messer both welding engineering and gas manufacturing. Messer's core business however, remained welding and cutting torches. Through this, he was able to build up a successful business for oxygen and acetylene production, which the Linde Company and Griesheim, after a long price battle, were able to squelch in 1929 through quota fixing.[129]

Marketing gases worldwide

Linde set his sights on internationalizing the marketing of oxygen, nitrogen and the installations for their production by founding in 1906 together with the French holders of the Fouché welding torch patent the Internationale Sauerstoffgesellschaft (International Oxygen Association, or ISG) in Berlin to promote and exploit the worldwide use of oxygen.[130] Linde founded or bought shares in businesses producing oxygen in England, Spain,[131] Switzerland,[132] Sweden,[133] Russia,[134] Italy,[135] Austria-Hungary,[136] and the United States.

Co-operation with the world's second largest company in the business, Georges Claude's Air Liquide, became particularly important. After Linde had won a patent suit in England against Claude in 1908, he used his favourable leverage to negotiate a co-trading agreement with Air Liquide.[137] In October 1908 Claude visited the Linde factory in Munich[138] under friendly circumstances. In his autobiography, which was published in the third year of the First World War in 1916, Linde wrote that Claude was, next to the Linde Company, the only innovative power in low-temperature engineering and that this was the reason for his co-operation with him. Such an argument appears rather uneconomical.[139] However, the co-operation with the technology-proficient Claude was a well-considered strategic decision. The two technological leaders had a good chance of preventing any other competitors from rising to their ranks.[140] Quotas were agreed upon for specific geographical areas and countries,[141] which the two companies gained in the face of tough competition from local subsidiary or partner businesses.[142]

Linde's co-operation with the English BOC was of an entirely different character. Founded in 1886 as Brins Oxygen Works, the company produced oxygen with the Baryt process and opened subsidiary plants in Germany. In 1896 Brins Oxygen Works took on the manufacture of Hampson air liquefiers.[143] Yet after Linde had won its patent suits against Hampson,[144] Brins Oxygen Works sought to co-operate with Linde.[145] The company was reorganized under the new name British Oxygen Co. (BOC) and Linde joined its board. By 1914 BOC had built 11 oxygen plants in England.[146] Air Liquide also participated in BOC as of 1908. Linde was therefore able, together with Air Liquide, Linde Air Products Co. and BOC to manufacture oxygen installations worldwide and sell oxygen.

Linde enters the American market: Founding of Linde Air Products

Following the turn of the century there was a tremendous field of untapped applications for oxygen and nitrogen waiting to be exploited in the United States. Low-temperature engineering offered the Linde Company more of an opportunity for expansion than refrigeration. The first step towards establishing an American subsidiary was to secure patent rights, which proved to be extremely difficult. Just a few days after his application for a patent in Germany, Linde's Berlin-based patent attorney sent a patent request to Washington in poorly written English. Later, Linde engaged two American attorneys who sought in vain for four years to get a patent for Linde's air liquefier. In the United States, the engineer Charles Tripler already had a patent from 1893 on his air liquefaction installation, the rights to which he defended rigorously. Linde, however, was blessed with luck in the summer of 1900. The famous electrical engineer Charles Brush[147] bought a Linde air liquefier for his personal laboratory in Cleveland. Charles Brush had sold his share of the Brush Electric Co. for 3 million dollars to the Thomson Houston Co. in 1889, which made him a wealthy man, albeit unemployed. Brush had his own well-equipped laboratory in Cleveland to keep himself busy as an independent inventor and businessman.[148] He bought the Linde liquefier to help in the work he had been conducting since 1896 on the discovery of new gases to support theories on aether.[149] Having heard of Linde's patent struggles, Brush offered in exchange for a 33 per cent share to fight for Linde's patent in the USA.[150] Linde accepted and Brush successfully won the full acknowledgment of Linde's patents by 1903.[151] Tripler's patents from 1893 were declared invalid after Brush left him standing[152] as the self-taught man attempted to 'explain' his liquid air perpetuum mobile motor during the patent hearing.[153] After having won this difficult battle, Linde briefly considered disposing of the patents.[154] However, after his application for rectification patents ran up against problems in the United States, he decided to establish his own company there.[155]

Building up oxygen production in Europe was a capital-intensive endeavour that forced the Linde Company to dip into its reserves. This in turn made the company all the more eager to develop potent business partnerships in America to tap into the new market overseas. Linde thought often of Fred Wolf in Chicago, Adolphus Busch – the largest brewer in America at the time, whose famous St Louis 'Budweiser Beer' he brewed with the help of Linde refrigeration machines – and Charles Brush. However, Linde heard nothing from Brush for three years after their patent battles had been won, leading Linde to doubt Brush's interest.[156] In January 1906 Linde told Brush of his spectacular success in Europe, which he used to convince Brush of his confidence in the American market.[157] Brush took the bait and put up 500,000 dollars in share capital, seeking American dominance amongst the shareholders while Linde sought German dominance.[158]

Linde needed an English-speaking man who would be loyal to him as general manager of the planned new company. He decided on Cecil Lightfoot, the son of T. B. Lightfoot, who had managed Linde British Refrigeration for 20 years. Linde had known the young Cecil since 1894, when he had arranged for an internship at the Linde associated company Sulzer Brothers in Winterthur for him. Cecil Lightfoot had been working since then at Linde British Refrigeration as an engineer. The elder Lightfoot sat in Linde's place on the Supervisory Board at BOC. At the end of the summer of 1906, Cecil Lightfoot went to America to seek out production locations, markets, and capital providers. He decided on Buffalo, New York as a location.[159]

In July 1906 Adolphus Busch and his son-in-law Hugo Reisinger, the German consul in New York, visited Linde in Munich and demonstrated interest in striking a business proposal. Reisinger was simultaneously one of the most important import/export businessmen in German–American trade and a patron of cultural exchange between the two countries.[160] In the autumn of 1906, Cecil Lightfoot travelled to America to see this old acquaintance of Linde's. Yet he already had his own ideas: Busch and Reisinger offered to buy all of the patents for 250,000 dollars.[161] Fred Wolf had begun to feel too old to start anew in this new field.[162] Even Brush had grown somewhat hesitant again. When Lightfoot telegraphed Linde on 30 November to this effect, Linde immediately left for America, arriving on December 15,[163] and in Cleveland negotiated personally with Charles Brush, who had succeeded in estabishing American dominance within the company. The majority of the shareholders were Brush's industrialist friends in Cleveland,[164] people whom Wolf had warned Linde about.[165] Linde insisted on Reisinger becoming a partner,[166] and then left in mid-January to return to Germany, leaving Brush with the legal power to finish founding the company.[167] The first problems arose with the issue of the French rights. Linde had made an agreement with his major competitor in Europe, Air Liquide, yet Brush refused Claude's attempts to become a partner in the American company. Air Liquide had nonetheless, at least indirectly, a stake in the endeavour thanks to the licensing of the valuable Fouché welding patent to Linde Air Products through the International Oxygen Association (ISG).[168] In a counter move, Linde tried to firmly install Reisinger and Lightfoot as loyal partners. Lightfoot was without doubt the right man to help build up the company. Less than six months after the start of construction, on Thanksgiving day 1907, he celebrated with 16 employees the opening of the first American oxygen plant in Buffalo. Linde Air Products produced not only oxygen, but acetylene and welding equipment as well – all that was needed for the entire process.[169] Nearly all of the devices came from Germany: the air liquefiers and rectification columns, even the welding devices and the first 5,000 pressurized bottles were shipped from Germany. Nobody in America had built containers for such high pressures at that point in time.

During the First World War, Linde Air Products outshone its parent company in Germany, growing to become the largest manufacturer of engineering gases.[170] The German Linde Company's influence on its American subsidiary had waned before the First World War. The exponential growth experienced by Linde Air Products consumed massive amounts of capital. The German Linde Company had increased its number of shares from 950 to 4,237 at 100 dollars from 1907 to 1913, but was unable to keep up with the capital increase, making its holdings in 1913 inconsequential. Carl von Linde was by 1912 no longer on the board of Linde Air Products. Following a severe row with the Linde Air Products management in 1914, Cecil Lightfoot was fired. Thanks to Linde's behind-the-scenes manoeuvring, a contract was signed making him a 'Consulting Engineer' for the Linde-Frank-Caro process.[171] Hugo Reisinger, the last man on the board loyal to Linde, died in the autumn of 1914 during a trip to Germany. The effort by the Americans to defeat the German attempt to exert influence on the company was unmistakable. As it became clear in 1916 that the USA was to enter the war, Linde tried to sell his shares and avoid expropriation. A bank in Chicago offered to buy up all of his shares at nominal value (1.75 million Marks), which was an acceptable price. Together with the manager of the Deutsche Bank, Linde considered transporting the shares in the U-boat *Deutschland* through the English blockade.[172] Attempts to sell the shares were obscured and may have failed entirely.[173] In 1918 the shares were officially on the books at an asset value of 0.[174]

During these years, the considerable hopes for the future of welding engineering had drawn the eager eyes of major gas companies towards smaller new companies just entering the market. At this time too the acetylene branch in the USA made the switch from luminary to burnable gas. The largest American producer of acetylene, the Union Carbide Co., bought shares in Linde Air Products in 1911 and founded in 1912 the Oxweld Acetylene Co., which immediately swallowed the Acetylen Apparatus Co.[175] Prest-O-lite, a manufacturer of acetylene bottles for automobile lighting, felt the weakness of the acetylene business and attempted to switch over to the booming welding market.[176] Union Carbide, which had produced electrodes, was also threatened by the demise of the acetylene lamp industry.[177] Linde Air Products, however, stood on the sunny side of things as an oxygen manufacturer. In November 1917 Union Carbide took over Linde Air Products and three other businesses in the branch: the National Carbon Co., Prest-O-lite and the Electro Metallurgical Co. The takeover appeared to be a merger; shares were newly divided following a code, the new name of the company was the Union Carbide and Carbon Corporation, and the subsidiaries remained as they were. The new company was one of the largest American businesses.[178] The defeat of Germany in the First World War had a significant impact on economic relationships within low-temperature engineering, shifting the dominance away from Germany. In 1920, Linde Air Products had a share capital of 15 million dollars and produced oxygen in 28 plants. For a while, the German Linde Company feared being taken over by the Americans, as in 1920 a mere 200,000 dollars would have sufficed.[179]

However, in terms of technical development, the Germans continued to be the leaders. Linde Air Products was one of the largest industrial consumers of low-temperature installations from Munich. The rectificators, above all the new columns for hydrogen and coke gas, came from Germany.[180] In the 1920s, Paulus Heylandt travelled six times to the USA to sell and explain liquid oxygen tanks to Linde Air Products.[181] Linde Air Products, however, slowly became independent in its own research. The old plant in Buffalo had stopped production with the First World War and was being used for apparatus engineering.[182] In 1923, the first true research employee was hired, Leo I. Dana, who had received his doctorate from Harvard and worked for a year at Kamerlingh-Onnes in Leiden.[183] In 1934, Dana's research department had 20 employees,[184] and in 1948 it moved from Buffalo to its own large complex in Tonawanda.

Linde Air Products dominated the American market up until the Second World War, when the businessman Leonard P. Pool began to establish a second company, Air Products Co., with an aggressive marketing strategy.[185] Pool eventually moved beyond distributing oxygen and began in 1941 to produce first welding equipment and then gas separators.[186] Whereas Linde Air Products sought to sell gas or large installations for gas production, Pool's Air Products Co. built small 'on-site' installations, which they either sold or leased to their customers.[187] Air Products made its breakthrough because of the war. The

American military was the largest customer of such small installations for years. After the war, Air Products had become so large it could no longer be threatened by Linde Air Products.[188]

This chapter has shown how Carl Linde transferred the organizational structure of his Wiesbaden development-oriented engineering company to the new field of gas separator and plant engineering. Linde sought and found a technicologically demanding field in which his company could out-compete with new developments instead of lower prices. This technical orientation provided the company the room to expand its ever-increasing scientific influence, which impacted the company leadership as well. However, it was possible for non-academic self-taught employees to rise through the ranks in the field of gas separation. The sudden increase in demand brought about by both world wars provided loopholes for market newcomers.

Gas liquefaction, gas separation and their applied technologies, particularly welding, but also steel coating, all came from Europe to the United States. In the period between the two world wars, the American Linde subsidiary outdid its German parent company first economically and then later in a number of areas of technology. However, following the Second World War, after European industry recovered its technological prowess, European low-temperature engineering sat once again at the top, both economically and technologically speaking. Despite significant research efforts in the USA since the Second World War, the European 'prime movers' – Air Liquide, BOC and Linde – have dominated the world in this field.[189]

Carl Linde's death and the generation change in company leadership

Carl von Linde died on 16 November 1934 at the age of 92. He was the deputy chairman of the Supervisory Board up until the very end. The company lost not only its founder, but also the one figure within the company who combined science, a passion for invention, and management talent all in one. The renowned refrigeration engineer Rudolf Plank boasted of Linde in his reflections from 1935:

[he] embodied an unusual synthesis of scientific detail with technical judgment and a developed sense of reality; these characteristics allowed him to establish a wonderful harmony between theory and practice early on in refrigeration engineering. He was ahead of his time, but also able to recognize limits so as to present solutions once a particular issue was ready to be put into practice.[190]

Carl von Linde had prepared for leadership transition long before by training his two sons Friedrich and Richard as well as his son-in-law Rudolf Wucherer to be the next generation to inherit his ideas.

Figure 2.3 The founder of the company and his family in 1911: Carl von Linde, his spouse Helene Linde, with his sons Friedrich (left) and Richard (3rd from left), daughters, sons-in-law, daughters-in-law

All three repeatedly emphasized, albeit at different moments, the need to continue the tradition of Carl Linde's business style. Fritz and Richard Linde wrote at the end of their careers in the Fifties: 'When our father passed on the leadership of the company to younger hands, we considered it our expressed duty to continue particular company traditions.' This included the 'axiom of holding absolutely to all contractual obligations with all partners and clients, regardless of any related costs'. This reliability was an important sales point in the particularly risky field of plant engineering. According to the third generation (Hermann Linde), one was to be able to depend on Linde so 'that every installation delivered by Linde was to be made to function, no matter the cost'. The high value placed on continuity and reliability was also evident internally. The company's second generation also sought to keep employees for the long term. As both sons reflected, 'we knew that satisfaction both on and off the job amongst as many employees as possible was the only way to ensure that the tasks at hand would be successfully completed'.[191] Both sons and son-in-law Rudolf Wucherer continued to position the company as a leader in technology. All three lived in Munich near the headquarters for low-temperature engineering research rather than Wiesbaden. They were all convinced that it was more important for the company to be a leader in technology than a market leader.

Following Wilhelm Schipper, Friedrich 'Fritz' Linde became in 1924 chair of the Management Board. Beginning in 1929 he also held the title of company general director, a move which was perhaps meant to strengthen his position against the powerful Supervisory Board. During his time as chair, Friedrich Linde was able to achieve a number of decisive business goals. The idea of establishing a cartel had long been discussed. In 1932, it was Friedrich who conducted the tedious negotiations with IG Farben to establish a cartel in the gas trade that favoured the Linde Company: the Linde Company and IG joined together as equal partners to form the Vereinigten Sauerstoff-Werke (VSW). In the absence of such an agreement, Linde would have had to fear being driven out of the gas business by the all-powerful IG Farben and its subsidiaries (e.g. Griesheim). Friedrich Linde was also quite successful in reassuring Linde's holding companies abroad during the difficult years between the two world wars. Within the company, Friedrich Linde had the final say in questions regarding business matters. Faith in his business ability and authority within the company was enormous. He left technical leadership in particular areas of the company – especially low-temperature engineering – to his brother Richard.

Figure 2.4   Board of Management and Supervisory Board of Linde AG in 1932. Seated, left to right: Dr Johannes Hess, Georg Proebst, Richard Buz, Dr Carl von Linde, Dr Otto Jung , Dr Bernhard Buhl, Heinrich Sedlmayr. Standing, left to right: Bruno Hippenmeyer, Rudolf Wucherer, Dr Friedrich Linde, Hugo Ombeck, Dr Richard Linde

Richard Linde was the man in charge of technological issues in Munich as of the 1920s. Looking back in 1966, Hermann Linde states:

the leading figure in apparatus engineering and construction was Dr Richard Linde, who led operations in a manner unthinkable today: patriarchal. Organization was for him secondary – he kept that in his head – but he did understand how to model and impress upon us a particular attitude that was so characteristic of him and those who followed him, that it continues to live on in the company today.[192]

Richard recruited valuable employees for the calculations and development departments and to shape the work style into one of research and development, which continues to be typical of the Munich departments. Of the three representatives of the second generation, he was the most musically talented and set a certain tone which carried on through the 1970s.[193] Many of those in the Munich technical leadership were not only musically inclined, but were also attracted to Linde through music. For Richard, who was also director of the erection department, work on air separation and refrigeration installations proved to be 'training tools for young employees whom he considered particularly capable'. In this manner 'he established quickly whether the men in question were capable of taking responsibility for their work and helping themselves through difficult situations'. The independence of project engineers was highly valued at Linde. This also meant that they were so-called 'punch bags', who were to take the blame 'for mistakes made by others'[194] just as they were expected to embody the company's emphasis on competence and quality.

Rudolf Wucherer, one of Carl von Linde's sons-in-law, was responsible for important technical innovations before the First World War. But after 1914 he switched to management and chaired the company's oxygen and acetylene plants for decades and became a board member in 1928. Between 1914 and 1929 he established and brought into operation 13 new oxygen plants and 25 new installations. By 1928 he had quadrupled oxygen production, tripled acetylene production and established regional monopolies in the container business. In doing so, Rudolf Wucherer was the actual founder of the company's gas department. This department continues today to provide the financial backbone of the company. While gas sales plunged and some factories closed due to the world economic crisis, Rudolf Wucherer, 'in anticipation of future developments, [used] the time to modernize operations'.[195] At Wucherer's suggestion, the department Ellira[196] for the production of electric welding equipment was founded in Höllriegelskreuth in 1937. Linde bought licences for submerged and shielded arc welding methods from the former Linde subsidiary Linde Air Products in New York. This strategic investment brought on the one hand improvements to the widely used method of gas welding and on the other increased argon sales. What in the 1950s was seen as a great success of the past, was considered only ten years later to be a dangerous diversification. The department of welding technology was sold as part of a move to focus on the company's core business.

Fritz Linde had a doctorate in physics and Rudolf Wucherer had a masters in engineering. However, Richard Linde was the one to take on the role of guiding research and development in the company. Working together closely with his father, Fritz developed the foundation for gas liquefaction and separation. However, after the First World War, he managed the company's business strategy. His greatest achievements were in securing Linde's position in various cartels for refrigeration machine manufacturing as well as in technical gases, which continues to be an international oligopoly.[197] Friedrich Linde was a talented businessman who bought other larger companies – Walb, Sürth, Güldner, Matra – guiding the group towards small refrigeration production, machine, and engine construction. Only Hans Meinhardt's acquisitions beginning in the 1970s were similar in strategic relevance. They were of course of a different financial standard, as the group had increased in size tenfold since Friedrich Linde's reign.

Rudolf Wucherer focused on corporate organization as head of Department C (gas works) and later as chairman of both the Executive and Supervisory Boards.[198] In contrast to Rudolf, Richard remained loyal to technological issues. After completing his dissertation at the leading thermodynamic institute at the Technical University Munich (which was co-founded by his father) he became director of the erecting department at Höllriegelskreuth. Until his departure from the Management Board, Richard Linde worked on improving the technology and physics of gas separation. His efforts ensured that Linde would remain the world's leader in this field. He also trained the third generation of the company's technical management. The young employees with the most potential began their careers in Richard's departments. Richard, however, grew increasingly withdrawn, exhibiting seemingly depressive behaviour. The horrors of the Nazi era, the death of three of his five sons and his increased loss of hearing led him to become a quiet and reserved man. Consequently, he did not enter the Supervisory Board after he stepped down from the Executive Board in 1950.

The older brother presented himself as the head of the company, as its organizer and strategist; the son-in-law as manager and watchman; the younger brother as inventor and researcher. All three saw themselves as carrying on the tradition of Carl Linde. In retrospect, it appears that Carl Linde's abilities and characteristics were split evenly amongst the second generation. This phenomenon may be interpreted in terms of family psychology without delving too deeply into Hellinger's theories of family constellations.[199] Rudolf Wucherer took over certain duties and characteristics from Wilhelm Schipper. For decades these three men worked together harmoniously and were highly competent in their respective duties. Hence their division of labour at management level gave them success. The exceptional combination that was Carl von Linde lived on in his three successors: the entrepreneur lived on in Fritz Linde, the manager in Rudolf Wucherer and the inventor in Richard Linde. The harmony showed quite literally when the families came together to make music. Fritz played first violin in the quartet, Richard the cello and Helene Linde the piano.

Their teamwork becomes all the clearer when we compare them to the third generation, to be explored in the fifth chapter. Fritz did not have children, which spelled the end of the 'entrepreneurial' side of the family. Of

Dr Friedrich Linde

Rudolf Wucherer

Figure 2.5 The second generation of Lindes on the board of management

Dr Richard Linde

Richard's five sons, three died during the Nazi era, two in the war. This was a tragedy not only for the family but also for the company management.[200] Richard placed much emphasis on education and family relations. All five of his sons were, like their father, musically talented and played music regularly. The two elder sons were already on their way up in the company before the war. Helmut, the eldest, held nationalist views to the dismay of his father.[201] As the eldest, he would have been likely to take over the strategic leadership from his uncle Fritz some day. Helmut had completed a dissertation in thermodynamics with Ernst Schmidt in 1937 in Danzig and was then sent to Wiesbaden as a Linde employee. Werner received his doctorate in chemistry in 1938. Gustav, the middle and most musical of the sons, committed suicide in the autumn of 1935 after his return from work service under the Nazi regime. Helmut and Werner died on the Eastern Front. Their deaths, particularly that of Helmut, left gaping holes that were not to be filled by the two younger sons who worked for the company after the war. Like his father, Hermann Linde was a respected inventor at the top of the company. However, even as a board member (as of 1961) and speaker of the board (as of 1972), he was less like his uncle Fritz the businessman and manager, taking more after his father. His brother Gert Linde only achieved a mid-level management position within the company. Rudolf's son, Johannes Wucherer, jumped in to fill this hole. With a PhD in engineering, he started his career in Richard Linde's department and enjoyed great success as director of calculations. Wucherer also applied his talent to plant engineering. He was the one who came up with the calculations for the expansion turbines built in Sürth.[202] However, as Johannes reached the position of chairman of the Executive Board in 1961, he took on the role of Fritz, as Hermann took over the management of plant engineering at the same time. Johannes Wucherer was a gentleman of diplomacy. Yet he was not a dynamic businessman. To make

Figure 2.6   Richard Linde performing a quintet with his five sons. From left to right: Helmut (first violin), Hermann (second violin), Werner (piano), Gert (page turner), Richard (cello), Gustav (viola).

matters worse, he disliked meetings as he suffered from low blood pressure. Consequently, the company and the family lacked a strategic-thinking dynamic individual interested in details of management in the Sixties. The rise of Hans Meinhardts to the top of the company in 1976, which spelled the end of a family tradition, was a result of the third generation's weaknesses in business and focus on technology. Hans Meinhardt fought for his position against Hermann Linde, not Johannes Wucherer. It is likely that during these battles, Johannes Wucherer himself was not convinced that family members were the best equipped to lead the company. This was also probably true, at least subconsciously, for Hermann Linde, who determinedly resisted his removal. An indication of this may be seen in the fact that the children of this third generation, that is, the fourth generation, were not systematically prepared to be part of the company. The writing was on the wall before Hermann Linde's resignation as spokesman for the Management Board in 1976. There are however, fourth and fifth generation descendents of Carl von Linde currently in the corporation.

3
Linde During National Socialism: 1933–45

Economic policy in the Third Reich

Economic crisis and the rise of the National Socialists

The rise of the National Socialists took place against the background of the world economic crisis of 1929–33; after a period of relative economic consolidation in the mid-1920s the worldwide economy collapsed. The German Reich – already economically weak during the Weimar Republic – was particularly impacted by this economic disturbance. Private assets were heavily hit by the inflation of 1919–23. There has been a long-term debate over the role that reparation payments played in amplifying the crisis.[1] The finance and sales crises at the end of the 1920s led to dramatic cuts in industrial production, resulting in mass unemployment, above all in the industrialized regions of Germany and particularly in the big cities.[2] There were almost 6 million people unemployed in 1932. Considering the limited unemployment benefits at the time, this amounted in practical terms to poverty for a large sector of the population.[3]

Linde was not spared by the economic crisis either, despite the fact that sales of investment goods such as mechanical and plant engineering were delayed in terms of their response to this crisis. The crisis impacted the Linde Company 'to its full extent' first in 1931; contracts dropped, profits in all departments decreased substantially and there was a shortage of work.

At the beginning of 1930, Department A (refrigeration units) could still compensate for declining domestic sales in large and small refrigeration units through an increase in its foreign sales. But towards the end of the year, 'in addition to inactivity in the domestic market, foreign buyers also' expressed 'increasingly stronger reservations'.[4] Furthermore, profit margins began to drop. In 1931, sales dropped more than 60 per cent below those of the previous year, and both layoffs and reductions in work hours could no longer be avoided. The material supplies of Department A had to be booked considerably lower 'because of the general fall of prices', which in turn prompted write-off losses. In the following year, sales dropped once again by nearly one-third of the previous year's turnover, and contracts dropped by about 20 per cent. Foreign contracts in particular decreased. The foreign share of Department A's contracts in 1932 hovered at only 30 per cent which was 'about on par with the average of the post-inflation years'.

By the end of the 1920s, cold houses and ice factories had more than enough business.[5] However, by 1930 they were suffering and in 1932 the cold store business reported 'altogether unsatisfactory' results. Furthermore, a 'senseless competition' in the ice market resulted in a 'drop in prices'. For the first half of 1931, the Höllriegelskreuth Department of Apparatus Engineering, as the plant construction division was called, 'had just barely enough business', while a major drop in contracts led to 'extensive layoffs and reduction in working hours'[6] during the second half of the year. In 1931 Richard Linde wrote in family letters (to his sisters) about Department B: 'gloomy times have – if only somewhat belatedly – arrived here in Höllriegelskreuth as well.'[7] In order to avoid layoffs, factory work was done only in two shifts – 26 hours per week each – which brought a corresponding drop in earnings for industrial labourers. Office employees also worked only 40 hours per week with a corresponding reduction in wages. The 'otherwise customary celebration' of the 25th anniversary of Richard Linde's services scheduled for 1 October 1931 'was cancelled due to the gloomy times',[8] as Richard informed his sisters in his letter. Linde succeeded in achieving satisfactory profits by being 'extremely thrifty'[9] at the oxygen and acetylene factories, despite their limited utilization. In contrast, at Departments C and D (Machine Factory Sürth and the Güldner Engine Works), losses were not to be avoided. Losses had already been declared in 1930 at Güldner, while Sürth could still profit from the 'important contracts for large-scale high-pressure compressors', which Section B had ordered.[10] Yet in 1931 the total turnover dropped by almost 30 per cent from the previous year, and in 1932 Linde's Ice Machines' business report declared that of all factories in the company, Department C was 'hit worst by the crisis' and 'despite all attempts to save' it could not generate enough to cover current costs.[11]

The National Socialists' seizure of power

The economic situation in the early 1930s was altogether catastrophic; with improvements to the situation nowhere in sight, the psychological impact of these dire circumstances was great. Richard Linde thus wrote in 1931, 'considering that almost everywhere the same difficulties exist[ed],' he believed 'no fundamental nor regime change was to come, even if one assume[d] that this would bring substantially more capable men into power'.[12] The National Socialists, who took control in March 1933, were certainly not economic politicians in the sense that their first priority would be the recovery of the German economy. They instead placed economic recovery in the context of another 'more important' goal of preparing for a war economy. Dietmar Petzina has characterized Hitler's crucial achievement in economic policy as a 'psychological moulding of economic policy', which 'moved job creation and the fight against the crisis beyond the sphere of economics, turning the upwards economic swing into a question of trust in the new regime and its policies'.[13] This to some extent is what allowed the National Socialists to succeed. Also to be noted is the extent to which the

Linde Company's perception of the economic situation changed. Thus the 1933 business report, which reported bleak numbers for almost all departments, nevertheless reads:

> The year bore signs of recovery in the German domestic market. Although we benefited only in a limited way from government job creation measures in the large refrigeration units sector, we nonetheless experienced a slight boom as a result of regained trust and a reawakened desire to conduct business domestically.[14]

Job creation as a political programme

The debate over the efficiency of the measures introduced by the National Socialists for job creation has been one of great controversy.[15] The reactivated world economy of 1933 was most significant in bringing an end to widespread unemployment. The National Socialist regime's job market measures were symbolically effective – it is one of the claims consciously spread and always repeated by the National Socialists that has survived through to the present day. Seen externally, National Socialist economic policy brought astonishing successes after 1933. There was a rapid rise of the economy in a country more strongly affected than all other countries by the worldwide economic crisis, and a change in its economic system towards a state-planned and controlled order, without resulting in private property losing its economic function.[16] Despite the 'socialist' character of the National Socialist Party's propaganda, with its model of a uniform and thus strictly hierarchically organized 'national community', the party supported businesses above all. The representative bodies for labourers and unions were abolished on 1 May 1933 – the 'Day of the German Labourer' – their properties being confiscated. They were replaced by the so-called 'German Work Front' (Deutsche Arbeitsfront – DAF), an organization compulsory for all employers and employees. With the wage freeze declared in 1933, which kept wages as low as they had been during the world crisis, nominal hourly wages increased until 1936 by only about 5 per cent. In terms of real buying power, they stagnated.[17] Property belonging to non-Jewish-owned businesses as well as their commercial and trade freedom remained untouched until the Second World War. Business incomes were not cut by the state and increased rapidly through the economic upswing and wage limits, which became an incentive for production expansion and investment.[18] Although the Nazi state could virtually dictate certain areas of the economy through its mass of armament contracts, demanding as it did larger production capacities for its military needs partially through contracts or direct subsidies and investment grants, the state did not take any measures, at least for the first three years after their rise to power, to 'militarize' the domestic economy's organizational structure. It was not until after 1936, when bottlenecks in production and materials created a need for regulation that the German economy came under comprehensive control.[19]

National Socialist armament policy and the four-year plan

References to armament were visible in the first job-creation measures after the seizure of power. The economic recovery seen in the Third Reich's early years was due in part to an armament-based economic trade. The share of armed forces expenses in the public sector's total expenses grew from 4 to 18 per cent in 1933–34, to 39 per cent in 1936 and finally 50 per cent in 1938. These expenditure figures point to a 'step by step degradation of economic policy into an instrument of armament policy'.[20] There were, however, various stumbling blocks to be eliminated. Until 1937 the Reichsbank and its president Hjalmar Schacht, who was responsible for financing armament and other 'important state' projects, pursued a monetary policy based on traditional criteria of stability. No opponent of Germany's forced armament, Schacht was, however, concerned about doing so on solid financing terms. In other words, he aimed to keep national debt manageable. This, however, made Schacht stand in 'strong opposition' to Hitler and Göring, which led him in turn to resign from his position as Minister of Economic Affairs in 1937.[21] Schacht was succeeded by Walter Funk, who was much more devoted and compliant to Hitler and did not attempt to limit the further expansion of national debt.

Large military projects were made possible by preliminary financing and a policy of non-stop policy credit, which in the long run placed a great burden on the national budget. In order to hide this essentially short-term national debt from the public, the Nazi regime made use of civil law regulations to create a dummy organization or limited liability company, the 'Metallurgischen Forschungsgesellschaft mbH' (Mefo). It was to appear to be a legitimate company dealing with the most important German armament businesses and its task was to provide a second 'valid' signature – obligatory under the rules of the Reichsbank – for the bills with which the Reich paid the arm supplier. The issuing bank could then cash these 'Mefo-bills' without violating the law – if only in letter and not the spirit of the law. Ultimately these tax credits amounted to the postponement of debt payments for the future. Not accidentally, the new finance plan was declared in June 1931 simultaneously with the Credit Policy Enabling Act. This act gave state leadership free rein over its credit and finance policy, which allowed for unhindered financing of the upcoming war.

As millions of previously unemployed people were reintegrated into the work force, their buying power increased, even amongst those whose income level remained comparatively low. This was in part due to the one-sided priorities of production and import policy, which kept the supply of consumer goods and food items down. 'An increasingly rapid military rearmament stood in conflict with the population's consumer necessities; satisfying both exceeded the possibilities of the German economy.'[22] Price-fixing, wage-freezes and 'Volk' savings accounts (for example for buying a Volkswagen) were the state's attempts at limiting public consumption. Formally, businesses still possessed a relatively high degree of freedom in decision making, while the state determined demand but also prescribed production capacity by controlling the allocation of raw materials. The question of the allocation of commodity production became increasingly important, meaning that from 1937–39 the raw-materials authorization paper for production became more important than actual financing. Alongside the increasingly critical shortage of workforce, finding materials presented the most difficult problem. In 1937 Richard Linde characterized the four-year plan for the Linde Company as 'a measure not well-thought out in its consequences, one which now demands all possible help and has led to circumstances, which are in part worse than those of the [First World] War'.[23]

Hermann Göring, who was appointed minister of the interior, Prussian prime minister and air commissioner, exploited quarrels between the army and the Ministry of Economy skilfully and strategically in order to extend his own sphere of influence over the armament economy. In May 1936 he instituted the 'Raw Materials and Foreign Exchange Staff' under the direction of Lieutenant Colonel Fritz Löb, which included leading personalities from the economy, including the IG Executive Board member Carl Krauch, who was in charge of all chemical issues. Löb and his colleagues succeeded in the following months to standardize and systematize their efforts, to make the German Reich independent of raw materials imports as much as possible. Löb's suggestions were not realized in this form, but later formed the basis for the planning of the four-year plan from October 1936.

As the Second World War began in September 1939, the aforementioned quarrels over authority and the lack of co-ordination between the four-year plan authorities under Hermann Göring, the Reich Ministry of Economy under Walther Funk and the Defence Economy and Armament Office of the Supreme Command of the army (OKW) under the leadership of Georg Thomas (1890–1946) were not resolved. Hence a radical restructuring of the economy for the needs of the war with armaments playing a central role failed to materialize at the beginning. At first, consumer goods industries continued to produce as they had, with few changes. Big companies in particular defended themselves against a centralized war economy. It was in fact Fritz Todt, who, after his appointment to the Ministry of Armament and Munitions in March 1940, made it his goal to tie industries already involved in armament production more strongly to the war economy. When Todt died in a plane crash in February 1942, it was his successor, the architect Albert Speer, who succeeded in expanding the ministry's authority and influence in the move towards a total war economy.

The policy of autarchy – synthetic petrol and rubber

When the National Socialists seized power, the goal of attaining economic independence from the import of raw materials became one of the Nazis' top economic priorities. The Linde AG was to benefit greatly from this goal. In the four-year plan of 1936, these goals were codified. Thus Fritz Todt, the director of the Office for Technology and inspector general for German roads, announced at the Nuremberg party conference in 1936 his justification for the major investment in autarchy: 'These three big areas: textile materials production, fuel for our motorization and creating what we lack in ores and metals, are along with providing food, the most important to us economically and politically.'[24]

Laws and regulations were thus introduced to keep industry and handicraft labourers from using imported raw materials so as to save on foreign currencies or make them available to other commercial businesses. The use of so-called synthetic 'domestic materials' was rather counterproductive for the national economy, since production costs were often considerably higher than those of the actual natural products. From the perspective of the National Socialists, its military-political interests justified this additional expenditure. For

> even where technical maturity and an advanced amortization of investment costs cannot lead to the price of domestic materials dropping to that of foreign products, one must remember to subtract from these additional expenses – which were necessary to prevent the use of cheaper foreign goods – that by which the entire national economy and therefore each individual has benefited, namely the fact that the specific domestic material was made by German labourers and engineers with German machines and within the German national economy.[25]

IG Farben played a central role in the National Socialist self-sufficient economy of autarchy. As the world's largest chemical company uniting Bayer, Hoechst, BASF, and other producers, IG Farben offered in 1925 a broad spectrum of research and production competencies for synthetic replacement materials. During the First World War, IG-Farben's predecessor, BASF, had found a replacement material for saltpetre using a high-pressure ammonia synthesis (the Haber-Boscher method), which made it possible to produce explosives despite the allied forces' naval blockade. This great technological breakthrough – for which Haber and Bosch received the Nobel prize after the First World War – was also a great success economically, even if some historians of technology argue over whether the focus on the synthesis of ammonia produced more disadvantages than advantages for the company. Thomas Hughes, the famous American historian of technology, is convinced that the technological-economic success of ammonia synthesis put IG Farben in the 1920s on the path towards synthetic petrol (and high-pressure synthesis). Hughes speaks of the company's crucial players being caught up in a 'Technological Momentum' that left them wanting a repeat success. Because of their policy for a self-sufficient autarchic economy, the National Socialists were highly interested advocates of petrol synthesis. This fact was, according to Hughes, one of the reasons why IG Farben financially supported the NSDAP, despite the large number of Jews in its management.[26]

The first ideas for replacing the very expensive natural raw material for rubber with a synthetic product already existed by 1906. The drastic decline of rubber imports during the First World War as a result of the blockade led to the construction of a pioneer installation in Leverkusen. However, this 'artificial rubber' was a failure, inferior as it was to natural rubber. It was not until the end of the 1920s, as researchers at IG Farbenindustrie AG developed the synthetic rubber Buna from *Bu*tadien with the use of sodium (*Na*trium) as the catalyst, that a synthetic material became available that approximated the natural material in its properties. However, the growing surplus of supplies of natural rubber on the world market kept prices so low that synthetic production became unprofitable. The manufacture of synthetic rubber more or less came to a halt by the autumn of 1930. Yet research continued and at the end of 1932 IG Farben introduced 'Buna N', a product that was even superior in its properties to natural rubber.[27] The rise of the National Socialists to power opened up new possibilities for these synthetic rubbers, as motor vehicle needs, armaments, and above all the goal of a self-sufficient autarchic economy in preparation for war created a great demand for 'domestic' synthetic rubber.

In 1936–37 the IG-Farben group established the Buna factory in Schkopau near Merseburg, following successful production in research facilities. In 1935 the research centre of the Schkopauer factory delivered only 2,400 tons; this meagre quantity could be later increased to 30,000 tons per year. IG Farben thus controlled the German production of synthetic rubber completely. From 1938 on in Germany, natural rubber was systematically replaced by Buna rubber; by 1939, 5 per cent had been replaced, in 1940, 61 per cent, 1942, 73 per cent, and in 1944, 97 per cent.[28]

The conversion of the solid fuel coal into liquid fuel was based on ammonia synthesis. This was a process which Fritz Haber had described in 1908 and Carl Bosch performed on a large technological scale in praxis. In this procedure ammonia would be synthetically manufactured at high temperatures and under very high pressures (200 At) from nitrogen and hydrogen. The first production of synthetic ammonia was in Oppau, Schlesien in 1913. In 1916, during the war, a large synthesis factory was built with the support of the state in Leuna (Sachsen). After the First World War Friedrich Bergius demonstrated that high-molecular carbon-hydrogen compounds in coal can take up hydrogen if they are subjected to high temperatures and very high pressures. The hydrogen needed for hydrogenation was obtained from conversion installations for water gas, coal, and coke. Liquid fuels could be obtained from coal through the described hydrogenation process, but also via synthetic methods. The synthesis of fuel according to Fischer-Tropsch could be implemented without pressure and at low temperatures. The synthetic gas would be converted into a fluid mix of carbon and hydrogen in the contact ovens. Fuel gases, petrol, and diesel type carbon-hydrogen products could thus be obtained from this mixture.

In the so-called Benzinvertrag (petrol-agreement),[29] which IG made with the Reich on 14 December 1933, and which was retroactively effective as of 1 December 1933, IG declared itself ready to triple the production of synthetic petrol by 1935. In return, the Reich made the commitment to purchase from the group those quantities of fuel non-sellable on the free market, and guaranteed altogether the reimbursement of actual production costs plus a 5 per cent profit margin. In order to increase the production numbers in so short a time, IG Farben had new plants built quickly through its subsidiary and partner businesses. Linde was the primary supplier of oxygen installations to these businesses.

Linde in the Third Reich

Linde and National Socialism

For an internationally active high-tech business with a global network such as Linde, National Socialism presented both an economic threat and an opportunity. It posed a threat because contacts with foreign markets suffered, restrictions were placed on trade (only some with the Soviet Union until 1939), travel was limited, employees were lost and the problem of foreign currencies made operations difficult. It brought opportunities, because the company came to occupy a privileged position in the ruling class of the National Socialist system, and because the political system itself increased demand for certain products (oxygen for instance). Thus, Linde profited considerably from the self-sufficiency autarchic policy of the National Socialist regime as the supplier for IG Farben's economically unprofitable synthetic petrol production. What was the Linde AG and its management's position towards the National Socialists?

When, in the summer of 1933, as a well-respected university professor, Carl von Linde received a written request to fill out a questionnaire regarding his ancestors according to the anti-Semitic 'law on the re-establishment of civil servant status', which removed Jewish civil servants from the governmental authorities, he rejected the request in a postcard and left it to the college to cross his name from the list of professors. This was certainly less an act of resistance than an expression of independent conviction and the confidence of the well-known and popular Munich business run by scientists.

As general director of the company, his son Friedrich Linde was in a different situation. Indeed, he took a stance far removed from that of National Socialist ideology and did not become a party member. However, as 'director of operations' and in order to safeguard the company's interests, he was unable to avoid National Socialist state institutions unlike his father and brother. This was true of him personally and also of his readiness to transfer responsibility to National Socialists within the company. In 1935 Friedrich Linde held the title of a 'Military Economy Leader'. With that he became at least nominally part of the National Socialists' so-called 'Military Economy Leader Corps'. This body was created in March 1935 on the suggestion of the Military Economics Staff and based on the legal foundation of supernumerary law.[30] The military economy was not an economical system, nor an economic procedure or organization. It was instead a form of voluntary commitment for participating businesses.[31] The acceptance of the title 'Military Economy Leader' was for Friedrich Linde less an expression of personal agreement with the aims and methods of the National Socialist state than an attempt to profit from armament through contracts and perhaps also an indication of his efforts to shield himself and the company against the authoritative infringements of the state.[32] Oetken, later chairman of the Supervisory Board, remembers that the title awarded Friedrich Linde 'respect among the powerful of the time'. Being awarded the title expressed the significance of the gas and refrigeration industry and the Linde Company. It was because of this title that Friedrich Linde was put in solitary confinement at

Stadelheim prison from July to November 1945 and was temporarily forbidden to be active in the company.

Friedrich's brother Richard had a different position in the company and could and wanted to express himself more unequivocally against the National Socialist regime. Furthermore, two of his closest colleagues were of Jewish origin, whereas Friedrich's closest colleague in Höllriegelskreuth was the operation leader Dr Alfred Hess, an uncle of Rudolf Hess, Hitler's deputy. Already in 1932 Richard Linde clearly expressed his thoughts on the NSDAP to his sisters; he saw in the National Socialists and their followers 'the great party of dissatisfaction, just like the Social Democrats had been before the [First World] War, and they [all] vote for Hitler for various reasons'. His presentiment was gloomy rather than optimistic: 'That is exactly what is dangerous about the National Socialists' he wrote in 1932, 'that they will surely – once they come to govern – like their ideal Mussolini – never let go of power again, even if they only have a small minority of the people behind themselves'.[33] Richard Linde decided in 1933 not to take any public office in National Socialist Germany and for this reason he even turned down the chairmanship of the German Association of Refrigeration Technology (DKV),[34] although after the war he took it over for a while. Rudolf Wucherer was also able to steer clear of any political involvement in the Third Reich; his only 'office' was that of the air-raid warden. Despite the Linde Company's success, Richard Linde became increasingly depressed and sullen under the National Socialist regime. As we have seen, he lost his two elder sons, Helmut and Werner, in the war to which he was strongly opposed. His third son, Hermann, became unfit for service because of heavy injuries in 1940 in France and survived the war, as did the fourth, Gert.[35]

While the Lindes had no direct contact with the National Socialist regime other than as leaders of an engineering business, they were close neighbours of Hitler. The 'Haus Baumgart' built by Carl von Linde on Obersalzberg was only 350 metres away from Adolf Hitler's 'Haus Wachenfeld', extended to 'Berghof' after 1933. The house was mainly used by Richard Linde and his family. In the first few years, Richard Linde found it merely unpleasant to have Hitler as a neighbour. After 1933 however, he was forbidden to drive his car on Bergstraße to his house. In addition, the friends and paladins of Hitler settled in the nearby houses. There were 'rivers of curious pilgrims on the Obersalzberg, in order to see the "Führer" or to at least take a picture with the door handle at the garden door of the Berghof in the hand'.[36] Increasingly large security zones were established surrounding Hitler's presence on the Obersalzberg. In December 1936, Richard Linde shared with his sisters, in a family letter, that the house and the plot of land on the Baumgart had to be sold and evacuated by February 1937.[37] The compensation was generous enough to prevent the Linde family from sustaining a financial loss, yet the displacement of the heritage of Carl von Linde, and particularly that of Richard Linde, is to be understood symbolically as well – it was not to be tolerated in the vicinity of the new rulers. After the war, the American army moved to the Obersalzberg, and the family could not buy its holiday residence again.

Figure 3.1   Morning allegiance in the workshop in Höllriegelskreuth during National Socialist times.

The two Linde brothers sharing responsibilities in Höllriegelskreuth at the time did in fact shape the atmosphere of operations.[38] However, there were also National Socialists present,[39] including Alfred Hess. Hess had been at Linde in Höllriegelskreuth since 1921 and had distinguished himself through his organizational measures, by introducing 'modern business in matters of contracts', and the notion of piecework in the middle of the 1920s, which had previously been considered untenable in Apparatus Engineering with its high share of handicraft and various production tasks.[40] A fundamental and lasting arrangement was the educational workshop organized by him in 1924, which trained mainly coppersmiths; a systematic training of apprentices did not exist at Linde beforehand.[41] In 1938 Friedrich Linde had appointed Hess as works manager and deputy plant manager.[42] Hess retired in 1942 after his nephew, Rudolf Hess, flew to England. At this point, he was already 70 years old. His successor was Ulrich Hailer, who had done outstanding work from 1928 until 1942 as assemblage and operations engineer. As operations director he had the difficult task of carrying production through all the perils and troubles of the late phases of the war and the years after the war; he died during an accident in the workshops in 1950.

Starting in 1936, Alfred Hess also directed Linde's newly founded business magazine. The department of 'magazine management' (Schriftwaltung – 'Redaktion' being scorned as a foreign word) was taken over by the advertising manager Janßen in Sürth. Department B had already in 1926 adopted the 'Altbayerische Werkswoche – the Weekly Magazine for the Community of the Bavarian Industrial Factories' and distributed it free of charge with good reaction among factory members. The factory magazine was published 'in collaboration with the Society for Labour Pedagogy [Gesellschaft für Arbeitspädagogik] in agreement with the German Work Front [Deutsche Arbeitsfront]'. The influence of Dr Carl Arnhold, the director of the German Work Front's office for Work Leadership and Vocational Schooling, was apparent here. To be sure, the magazine reported mainly from the company's factories and about individual 'Gefolgschaftsmitglieder' (the National Socialist expression for blue-collar labourers), yet National Socialist propaganda – above all that of the German Work Front – was also printed.[43] However if Friedrich Linde or Rudolf Wucherer ever took positions in the magazine, they did so in a neutral manner as far as party politics was concerned.

Jewish employees at Linde

Having seized power, the National Socialists engaged in an unprecedented smear campaign against Germany's Jewish citizens. This campaign robbed the Jewish population step by step of its rights, until finally deporting and murdering them. Already in the first weeks after Hitler's government took office on 30 January 1933, new legislation made it apparent that National Socialist terror was not only a temporary symptom of a 'revolution', but that it had set the goal of removing the Jews from all areas of public life, denying them every future commercial and social activity. The regime soon began to put pressure on non-Jewish partners or owners of businesses in order to force the removal of Jewish members from companies. At the NSDAP's Reich Party Conference in 1935, the 'Nuremberg Laws' were announced, which were directed above all against German citizens of Jewish faith.[44] This legislation denied Jews their civil rights; by decree of these laws, the 'members of the state' were to be distinguished from 'Reich citizens'. Jews (or people with parents and grandparents of Jewish origins), by definition, were not to be amongst Reich citizens; so-called 'mixed marriages' between Jews and 'Aryans' were forbidden from then on. Existing 'mixed marriages' were tolerated for a while, but the state's pressure on 'Aryan' partners to annul their marriages intensified increasingly. Exceptions to these rules (for example those regarding former 'front fighters', German-Jewish soldiers who had fought in the First World War) were increasingly restricted or abolished through 13 amendments in the ensuing years in order 'to protect German blood and German honour'. On 'Reichskristallnacht' – the night of 9–10 November 1938 – when SA members and National Socialist party supporters systematically plundered, destroyed, and partially burned Jewish synagogues, establishments, and businesses, 260,00 male German-Jewish citizens were arrested. In addition to the many instances of harassment and attacks against Jews, the National Socialist regime also issued an order on 12 November 1938 to raise 1 billion Reichsmark in 'atonement funds' for the damages incurred during Reichskristallnacht. The money was to be provided through private assets, since insurance companies were instructed not to regulate the damages.[45] The exclusion of Jews from German economic life progressed rapidly; stocks, real estate, and business shares were already forcibly 'aryanized' – that is, they were transferred to Germans, often for ridiculously small compensation payments. A Jew seeking to leave Germany in 1938–39 was forced to pay a 'Reich Escape Tax', which could amount to 50 per cent or more of his or her capital. Those Jews able to leave Germany after 1938 and find admission abroad (many countries limited their admission of Jews or closed their borders to them altogether), could rarely rescue any of their possessions other than the clothes on their backs. Good connections were more important than money at the time – both in order to leave Germany and to find citizens in the country of exile who would commit to bearing the financial burdens of residency or provide a workplace for the exiled person.

Some of Richard Linde's Höllriegelskreuther employees were classified by the National Socialists as Jewish and were therefore persecuted. This included two people with crucial influence on the company's development at the time: the engineer Philipp Borchardt, director of the Assemblage Department, and the chemist Franz Pollitzer, an early assistant to the Nobel Laureate Walther Nernst and director of the Chemical Labour. Both of these men worked within the company in their leading positions until they were arrested on 'Reichskristallnacht' and were taken away to the concentration camp[46] in Dachau. In a family letter to his sisters on 7 December 1938 Richard Linde writes:

In the last weeks I have had plenty to do, because my two closest colleagues were arrested as non-Aryans. One was released 10 days ago, but must first recuperate somewhat and is not allowed to return to the office, but he can work for us at home. The other will hopefully come tomorrow. One cannot write or speak of this matter, without vomiting.[47]

Great efforts were required to have them released and succeeded 'only due to intervention by some influential National Socialists and the reference to the fact that these Jewish employees were responsible for the fulfilment of important state contracts related to the production of synthetic petrol'.[48] However, both men did not resume their work again after their release from the concentration camp. It had become abundantly clear that the lives of Pollitzer, Borchardt, and four other Jewish employees were endangered by continuing to stay in Germany. Richard Linde personally arranged to have his colleagues sent to the competitors, L'Air Liquide in Paris and the British Oxygen Co. in London. Linde's traditionally good connections with both important European rivals paid off.[49] At the beginning of 1939, Borchardt left for England and Pollitzer for France. In contrast to the majority of German Jews they did not lack capital for emigration.[50] Philipp Borchardt found a relatively happy solution in 1939: he was entrusted with bringing an ammonia synthesis installation set up in England into operation and was also made operation leader of this installation.[51] However, after the outbreak of war in 1939, Franz Pollitzer was confined as a German civil prisoner, caught by the Gestapo after German troops marched into

France in 1940, and was gassed in 1942 at the concentration camp of Auschwitz.[52] Philipp Borchardt survived in England and returned to his old workplace in Höllriegelskreuth after the war. In 1946 Richard Linde unsuccessfully mooted Borchardt his successor on the Executive Board of the Linde Company. Borchardt remained in the second tier of leadership until his retirement. Two other Jewish employees, Dr Lothar Meyer and Dr Paul Schuftan managed to avoid the concentration camps. Lothar Meyer (in the Chemistry Department from 1932) was able to elude the persecution just in time in 1938, and Paul Schuftan had already immigrated to England in 1936 to assume a position at the British Oxygen Company. After the war he visited Linde as an English officer. His interest was above all in the successes accomplished until then in the area of ethylene production.[53]

In 1931, two years before the rise of the National Socialists to power, Linde had acquired half of the shares of the Marx & Traube GmbH[54] in Frankfurt am Main, 'in order to give its factories the additional business of supplying this company'.[55] Erich Marx kept the other half of the shares and led the business after becoming general manager. Marx had become a market leader in tools and tool machines for motor vehicle maintenance by constructing and compiling special equipment for cylinders and other automobile engine parts. The company delivered tools and equipment to most of the major automobile and truck manufacturers.[56] However, the company was on the verge of bankruptcy due to the economic crisis. The company was interesting to Linde because Marx himself built only a tiny part of the tools and machines, and – like Linde earlier – manufactured under licence, which he had acquired largely in the USA. After Linde's takeover, the under-operated Güldner division became involved in the licenced production of special tools, equipment, and machines.[57]

Marx was of Jewish background and when 'as a result of the political circumstances an Aryanization of the Matra factories became necessary', he offered the remaining 50 per cent of his business shares in the company to Linde. Linde bought these shares on 4 November 1935 for another 150,000 RM. Erich Marx left the company as partner and manager[58] and Linde appointed from the staff Kohlmeyer and Lauer as new managers. At this time, with 73 employees and an annual turnover of roughly 1 million RM, the company was a medium-sized business. A law in 1937 demanded the names of (former) Jewish owners of the company be crossed from the indices, prompting Linde to change the name of the company on 20 January 1938 from Marx & Traube to Matra Works. In this way the names of the founders were preserved as an anagram. Marx successfully fled Germany and fought a legal battle with the Linde Company after the war over the rights to regain his business shares or appropriate compensation.

Business development in the 1930s

'For the Linde Company, the years between 1936 and the outbreak of the Second World War produced increasing turnover and profits.'[59] This however, did not mean that all departments turned a profit. The total losses for Department A in 1936 amounted to 1,869,440.29 Reichsmarks (RM); commercial refrigeration, as well as cold houses and the stadium all showed losses, just like Department D (loss: 544,331.95 RM) and the headquarters in Wiesbaden (loss: 605,063.54 RM); a positive balance could be shown only by offsetting the negative balance with assets and reserves (net earnings of the balance in 1936: 1,864,781.08 RM, gross earnings including write-offs and reserves 11,310,204.46 RM). Department C made a modest gain of 14,373.63 RM; the burden of the whole enterprise was carried by Plant Engineering alone, which achieved a net profit of 4,869,243.23 RM; together with the special amortization and the reserves the gross earnings of section B were 9,801,363.52 RM.[60]

Refrigeration units Department A: cold house success

Starting in 1933, Refrigeration Engineering slowly began to rise out of the crisis and domestic business exhibited signs of development once again. The number of contracts had increased by one-third above that of the previous year, almost re-obtaining the amount in 1931. In the following year, 1934, turnover was about 25 per cent higher than in 1933, but still around 50 per cent below the best values of the late 1920s. Despite the recovery, the Commercial Refrigeration department incurred further losses until the beginning of the Second World War. This was due to the fact that production capacity and workforce numbers had dropped, making the department unprofitable and that prices never once topped expenses 'as a consequence of exaggerated competition'. Linde had not yet learned to survive in a market of competition focused more on price than innovation – that is a market in which optimizing production was crucial. This continues to be an issue for the engineering-oriented Linde Company.

In the 1930s, the domestic refrigeration factory in Mainz-Kostheim produced refrigerators, refrigeration installations for confectioneries, display cases, cold shelves, ice creameries, and cold cans; in addition the factory worked for Matra. For the factory in Sürth, Kostheim delivered equipment for the rapidly growing freezer economy.'[61] The focus here was on freezer systems for storing army provisions and food supplies (frozen fruits, vegetables, fish) – the demand for frozen consumer goods had yet to develop in the Germany of the Thirties and Forties. The majority of the population did not own refrigerators, using instead traditional storage and conservation techniques to keep food supplies fresh.[62]

The frozen storage of food supplies was a new area of refrigeration technology; the selection of food items in question as well as economical freezing and defrosting techniques had to be researched first. In 1936, Professor Rudolf Plank, the most famous refrigeration technology expert between the wars, founded the 'Reich Institute for Foodstuff Preservation' (Reichsinstitut für Lebensmittelfrischaltung) in Karlsruhe. Together with food supply and packaging industry companies, the army established in Munich the 'Institute for Foodstuff Research' (Institut für Lebensmittelforschung). Linde provided testing facilities for these institutes.[63]

Figure 3.2 Refrigerator manufacturing is still manual work: the wood shop in Mainz-Kostheim, 1926

The cold houses belonging to Linde's Ice Machines had in 1933 shown increasingly unfavourable economic results compared to 1932. It was not until 1934 that turnover began to improve due to strategically enacted 'Reich measures to preserve agricultural products'.[64] In the following year things rapidly began to improve for cold houses as a 'consequence of Reich measures to secure the population's food supply'.[65] Linde's cold houses were an important link 'in the chain of Reich measures to secure food supplies for the people'.[66] Thus, after a long pause, Linde began in 1934 to build new cold houses and as of 1935, skating rink production began to boom. The first ice rink opened in Nuremberg at the end of 1935.[67] In 1936 Linde sold similar installations to Hamburg, Munich, Krefeld, Dortmund, and Cologne as the Winter Olympics made ice sports more popular. To be sure, Linde's own installation in Nuremberg was no economic success, but it nonetheless served as a model for cold houses, and became a profitable business sector for Department A.

Beneficiaries of the National Socialist policy of autarchy in Department B

Just as there was a delay in the economic crisis reaching the department of Plant Engineering and Construction due to the extended amount of time needed for such work, economic recovery also lagged behind. As of 1934 in Höllriegelskreuth, there was 'no sign of an economic recovery other than an increase in inquiries'. The primary business for the future was already emerging, namely 'larger factory projects for the manufacture of artificial petrol, a pet issue for the government'.[68] Central to the issue of construction was the need for new gas separators, 'which we build according to a new procedure, for which we alone have the licence in Germany'.[69] Over the next few years, Apparatus Engineering in Höllriegelskreuth was thus entirely devoted to oxygen installation units used for synthetic fuel[70] and artificial rubber (Buna) manufacturing. Thus Richard Linde reports to the Supervisory Board in 1938: 'The installation delivery business of Department B continues at a capacity that can hardly be exceeded keeping us busy with 52-hour working weeks in the first quarter of 1940.'[71] In a family letter to his sisters, he sarcastically stated: 'We are doing as well as other companies, the director of one having said to us recently, "The amount of work is absurd, but this is alleviated to some extent by the lack of materials."'[72] As a central figure in Höllriegelskreuth, Richard Linde gathered in the 1930s and 1940s a host of competent young scientists, together with whom he developed new methods. Thus Dr Rudolf Schlatterer (1936), Dr Johannes Wucherer (1937) and Dr Bruno Hippenmeyer (1943),[73] and to complete Pollitzer's team in the Chemistry Department, Dr Fritz Rottmayr and Dr Josef Weishaupt, began working for the company. In the years 1933–43 the number of salaried employees grew from 108 to 255 (and the number of industrial labourers from 168 to 517).

Linde built the first four oxygen installation units for manufacturing fuel from coal in 1935 for the Braunkohle-Benzin AG, Berlin, in Böhlen near Leipzig; and four more in Magdeburg with 2,600 cubic metres oxygen output per hour each. Four installations of the same size for the Mineralöl-Baugesellschaft in Zeitz followed in 1937, three installations for the Aktiengesellschaft Sächsische factories in Böhlen near Leipzig in 1938, two installations for the Braunkohle-Benzin AG in Zeitz, a small installation for the Wintershall AG in Lützkendorf, then in 1939 alone, six installations for the Mineralöl-Baugesellschaft in Brüx, and an installation for the Braunkohle-Benzin AG in Böhlen; in 1940 four installations for the Sudetenländischen Treibstoffwerke in Brüx, three more installations for the Wintershall AG in Lützkendorf and two installations for the Aktiengesellschaft Sächsische factories in Böhlen and in 1941 two installations for the Sudetenländischen Treibstoffwerke AG in Brüx. This was all essentially a gigantic construction programme aimed at preparing Germany for war. The biggest single client was IG Farben and its subsidiary, Ammoniakwerke Merseburg GmbH, to which Linde supplied seven oxygen installations for Leuna and six for Heydebreck in Oberschlesien (with the stamp 'for Luftfahrtanlagen GmbH, Berlin'). The contracts were a lucrative business for Department B, not only because a large number of installations were to be built according to the existing system developed by Linde, but also because, with the introduction of hydrogenation, new types of low-temperature technology had to be developed. The production of synthetic petrol was at least for a while a great success for IG Farben too, not only because of guaranteed state purchases, but also because of significant income from royalties abroad, particularly from the United States, IG Farben's most important licencee.

Due to the reasonable price of oil, the production of synthetic petrol nearly stopped during the peace economy in the years after the war. Only during the oil crisis of the 1970s did it experience a brief comeback in the areas of research and development. During the coal chemical industry era, Linde developed competencies in plant engineering and construction, which it could later apply to petrochemistry, such as ethylene plants, which will be discussed in the fifth and sixth chapters. The same was true also for the German chemical companies, which adjusted themselves astonishingly quickly to petrochemistry after the war.[74] The prevailing mood in the German chemicals and gas industry was that everything seemed easier with oil.[75]

Department C: guaranteed profits for gas factories

Of all departments at Linde, Department C (gas factories) profited the most from the economic upswing after 1933. The sale of condensed gases 'for which we have distributed 24 installations across Germany' was by 1934 'no longer so far away from the good years of 1928 and 1929' (with oxygen). Indeed, as was the case with acetylene, sales in 1934 even 'surpassed them by far',[76] rendering the financial results of this 'most important business branch' at Linde quite satisfactory.[77] Sales continued to increase in 1935, although production capacity at the factories had yet to be exhausted. Bottling stations had to be expanded so as to keep up with deliveries to all customers. The annual yield of this division within the company was limited due to intervention on the part of the Reich Ministry of Economics, which 'suggested' in the autumn of 1935 that Linde lower its price of oxygen, as is cautiously reported in the business report of that year.[78]

Because sales for condensed gases continued to grow, the oxygen factory in Nuremberg was moved in 1936 from its location next to the cold house and ice factory to newly acquired premises. The same action was to be taken in Dresden in 1937. In 1938 the oxygen factory in Hamburg-Wilhelmsburg – shut down since 1927 – and the oxygen factory in Mühlheim-Ruhr – abandoned since 1931 – were brought back into operation again. Factories in Untermaubach, Reisholz, and Braunschweig were expanded; and production at the remaining factories was high 'in some cases pushed to the limit of their capacity'. Although having expanded, the bottling station could not continue to do so in 1938 due to the steel shortage which disrupted 'smooth service to clients'.[79]

In the 1930s[80] Linde made an astonishing variety of diverse restricting mutual agreements with various domestic and foreign gas manufacturers, for the production and delivery of oxygen, nitrogen, rare gases, and carbon dioxide, as well the construction of gas separator installations. This was made public after the war when the occupying forces sought to break apart market-controlling cartels in Germany.[81] The buyers of Linde oxygen installations were thus forbidden from purchasing surplus oxygen, licensees were forced to agree to delivery restrictions, and Linde tried to reach agreements with competitors on sales, regions, prices, and deliveries, so as to keep the price of industrial gases stable and able to turn a profit. The most important organization here was the Vereinigte Sauerstoffwerke GmbH in Berlin, in which Linde and the IG Farben each had half of the shares. By means of this enterprise, the two gas manufacturers would come to agreement on the sales volume and prices.[82]

Machine engineering in Güldner D: losses despite growth

After 1933, Linde's subsidiary, Güldner-Motoren-Werke, produced primarily diesel engines, including large engines for ships, small ones for construction machines, locomotives, and tractors (for example Kramer tractors). Compressor parts were also manufactured. Although turnover increased after 1933, this business segment in the company experienced losses after the economic crisis, as production was simply too expensive. The hope for a growing and profitable business segment was linked to the new contracts for constructing small diesel engines. Above all, after 1937 the domestic sale of diesel engines for airplanes increased considerably, and it became necessary to enlarge the plants significantly. The 'progressive motorization of farming and the lively farming activities' led in 1938 and 1939 to an even faster increase in small diesel engine sales, as did those of large ship diesel engines.[63] From 1938 on, leaving the area of diesel engines, Güldner began in 1938 to produce its own farm tractor with a standing one-cylinder, four-stroke engine with 20 PS power. This became in the 1950s the most profitable division within the company for some years.

The other side of domestic business: increasing export problems

In order to keep the isolated domestic economy of the German Reich going, the economy's relation to the outside world had to be fully restructured.

> The most important measures were the bilateralization of foreign trade, the quantitative restriction of imports 'vital' to the national economy, in other words, imports needed for the armament economy, the targeted support of exports with a complicated system of Reich Mark devaluation against other important currencies, and finally redirecting trade flows from Western Europe and North America to Eastern and Southern Europe and South America.[84]

Rather than retreating, Germany's interaction with the world economy merely shifted regions, as the German economy focused on particular products or production groups. In 1937, the German engineering industry succeeded for the first time in surpassing England and the USA in exports. Germany's share of the world market was at 37.6 per cent and stood ahead of the United States and England, each with 35.6 and 26.8 per cent respectively. In terms of value, machine exports made up a quarter of machine manufacturing in Germany. 'This export is especially significant for the German economy, precisely because the machine industry brings in more foreign currency than it spends.'[85]

Within the hierarchy of the National Socialist's economic goals, exports played a crucial role until the beginning of the Second World War.[86] However, because of competition for materials allocation, armament held the greatest leverage over exports, despite the government's otherwise-sounding declarations of intent.[87] Furthermore, the National Socialists' aggressive racial and foreign policies led particularly in the latter half of the 1930s to international criticism and boycotts. The circle of countries for export grew increasingly small.

The world economic crisis had led to a massive decline of foreign contracts in all Linde departments; Hitler's economic policies hindered the resumption of normal relations. Because of the 'efforts intensifying everywhere towards protecting domestic industries', Department A's foreign trade had strongly decreased from about 50 per cent in 1931 to 25 per cent of an already decreasing turnover in 1933.[88] From 1935 on, exports were on the rise again. However, losses were sustained due to price competition and Refrigeration Engineering adjusted poorly to such circumstances.

Department B experienced a recovery in installations exports in the mid-1930s. Richard Linde wrote in 1936: 'We have at the moment a relatively high number of foreign contracts, in particular with Russia and Japan. There are currently a number of potential contracts in Japan, so many, that Mr. Borchardt has been sent there for negotiations.'[89] At the same time, the department suffered from externally induced and poorly co-ordinated intervention. Partially frustrated, partially bemused, Richard Linde writes: 'About one-half are foreign contracts, which are always a matter of priority – also for the German government – the other half are for the four-year plan, where one thing is always purported to be more urgent than that ordered right before.'[90] Yet after 1937, the number of foreign contracts dropped significantly.[91] Richard Linde would have preferred to produce primarily for export, and not only because he had relatively little difficulty in obtaining materials thanks to the incoming foreign currency through foreign contracts.[92] However, Apparatus Engineering had become so occupied with oxygen installations for synthetic fuel and rubber that it was no longer possible to handle export contracts in a timely manner. Linde certainly would have preferred to confirm his company's international reputation by delivering on time, but Apparatus Engineering was unable to fulfil even its domestic contracts in a timely manner. Because domestic contracts were of a higher priority, Linde remained behind schedule on its foreign contracts.[93]

The Linde Company's other domestic factories and departments manufactured on a much smaller scale for export in comparison to Departments A and B. Güldner, for example, manufactured stationary motors and profits remained narrow despite all efforts.

Because a large number of qualified young coppersmiths were conscripted into the army at the start of the war in September 1939, Department B's exports were hindered. In order to at least partially compensate for the drop in production resulting from the transfer of skilled German labourers into the military, the industrial capacities of those countries allied with Germany or those invaded by German troops were incorporated into the German war economy. With its delivery span of up to two and a half years which hindered the expedition of industrial production, the Linde Company was forced to employ a French subcontractor: 'As reported already in the last meeting, we had – indeed as decreed from above – to subcontract parts of large oxygen installations to Fa. L'Air Liquide in Paris.'[94] Linde wanted to render the further transfer of contracts to L'Air Liquide in occupied France independent of 'certain services to be given in return [technology transfers]'. With regards to gas separators however, Linde was reluctant to lay its cards on the table as long as the company had no financial investment in L'Air Liquide. The Reich Ministry of the Economy advocated an investment on the part of Linde: 'The R.W.M. has inquired through the business segment of Machine Engineering whether the question of an involvement is already being considered from our side; such participation would certainly be welcomed by R.W.M.' In Paris there was little enthusiasm for this idea. In the end, the partnership failed due to high share prices: 'At least 6,000,000 RM would be necessary to have a 25 per cent partnership, and the question is whether we should consider that.'[95] However, L'Air Liquide was 'itself absolutely ready to grant us a re-involvement in the Italian oxygen company S.I.O.', in which Linde had been a partner before the First World War.[96] The partnership was expropriated after the war, and L'Air Liquide held 75 per cent of the capital. The acquisition of these shares failed because of an Italian government law forcing the elimination of all foreign investment in Italian companies, regardless of the German–Italian 'pact of the axis forces'. Wanting to alleviate the load on Apparatus Engineering in Höllriegelskreuth, Linde had sought to establish an Italian subsidy, yet this too failed to come to fruition.[97]

Linde during the war

The onset of the Second World War brought significant changes to production at the Linde Company once again. Contracts increased to where they could no longer be properly handled, while Linde was continuously losing employees to the military. Linde resorted to using forced labour.[98] Thirdly, Linde had to confront the question of how to deal with operations or establish new manufacturing facilities in the conquered territories. And then, by 1942, Linde was increasingly faced with the destruction of its production facilities through air raids and the loss of its production stations in the east.

All company branches were in one form or another tied to armament production. The oxygen factories in Hamburg-Wilhelmsburg, Hiltrup and Illertissen each delivered 200 cbm oxygen per hour in fluid form to the Air Force,[99] while the refrigerator cabinet factory in Mainz-Kostheim produced and assembled mobile repair units for the military. The new welding department in Höllriegelskreuth, Ellira, 'during the war manufactured welding equipment for the armaments industry exclusively and in 1944 was involved with U-boat hull welding, although the boats were not used in battle'.[100]

As war production in Germany ran full speed ahead, the business report of 1943 reads: 'All our factories in the accounted business year ran at full production capacity.'[101] Therefore, considering the constant labour and materials shortage at Department B, Richard Linde notes: 'The prospects are satisfactory in the sense that we are provided contracts for this year and the next. How these are supposed to be conducted on time, we cannot tell.'[102] Starting in 1943 allied forces conducted heavy air raids, damaging almost all of Linde's factories and cold stores. By the summer of 1943, Linde had already built an alternate operation in Schalchen at Trostberg.[103]

Linde and the National Socialist rocket programme

The Linde Company's most spectacular contribution to the National Socialist regime was its participation in the rocket programme, which was conducted through a subsidiary, the Heylandt-Gesellschaft für Apparatebau mbH in Berlin-Britz. Linde's involvement through Heylandt could have been a cautionary measure planning for the post-war future. More likely however, the rocket experiment management led by Wernher von Brauns turned to Heylandt because of its expertise as a world leader in liquid oxygen transport vessels, which Linde was not.[104] Heylandt was also probably favoured because of its long-term collaboration with the rocket pioneer Max Vallier, with whom Heylandt worked on rocket engines until his death on the Berlin Avus race track. In the early 1940s, the company received major contracts as the central partner in manufacturing, storing and filling gases as part of the V2/A4 rocket weapons programme.[105] Heylandt planned and manufactured the liquid oxygen installation units for the fuel supply of this weapon, given the misnomer 'Retaliatory Weapon (Vergeltungswaffe)' (or V-Weapon), in Friedrichshafen, Peenemünde, Oberraderach, and in Eperlecques[106] on the French Atlantic coast. The fuel was 98 per cent pure alcohol, distilled from potatoes. Each plant needed its own large oxygen factory, as each rocket start required some eight tons of oxygen, from which at least 1 per cent per hour would evaporate. At the end of 1939, construction was begun on the first ever large-scale industrial liquid oxygen plant as part of the programme to expand research institutes and start production-line manufacturing in Peenemünde on the Baltic island of Usedom. The plant was put into operation on 27 July 1942. The production of oxygen using the Linde process ran around the clock in three shifts, producing 13 tons daily. The factory consumed an enormous amount of energy: 22000 kW. A second plant of the same size was built in 1942–43 in Oberraderach at Friedrichshafen. Because of the project's urgent nature, engineers and designers were transferred from Höllriegelskreuth as well.[107] Because operations at Heylandt ran less smoothly and were less co-ordinated than Richard Linde had expected, he sent Walter Ruckdeschel to Berlin on 18 May 1944, as he was 'one of our most competent engineers, who spent the last years working mostly in Italy, as manager of Heylandt'.[108] Ruckdeschel now led Assembly and Operations of the oxygen supply centres produced at the Heylandt-Gesellschaft für Apparatebau in Germany, in Austria, and France.[109] Some 65 per cent of the preparation work for the liquid gas aspect of the rocket programme was done ultimately at Linde in Höllriegelskreuth and Schalchen, since the tight deadlines of the military weapon office would not have otherwise been met.[110] Shortly after the end of the war, Americans came to the factory in Höllriegelskreuth to take the documents of the rocket programme with them. Paulus Heylandt was taken to the USSR in 1946 by the Soviet military, where he died in 1946 or 1947.

Matra's tool sets

Both National Socialist armament and the war provided the Matra factories with a rapidly increasing turnover, growing from barely 1 million RM in 1935 to 8 million RM in 1935, 19.5 million in 1941 and finally 58 million RM in 1943.[111] The military and air force had become primary clients, using Matra's tool sets and complete mobile repair units on trucks for the repair and maintenance of motor vehicles, tanks, and airplanes. Matra developed 'in closest collaboration with the official authorities' a unit construction system for all aircraft and aircraft engines used by the military. The company enjoyed a kind of monopoly position with the army in this field.[112] The manufacture of special tools as well as tool sets for aircraft and motor vehicle repair did not in fact constitute actual armament production, but the demand for such items was a consequence of the war. Indeed, extensive military contracts prior to the war had prompted the construction of an additional manufacturing facility in Kahl am Main.

Stationary engines and wood gas tractors from Güldner

Güldner also produced for Matra; as mentioned before, Linde had decided to purchase Matra primarily for the purpose of keeping Güldner in operation. Among other things, Güldner produced cylinder drills, valve and point grinding machines, as well as hydraulic presses with the Matra label. The state programme to reduce model variety, which was put into effect after the start of the Second World War, forced Güldner to begin producing large engines.[113] Güldner thus produced diesel engines for vehicles and factories. Deliveries of engines for the construction industry and farm tractors increased considerably, as did the manufacture of stationary and moveable diesel dynamos. Tractor production itself, taken up in the same year, also grew rapidly.[114] In order to increase production during the war, the number of engine models had to be reduced. Hence Güldner developed a standard diesel engine for tractors. As it became clear that the Reich's fuel production capacities were overwhelmed by the needs of motorized warfare and an air war, all vehicle engines not directly relevant to the war were adjusted to gas and wood gas fuel. Güldner developed a tractor equipped with a generator to carbonise wood.[115] Thus, until the heavy air raids of 1944, the Güldner engine factories were in full operation.

Oxygen installations for IG Farben at Auschwitz

The most problematic example of the Linde Company's involvement with the National Socialist system was its participation in building production facilities near the extermination camp Auschwitz. In 1941, Apparatus Engineering at Department B built four oxygen installations[116] and two helium installations[117] in Auschwitz-Monowitz. Only a few kilometres east of the extermination camp Auschwitz, IG Farben started building there in 1941 its fourth factory for synthetic rubber (Buna), using inmates from Auschwitz II, which IG Farben itself had built. 'From the beginning, management put prisoners to work on building its factory and in the fall of 1942 opened its own concentration camp on the ruins of the village Monowitz.'[118] The lives and health of the prisoners at Monowitz was not a matter of concern;[119] those no longer able to work would be gassed at Auschwitz-Birkenau. Three further installations [TR 78–80], which were ordered from IG Farben Ludwigshafen for Auschwitz between 1941 and

1944, were not completed due to war circumstances.[120] It is unlikely that the use of camp prisoners in building the factories was meant to be kept secret from Linde's erectors, as it took months to erect several large oxygen installations on the factory premises. Because erectors were required to send to Höllriegelskreuth detailed reports from non-local locations about their progress and problems with participating construction companies and their labourers, management must have been aware of camp prisoners' working and living conditions.

Linde's forced labourers

The German war economy was increasingly dependent on forced and foreign labour, for whose recruitment the General Plenipotentiary for Labour Allocation, Fritz Sauckel, was responsible. Starting in 1942, the German territory – the so-called 'Reichsgebiet' – was 'raked' for labourers fit for military service in regions not directly crucial to the war effort. Women were thus employed to partially compensate for this labour loss as were foreign and forced labourers. At the end of 1944, there were more than 7.5 million foreign and forced labourers – a good 20 per cent of total employment – in almost all areas of the German economy.[121] They worked primarily in armament operations though nearly a half worked in agriculture. Foreign civilian labourers were referred to as foreign labourers, *Fremdarbeiter*, by contemporaries and in the older historical literature. Today's literature refers to these civilian labourers, prisoners of war, and inmates together as 'forced labourers' and a distinction is made between them and voluntary civilian labourers.

Marc Spoerer distinguishes four groups of foreign labourers used by the National Socialist regime in terms of the degree of their freedom of action: voluntary foreign civilian labourers (labourers from states allied with Germany at the time, including Bulgaria, Italy (until 1943), Croatia, Rumania, Slovakia, and Hungary, and those from neutral Spain and occupied Denmark as well as many labourers from Western and Southern Europe), forced labourers subject to conscription (civilian labourers from the occupied regions within Poland and the Soviet Union, and also prisoners of war from Belgium, France, Great Britain, and Yugoslavia), forced labourers from Poland and the Soviet Union as well as Polish Non-Jewish and Italian prisoners of war and the fourth group, forced labourers without any power over their existence whatsoever (Polish-Jewish prisoners of war, concentration and extermination camp inmates, and inmates of forced labour camps and ghettos).[122]

According to current knowledge, all of Linde's factories used foreign labourers, prisoners of war and foreign civilian labourers at least for a short period of time between 1943 and 1945. However, according to still available staff documents and witness testimony, no camp inmates worked there. The precise number of all foreign labourers working for Linde during the Second World War is difficult to establish due to the disparate written record.[123] In connection with the foundation for forced labour recompense 'Remembrance, Responsibility and Future' a 1998 joint initiative of German industry and the German federal government, the Executive Board at Linde asked its former employees about its forced labourers and found the following results: at Höllriegelskreuth (records 31.7.44): 58 Eastern labourers, 15 prisoners of war and 103 other foreigners. Sürth: 20–30 prisoners of war, 30 forced labourers; Mainz-Kostheim: an unknown number of prisoners of war; 1943 Matra: 250 (mostly Belgian and French); STILL: 30 forced labourers, altogether 400–500.[124] Since then, more precise numbers for Höllriegelskreuth and Schalchen from July 1944 have been found in staff documents.[125] Older personnel documents were presumably destroyed during a major bomb attack of 19 July 1944. According to these figures (see below) the share of the foreign labour force in September 1944 made for nearly one-fifth of the total staff. The prisoners of war were not paid for their work. The foreign labour force – including 14 women – came from the following countries: Belgium, France, Greece, Holland, Italy, Yugoslavia, Croatia, Latvia, Czechoslovakia, and Ukraine.

Table 3 Foreign and Eastern labourers in Höllriegelskreuth and Schalchen

(Numbers excluding Höllriegelskreuth Schalchen Prisoners of War)	Eastern Labourers	Other	Others only
31 July 1944	70	119	15
31 August 1944	64	120	15
30 September 1944	64	120	13
31 October 1944	64	127	16
30 November 1944	64	126	16
31 December 1944	64	122	16
31 January 1945	62	122	15
28 February 1945	62	123	15
31 March 1945	62	122	16
30 April 1945	60	70	16
31 May 1945	2	–	–
30 June 1945	2	–	–
31 July 1945	–	–	–

Source: Linde AG Archives (Höllriegelskreuth) Ein- und Ausstellungs- buch vom 1.7.1944–31.12.1947.

Richard Linde wrote in 1941: 'We have currently 20 foreign labourers, almost all of which are uneducated, but generally speaking useful labourers.'[126] In May 1943 his report states:

Regarding foreigners, we have so far: 76 Eastern labourers, 17 French, 8 Belgian, 19 Other. The French came partially from L'Air Liquide in Paris. There are however, only very few comparable to our German labourers. Generally, their work tempo is half [of that of the Germans]. Nonetheless, they have at least some skills, while the rest of the foreigners, at best, can be considered to be skilled after some time.[127]

Assessing these statements from a contemporary perspective, it is important to be reminded of the fact that Linde wrote this in an effort to limit the number of his German labourers being sent to war with the military. In December of 1943 – obviously in relation to work on the V2 programme – the highest record of foreign labourers was reached: 'We currently have a staff of 1054 people, including 132 draftees, 12 liable to military service, 232 foreigners. We have set up barracks to house the latter.'[128] In June 1944 the number clearly dropped once again: 'The staff at Höllriegelskreuth and the nearby factory (Schalchen) together is currently at 915 persons (excluding those taken by the military), of which 205 are foreigners.'[129]

Most foreign and forced labourers were placed in barracks, which were set up by the Linde Company. Franz Bäumler, who started in 1943 as an apprentice in the copper-smithy, reported in an interview on foreign labourers' living conditions at Linde:

The French were put up in a barrack left to the canteen. They enjoyed certain freedoms; they could go outside. The Russians, on the other hand, were always under surveillance. The premises however, were not guarded particularly strictly. On top of the rail-station there was a military post, which was set up only after the attack. On our own premises I never saw a soldier, who would have [been responsible for] watch[ing] foreign labourers. Watching the Russians, as far as I remember, was in the hands of Linde's factory security. They were armed.

Franz Bäumler could not or did not want to remember Eastern labourers being harassed: 'I can with certainty exclude Linde [from such behaviour] – that is at least my impression. The French were better off regarding food, they all looked good – in contrast to the Germans.'[130] A letter addressed to 'Monsieur le Directeur' Richard Linde, signed by 57 foreign labourers with mainly French and Italian names, is dated 3 January 1945. In this petition the foreign labourers complain about the 'injustices se passant au'camp de la Gesellschaft Linde's Eismaschinen'. The complaints were of a relatively minor nature, namely shortened leave times away from camp and delays in the distribution of cigarette rations. This group of foreign labourers sued over equal treatment rights: 'Nous sommes employés depuis un an, deux ans, à la 'Gesellschaft Linde's Eismaschinen' et nous n'avons eu qu'a nous louer d'être ici; car une fois à l'atelier ou au bureau, nous sentons que le français et l'étranger n'sont pas détestés et sont traités comme l'ouvrier allemand: nous vous en remercions.'[131] After German capitulation on 8 May 1945, all but two foreign labourers left the company.

Güldner in Aschaffenburg also had Russian prisoners of war and French civil labourers. A report to the Supervisory Board about Department D in Aschaffenburg on 5 June 1946, tells of the air raid on the factory on 21 January 1945: 'Under the weight of the 1½ m strong concrete wall of the air-raid shelter, 8 Russian prisoners of war and 3 French civil labourers lost their lives, 26 more were injured. Since the attack happened on a Sunday, only a few German personnel were present, of which no one was hurt.'[132] Sürth and Kostheim also had foreign labourers, as a letter from director Otto Hippenmeyer from 26 September 1944 indicates:

I have heard that foreign labourers are to be taken away from the factory in Sürth, as they are from the factory in Mainz too. We are in any case to count on a transport of Russian men and women in the next days. It is possible that the French – amongst whom are some excellent and competent skilled labourers – can remain, but this question is not yet decided.[133]

There were a number of foreign labourers at the Linde subsidiary Heylandt in the rocket programme. Ruckdeschel stated in his memoirs that he, as the new managing director at Heylandt in June 1944, found a number of French foreign labourers at its alternate factory in Gassen that were living in poor conditions and for whom he instituted proper work and food supplies. Ruckdeschel also described manufacturing facilities in Saalfeld for the A4 rocket engine works at which concentration camp prisoners were forced to work under inhumane conditions. 'The prisoners in Austria were better off ... we were able to use them even as apparatus watchmen and servicemen.'[134]

Planning for the post-war future

Throughout the Thirties and also during the war, the project of large oxygen installation units for the iron and steel industries stood at the very top of the agenda at Linde. Before the First World War, Hugo Stinnes had, as already discussed, ordered oxygen installations for air blowing in steel manufacturing. It was not until the pioneering inventions of Matthias Fränkl, which brought about a drop in the price of oxygen manufacturing, that the method of oxygen blowing for steel manufacturing could be introduced to Austrian steel factories. This project was highly significant for the future development of installation construction and the sale of oxygen for what is today referred to as on-site installations. Another consumer in the iron and steel industries was the blast furnace user. The first experiment facility for oxygen enrichment of the blast furnace blow was ordered from and delivered by Linde and the L'Air Liquide jointly for the Gute-Hoffnungs-Hütte (GHH) in 1938.[135] For Linde, the additional follow-up of a regular installation was most important, 'because the entire iron and steel industry has always waited for GHH to order a large installation'.[136] Following the first order from GHH, other steel factories like Krupp and Hoesch eventually requested similar projects from Linde. Oxygen installations for the enrichment of the blast furnace blow brought about a new order of magnitude to installation and plant construction. Already counting on similar contracts after the war for Apparatus Engineering in Höllriegelskreuth, Richard Linde wrote to the Supervisory Board in 1941: 'In Höllriegelskreuth we face the question, of whether we should not reckon with a high level of production capacity after the war's end, which would be likely to require the enlargement of our manufacturing facilities.' For this reason 'the oxygen manufacturing installation, which has no organic relation to Apparatus Engineering management' was to be transferred 'to some industrial region near Munich'.[137] Indeed the war hindered the construction of the Hüttenwind oxygen installations as well as the transfer of the oxygen installations. Because of the destruction of the oxygen facilities in the bomb attack of July 1944 the transfer became obsolete; the transfer of oxygen production and bottling from Höllriegelskreuth to the newly built factory in Lohhof north of Munich materialized only during the late 1950s. Yet on-site oxygen installations for the iron and steel industries became one of the most profitable branches at Linde after the Second World War.

The extraction of methane as fuel for vehicles was another interesting project. Berliner Gaswerke contracted Linde's Department B to develop an installation, with which a gas rich in methane could be manufactured from illuminating gas (carburetted hydrogen). This gas – bottled at high pressure in steel containers – was to serve as fuel for the municipal transport fleet. The installation worked according to a patent developed by Rudolf Becker, DRP 762 787. Because of the difficulties in obtaining materials, its completion was repeatedly postponed; shortly after its erection, the installation was destroyed in a bombing attack. Since the 1970s, Linde has again been involved in research on the use of gases as a vehicle fuel; today hydrogen technology stands at the core of this endeavour. A project involving the ship transport of large quantities of liquid gases was not carried out. In March 1944, the Linde Company received an inquiry from 'Reich Factories Hermann Göring' (Salzgitter) for an installation for extracting liquid methane, which was supposed to be brought to Berlin by barge across the Mittelland canal. Because of war conditions, this failed to come to fruition.[138] This idea was to be taken up once more decades later.

Factories destroyed

Armament production in those regions not yet occupied by the Allies was, until the beginning of 1945, running at full speed, while the manufacturing of goods for the general population become increasingly limited from 1940. The bombing raids of the allied forces had, since 1943, incurred heavy damages in industrial centres, making consistent production impossible. Production facilities vital to the war effort were often transferred to regions less threatened by war.

By 1943, Linde's factories and cold storages were being hit by sporadic heavy air raids. The old cold store building in Munich was badly burned in September 1943 in a bombing attack;[139] only the new, less damaged building could be brought into operation quickly, since the refrigeration machine installation unit remained unharmed. Berlin-Britz, the factory of Linde's satellite company Heylandt-Gesellschaft für Apparatebau, had already suffered several air attacks and major damage during that year. Ruckdeschel, who had been sent from Höllriegelskreuth to Heylandt as management director, oversaw the transfer of this operation to Gassen in the Niederlausitz. Another set of provisional production was started, yet was abandoned as the Soviet troops advanced. Equipment was lost; Ruckdeschel and part of the permanent staff avoided the front 'and in an adventurous escape reached Schalchen at Trostberg, where operations were transferred from Höllriegelskreuth to rented rooms at the south German calcium nitrate factory in Trostberg'.[140]

Like its neighbouring electrochemical factories belonging to Pietsch (now Peroxid), the Linde factory in Höllriegelskreuth was vital to the war effort.

Höllriegelskreuth supplied, among other things, oxygen for V2 weaponry and Pietsch produced hydrogen peroxide for U-boats and it was therefore considered to be particularly vulnerable. On 19 July 1944, a heavy air raid hit the factory in Höllriegelskreuth and destroyed large parts of the factory facilities. The oxygen factory, the workshops, the shipping hall, the operation's office and two residential houses were totally destroyed; the large erection halls, the laboratory building and another house were damaged.

Figure 3.3   Destroyed factories in Höllriegelskreuth at the end of the war in 1945

As only one-third of the erection hall was damaged, work could be continued immediately following the construction of a partition. As a result, the factory ran until the end of the war at about 40 per cent of its working capacity previous to the air raid. The external, seemingly ruined condition of the factory was left unchanged so as to deceive allied aerial reconnaissance operations. The destroyed oxygen factory was never rebuilt; a provisional bottling station, which received deliveries from one of the air separators of the Süddeutschen Kalkstickstoffwerke in Trostberg built by Linde, secured oxygen supply in Höllriegelskreuth.

In 1944, the Güldner engine factories in Aschaffenburg were also heavily damaged. In November 1944 they experienced their first heavy air raid, and then shortly before the end of the war, on 21 January 1945, they suffered a second major attack, in which they were nearly destroyed.[141] Since the manufacture of tractor engines for wood gas fuel had been classified as vital to the war effort, the necessary production facilities and manufacturing barracks were transferred to premises owned by the agricultural machine manufacturer Fahr (later to become a co-operation partner in tractor manufacturing) in Gottmadingen in Baden, where production continued until mid-April 1945.[142] The same was true for the Matra factories. They were rather heavily damaged by bombs, but the alternate factory in Kahl remained unharmed and quickly resumed production after the end of the Second World War. These relocations to Schalchen, Gottmadingen, and Kahl were extremely helpful in making a new start after the war.

Many cold stores and oxygen factories were also damaged. The cold store in Magdeburg was so heavily damaged by one of the major attacks in 1944, that it could not operate for a good while. The oxygen factory in Bielefeld was destroyed on 7 December 1944; equipment that was still usable was to be transferred to Porta at the beginning of 1945 to supply oxygen to a company there manufacturing radio tubes.[143] 'It never came to that, and the installation parts used there were also substantially damaged.'[144] The cold store in Dresden was badly damaged in 1945 during a heavy bombing raid which nearly obliterated Dresden, while the cold store in Nuremberg remained essentially undamaged; the air raids of 1945 caused minor damage to the Linde stadium in Nuremberg. The management of Department A, housed in Wiesbaden, suffered heavy damage in February 1945 from an air attack. The entire Wiesbaden design archive was lost in this raid – a loss that surely weighed heavier than the destruction of buildings and equipment.

Further losses were incurred as a result of Allied recapture of German-occupied regions. Linde, sceptical as it was of the eastern conquest, did not build factories there, unlike other companies such as AGA, which had set up gas factories in the so-called 'General Government'[145] ('Generalgouvernement') and had to retreat hurriedly in 1944–45 leaving valuable installations behind. In 1940, Linde had set up a few gas factories in France, but only in Alsace and Lothringen, leaving the rest of France to L'Air Liquide as a regional monopoly, in post-war terms. There were, however, cases where Linde was concerned that its installations may fall into competitors' hands. For this reason, the gas liquefaction department sent a telex message on 7 March 1945 to the responsible chief engineer at the Reich Air Ministry in Berlin stating that in Ajka, 30 km north of Lake Balaton in Hungary, there was 'an installation for obtaining krypton, built by us, operated by the Hungarians'. Linde requested its 'complete destruction, especially of the separators in case of a possible retreat, as the Hungarians and Russians have a great interest in reconstructing the Krypton installation'.[146]

Allied troop occupation ran unspectacularly in most factories. In Höllriegelskreuth, documents were burned before occupation. Once occupied, technical specialists followed American troops into the factories, especially Höllriegelskreuth. They were amazed by the small size of the workshops at this 'global company'. They were particularly surprised to find no experimental installations, especially since Linde traditionally tested new installations one-to-one with its clients. The emigrant chemist Dr Paul Schuftan, who had worked at BOC since 1939, led an English–American commission which confiscated all of the documents and engineering drawings at Höllriegelskreuth. These documents were passed on to the American competitors, making Linde's international competition tougher.[147]

Summary

The fact that Linde was not directly involved in armament production turned out to be an important advantage, particularly after the defeat of 1945. Linde did, however, produce gas separators for synthesis factories and the V2, which were vital to the war effort; it supplied industrial gases needed for armament production, manufactured tool sets and mobile units for the repair of military vehicles and airplanes, delivered diesel engines for emergency power generators, and operated cold stores, which supplied both the population and military. It is therefore fair to claim that Linde was extensively involved in the National Socialist armament programme and its war preparations, and that Linde profited from the Second World War considerably in financial terms as well as in terms of expanding the production of all its factories. One can also say fairly that Linde already had the know-how, factories, facilities, and expert personnel in the relevant branches before the National Socialist regime and did not alter its production programme for the war, did not directly manufacture any armament products, and that the company leadership sympathized neither with the political nor with the military aims of the National Socialists. One should also take into account that commercial freedom was heavily restricted in the National Socialist system, particularly once the war began. However, it must also be stated that there is no indication whatsoever in the company's business documentation that company leadership would have, either for ethical or other non-economic reasons, refused in any way the production demands and wishes of the National Socialist regime. The Linde Company was entangled with the National Socialists' war crimes and crimes against humanity in various ways. Apparatus Engineering built oxygen installations in Auschwitz on contract with IG Farben; because much of the construction work of IG factory installations there was done by extermination camp inmates, it is to be presumed that the company leadership had knowledge of this fact. From 1943 onwards, foreign labourers and prisoners of war were used at various Linde factories; preserved documents do not record any ethical considerations on the part of the company leadership on this matter. According to existing documents, the foreign labourers at Linde were well treated, given the circumstances; concentration and extermination camp inmates were not used at Linde. Thus, one could summarize the company's involvement in the National Socialist economic system and the Second World War by stating that company decisions were motivated rather by commercial, than by ethical or moral, considerations. Linde did, however – and this was likely to be in large part thanks to Richard and Friedrich Linde – make an effort to maintain traditionally respected value standards in the company, even under the dictatorship of National Socialism. After the war, the Linde Company made a concerted effort – particularly under the direction of Hermann Linde – to not accept military contracts.

Linde withstood the National Socialist era in terms of economics comparably well. While surely dramatic, the destruction of its factories was not unusual, as all industrial companies suffered the same losses. The company grew significantly as a result of armament and war. Company leadership had kept its eyes open and was far-sighted about the process of expansion. Linde, for example founded no subsidiaries in the occupied eastern territories. In the west, Linde did indeed set up branch factories in Alsace and Lorraine, but not in central France and the company reacted to suggestions for a (partial) takeover of L'Air Liquide with restraint. Linde's co-operation with the sensitive rocket industry was conducted through the 100 per cent subsidiary Heylandt. Linde was also fortunate about the location of its production facilities, which apart from a few oxygen factories and cold stores, were all in the three western zones. For the most part, the leadership team survived the war and was at the company's disposal again in 1945. The Linde Company's products certainly enjoyed greater demand during the war, but were also suited to a peacetime economy. Linde did not need to adjust its production after 1945, but rather to begin it again. Linde did not participate in confiscating its international competitors' facilities during the war, which also facilitated Linde's re-entry into the international community thereafter.

4
Linde During the Great Economic Boom, 1945–66

Lean years in Germany, 1945–49

1945: zero hour?

The end of the Second World War brought the destruction of cities and houses and much of industry. Most of all, it put an end to National Socialist tyranny and the controlled economy. The German populace experienced these events as a central rupture, as a collapse, and at the same time as true liberation. After the initial shock, the dominant emotion among most survivors was the feeling of having narrowly escaped once again. The extent to which survivors felt liberated varied, of course, depending on the losses they had personally suffered: of time (for some, also through post-war incarceration as a POW), family members, homeland, social status, and prosperity. Many people had experienced such horrors that they were unable to feel that they were beginning anew, nor that 'life was starting again', nor even that 'life went on'.

This experience of collapse and, implicitly, of a new beginning was one reason that the term 'zero hour' came into use despite the tremendous continuity of persons, structures, experiences, and property ownership in post-war Germany. This continuity – particularly in West German society, and above all in politics and the economy – was later criticized by the left and by the East German leadership, and rightly so. In the direct post-war period and the early Federal Republic of Germany, the term 'zero hour' was used gladly in part because it emphasized the complete destruction of the old system and thus the situation of a new beginning and successful reconstruction. Moreover, 'zero hour' drew a sharp line between past and present. The clocks were set back – to zero. History prior to the zero hour was thus suppressed. A veil of silence cast over the crimes of the National Socialist system was a widespread cultural phenomenon during the years of reconstruction, up to the cultural turning point of 1968. Those who had been oppressed and persecuted by that system scarcely found a sounding board for their memories or accusations, even after 1945.[1]

With its unconditional surrender, the National Socialist regime was removed from power. Even before the end of the war, the Allies had announced a restrictive denazification and occupation policy, which however was administered ever more liberally with the advent of the Cold War between the Western powers and the Soviet Union. This was because the technical skills of the followers and supporters of the National Socialism system were very useful in establishing stable administrative structures.

Compared to the Western occupation zones, the areas east of the Elbe River occupied by the Soviets were much harder hit by losses from reparations and dismantlement.[2] However, destruction from the air war had been greater in the west. Due to shortages of supplies and labour as well as the dismantling of industry by the occupying powers, industrial production fell in the second half of 1945 to about 10 to 20 per cent of its pre-war level. In both the west and the east, the Allies removed large quantities of the available documents and equipment from their respective occupation zones in order to analyse them. Moreover, key experts were persuaded – more or less voluntarily – to continue their research in strict secrecy in the US or Soviet Union in order to maximally exploit the progress achieved in Germany before 1945.[3] These 'intellectual reparations' were accompanied by the withdrawal of trademark and industrial property rights; important patents and designs were provided free of cost to interested foreign companies.[4] This did not occur only in the high-technology area; it affected the entire span of technical products.

Along with the military loss came the loss of foreign holdings and industrial property rights. Most German subsidiaries and production sites in Soviet-dominated Eastern and Southeastern Europe were expropriated without compensation.[5] This did not occur in comparable fashion in the West, but German companies were forced to sell their foreign holdings. On top of this came the policy, pursued mainly by the Americans, of breaking up conglomerates such as IG Farben and the Deutsche Bank.

Economic aid, currency reform, and reconstruction in West Germany

The economic principles agreed to at the Potsdam Conference, and especially their emphasis on the development of agriculture and peacetime industry, gave hope that Germany would not sink into economic chaos despite all the reparation demands and limits on production. In late July 1945, the Allied Control Council was constituted, but as yet there was no central German authority.

The victorious Allies bore sole responsibility for the fate of the vanquished, but already by the end of 1945 all four occupying powers were forced to recognize how difficult it would be to translate the Potsdam economic reconstruction guidelines into regular practices that would assure the survival of the populace. The influx of ethnic German refugees from the eastern territories previously owned by the German Reich (but now partitioned off from it) initially exacerbated food and housing shortages in all the occupation zones.[6] Since the reparations question proved to be unsolvable and the Western and Eastern occupying powers increasingly cut off their spheres of influence from one another, the economy actually contracted in 1946 and 1947 compared to 1945. Industrial production was impeded not only by wartime destruction, material shortages, and transportation problems, but also by a scarcity of skilled workers. The worker shortage, in turn, resulted in part from the lack of habitable housing in direct proximity to workplaces, for public transit systems were also crippled and totally overloaded due to fuel rationing and the destruction and seizure of vehicles. For these reasons, but also due to the already tense food and supply situation, freedom of movement was restricted, particularly for would-be immigrants to the larger cities. After 1945, German agriculture produced barely half the food needed, and forcible imports from occupied territories ended after the war.[7] The supply situation was accordingly critical. Many people went hungry between 1945 and 1948, especially in the cities.[8] The foodstuffs crisis also put a strain on industrial production.[9] Without enough to eat, the coal essential to all production and mass transport could not be mined, so that even modest economic reconstruction was doomed to stagnation.

With the deepening rift between East and West, the economic crisis in the two zones took on a political dimension. By 1947 at the latest, both the Soviet Union and the Western powers were increasingly interested in creating an economically stabilized bulwark against the other system. This attitude toward their zones prevailed in Washington and London by mid-1947. In the west, the economic integration of the American and British zones had already begun in January 1947 with the so-called Bizonia under joint German professional administration.[10] In July 1947 US secretary of state George Marshall presented his plan for promoting the reconstruction of the European economy with American aid in the form of cash subsidies and goods.[11] In January 1948, the joint administration of 'Bizonia' was reorganized.[12] While the Marshall Plan began to bear its first fruits in Western Europe, the German economy continued to stagnate. The Soviet zone was, of course, ineligible for Marshall Plan aid. But the breakthrough to real economic recovery in the west was achieved only with currency reform in June 1948. Each resident of the western zones received 40 Deutschmarks as a per-person allowance in a one-to-one exchange against the old Reichsmarks; savings and stocks were converted in a 1:10 ratio.[13] The currency reform represents an important caesura in West German post-war history. For the populace, which had been suffering acutely from the shortages of provisions and supplies and which had little purchasing power with the old Reichsmark, a true 'economic miracle' occurred: on the day after the currency reform, the shop windows were once again full; groceries and other goods were – at least in part – no longer just distributed but presented to the buyer – to the extent that he or she could afford them.[14] From about 1949 on, repair and reconstruction work had generally progressed to the point that regular production could resume. Joachim Radkau and others have pointed out that the reconstruction, as well as the economic miracle with its success in exports, occurred on the basis of pre-war technology, without new research and development efforts.[15]

Then, in 1949, the partition of Germany was set in stone for 40 years with the founding of two states on German territory. A short time later, the German Democratic Republic (GDR) was admitted into the new Council for Mutual Economic Assistance (Comecon) of the Warsaw Pact countries. Within their respective political systems, both German states now entered into an increasingly tight network of trade relationships, while German–German economic contacts dwindled significantly.[16] Economic recovery in the GDR began later than in the Federal Republic. The first five-year plan took effect in 1951, which brought initial successes under tough economic conditions. Only in 1958 was the rationing of foodstuffs with ration cards abolished and the pre-war level of provisioning achieved once more.[17]

Linde in 1945: hard hit, but not fundamentally threatened

The production sites of the Linde Company were certainly damaged relatively badly in the air war, as described above. According to the company's own estimates, 75 per cent of its buildings and production facilities were destroyed in the Second World War.[18] This may be somewhat overstated. In any event, reconstruction of damaged plants was nearly completed by 1949 despite the general shortage of construction materials and skilled workers.[19]

As already described, Linde's plants lay predominantly in the western sectors. These were favourable starting conditions. The factory of the Linde subsidiary, the Heylandt Gesellschaft für Apparatebau, in the Britz district of Berlin, was almost entirely demolished. After multiple air attacks, the buildings and plants were largely destroyed; the company's subsequent relocation to Gassen in the Niederlausitz (which was occupied post-war by the Soviets) then led to the complete loss of its machinery. Most of its employees migrated to the west and were hired by the Linde factory that had relocated during the war to Schalchen, near Munich. Paulus Heylandt emigrated to the Soviet Union in 1945 under unexplained circumstances, where he was supposed to conduct rocket experiments. But he became seriously ill and died in 1946 or 1947 in the Soviet Union.[20]

To be sure, Linde had supplied products and plants that were essential to the war effort, but none of its divisions had been directly involved in the production of weapons and military equipment. This was a decisive point, for companies that had been largely or completely converted to weapons and ammunition production were expropriated without compensation and often totally dismantled. But only in the autumn of 1947 was the final decision made that Linde, as a whole concern, would be spared dismantlement in the western zones.[21]

But Linde had profited from the boom of the 1930s and the war years, and economically it was much better off in 1945 than in 1933. Its plants were indeed damaged but not wholly destroyed or dismantled, and its products were in great demand during the early post-war period. As a result, in 1946 Linde earned a profit of 1,706,506.44 Reichsmark (RM), which however was offset by a loss of 3,678,275.30 RM carried forward from the previous year. The remaining loss of 1,971,768.86 RM was carried forward to 1947.[22] Thus, 1947 was Linde's last year 'in the red'. But what in retrospect looks like the prelude to a successful economic upswing was perceived quite differently by contemporaries in 1947. Philipp Borchardt, who returned to Höllriegelskreuth from exile in England, noted on 7 March 1947: 'Export opportunities have been our main worry the entire time. Whether it will ever go anywhere is highly doubtful.' Although Linde was allowed by then to correspond directly with foreign customers, all deals still had to be concluded through the Export–Import Agency in Minden.[23] Borchardt suspected that the Allies would keep Linde 'under a glass cover, and might even bar us from going abroad at all'. He quoted Richard Linde, who believed that 'the Allies' actual intent is not to allow us any exports, with the exception of raw materials such as coal', and who was convinced that 'it would be cheaper for the Allies to keep us only just alive by delivering the bare necessities, but to stop our exports entirely in order to protect their own exports from our competition'.[24] Richard Linde was wrong about this, however. The costs of feeding the population, setting up administrative units, and reorganizing the economy and transportation in the western zones placed such a heavy burden on the occupying powers that starting in 1946 and especially from 1948 onward they promoted German companies' production and exports. Of course, this effort remained internally inconsistent and was also thwarted by countervailing regulations. Communication with foreign companies thus remained difficult at first:

> Postal service to foreign countries continues to be unspeakable. A letter takes 14 to 18 days! The customer just doesn't receive an answer to his inquiry and he can – no, he must – place his order with one of our competitors abroad if he doesn't want to lose an endless amount of time. He can place the order with our competitor before we have received his inquiry.[25]

And 'abroad' began in Germany, for co-operation was problematic not only between the Western occupying powers and the Soviet Union, but also among the Western Allies. Even the delivery of equipment within the American occupation zone, which included Bavaria, was not trouble-free, as Richard Linde reported from Höllriegelskreuth to the Supervisory Board in Wiesbaden in July 1946:

> Till now, relatively few orders have arrived from the American zone, which is detrimental in that it further complicates material allocation. For example, orders destined for the English zone are not provided for by releasing material quasi in processing traffic, but must be covered from the 'Bavaria contingent', and the Bavarian State Office for Iron and Metal understandably refuses to allow the quantities of iron that come in (which are very small anyway) to return to the English zone.[26]

Yet the outlook for the division of Apparatus Engineering in Höllriegelskreuth was 'not unfavourable', for 'promising inquiries about relatively large equipment' were coming again not just from within Germany but also from abroad. Some were from 'old customers from Norway and Belgium', but 'inquiries have come from England and the US, mostly through visits by commissions, many of which have called on us'.[27]

Due to the currency conversion from the Reichsmark to the Deutschmark, 1948 was a shortened business year; in exchange, the following business year ran from the currency reform in June 1948 until December 1949. In the initial balance sheet denominated in Deutschmarks on 21 June 1948, Linde's stock capital was converted in a 1:1 ratio to 34,266,000 DM (34,000,000 DM in bearer shares and 266,000 DM in personal shares with 20-fold voting rights); at the same time, reserves of almost 22 million DM were created.[28] For five years, from 1944 through 1948, the company had not distributed any dividends, but it resumed paying them in 1949. Common shares paid 4 per cent, while dividends for preferred shares were set at 3 per cent by Linde's articles of incorporation, but had not been paid since 1944; these were now paid retroactively, but devalued 10:1 for the time prior to the currency reform, as was usual. Indeed, the currency reform in the western occupation zones laid the long-term foundation for economic recovery, but in the short term it led to losses through the devaluation of bank deposits and outstanding debts.[29] There was also a painful lack of liquidity, which hindered investment. All investments and repairs over 2,000 DM had to be approved in writing by the CEO.[30]

The years following the currency reform were also a period of initial reorganization at Linde. A much-debated question was the necessity of establishing an administrative centre. In the preceding 20 years, the company had grown significantly and acquired production sites in various locales. In the view of upper management, this called for the establishment of a true central administration, or headquarters. Wiesbaden had always been regarded as head office, but until now it primarily had the character of a central engineering office for Division A. Central staff offices, by contrast, hardly existed in Wiesbaden. In 1949, a headquarters building was erected in Wiesbaden next to the (partially destroyed) villas at Hildastrasse 4–10, on the same plots. But by today's standards, this centre remained very modest in its scope. Only in the 1960s would it assume a central management and control function for all segments of the Linde Group, with divisions for organization, planning, marketing, and later controlling. The business segments still managed themselves almost autonomously.

Another question was Linde's manufacturing structure. In the preceding decades, Linde had grown from an engineering office into a manufacturing company (with strengths in development and planning). In-house production not only made money; above all, it guaranteed that quality standards would be upheld, and it simplified – or made possible in the first place – the production of complex equipment. It was thus no surprise that a systematic study of plant safety and potential operational simplifications at all of the Linde Group's production operations (commissioned by the Executive Board and carried out by director Ilzhöfer) recommended further increasing the vertical range of manufacture – that is, manufacturing certain goods at Linde that had previously been purchased from other companies.[31] In more recent decades, efforts have been made to enhance efficiency primarily by reducing the vertical range of manufacture. But in the economy of (material) scarcity during the late 1940s, a high vertical range of manufacture was regarded, on the contrary, as desirable. It made production more predictable and robust. In a sense, this is still true today. When auto-makers nowadays force their suppliers to relocate in immediate proximity to their factory grounds, this is in fact an (outsourced) increase in the factory's vertical range of manufacture.

Let us now look at the development of production in the individual divisions. As early as 1946, and through 1948, Division A's commercial refrigeration business was distinguished by 'brisk demand, which can only be inadequately satisfied, however, owing to shortages of materials and skilled workers'.[32] Only in 1949 were normal work conditions achieved again; from then on, the commercial refrigeration business developed 'well and led to growing sales abroad'.[33] For the household market, production was resumed at the Kostheim plant in 1947 after its production facilities had been rebuilt; by 1948, the plant was already 'quite busy' again.[34] The Nuremberg Ice Factory and Cold Storage exhibited no significant damage in 1945, while the cold store in Munich sustained considerable war-related damage to its old building in 1944. The complete rebuilding of the Munich cold store 'did not make the desired progress due to the generally known problems' – that is, manpower, material, and means of transportation were all lacking – but it was finished in 1949. In the cold store market, Linde profited particularly from the needs of the occupying powers. The Americans supplied their troops to a great extent via refrigerator ships from the US. The cold stores were thus 'predominantly used by the occupation force' and were 'full'.[35] The Linde stadium in Nuremberg was seized by the American Red Cross after US occupation troops marched in and was vacated again only in 1952.[36]

In Höllriegelskreuth, 'normal production restarted on a small scale' in 1945 after the initial post-capitulation clean-up.[37] Of course, the clean-up and repair work continued in Höllriegelskreuth from 1946 to 1949, but as early as 1946 production resumed of air and gas separation plants, 'for which we had so many orders domestically and also, toward the end of the year, from abroad'.[38] Time and again, the sources report a shortage not only of material but also of manpower. By 1948, the number of foreign orders increased to 70 per cent of Linde's order volume; this resulted in part from the comparatively low German wages and prices before the currency reform.[39] Yet production rose even faster in 1949 after the currency reform 'due to both improved employee performance and new hiring'.[40] Most of the larger foreign orders for plant engineering were earmarked for the nitrogen industry, 'both pure nitrogen plants (as for Norway) and some connected with coke oven gas separation plants (especially for Belgium, Spain, and Italy)'.[41] In certain areas, Linde was still recognized 'as by far the most capable company', while its most important international competitor, L'Air Liquide, had 'relatively few' operations outside France at that time. Good profits could be made in exports; foreign prices were about 40 per cent higher than domestic ones! The situation looked different for oxygen plants; here, strong competition from American companies was becoming evident.[42] During the war and its aftermath, 'downright turbulent growth' had occurred in the US 'in the construction of very large oxygen plants for gas generation, methane cracking, and the steel and iron industry'. At least five companies had large plants on order, including the market leader, Linde Air Products (LAPC), which had belonged to Union Carbide since 1917. Since Linde in Höllriegelskreuth was not very competitive in the US due to high freight and customs expenses and its lack of an American subsidiary, it pursued collaboration with LAPC, at least in the form of 'engineering help' (assumption of planning services). To this end, Philipp Borchardt travelled to New York in late 1948 at the invitation of LAPC, but he did not accomplish much. LAPC rejected the engineering collaboration, allegedly due to political intervention, and moreover wanted 'to no longer be seen as associated with us'.[43] But Linde had become quite self-assured about technology. In England, too, Linde initially sought closer contact with British Oxygen (BOC), 'which even approached us to offer their financial assistance' – an offer that Linde 'naturally rejected' because 'we don't want to be dependent on foreigners', as Richard Linde self-confidently wrote to the Supervisory Board.[44]

The gas business, previously conducted by a sales company jointly operated with IG Farben, was taken over by Division B after 1945. The oxygen and acetylene plants produced 'primarily for the foreign troops' after the Allies marched in; as early as the end of 1945, output in the remaining plants in the west attained a level that corresponded 'approximately to that of all German plants in 1933'.[45] The plants in the Soviet zone were still kept on the books for several more years, even when they were sequestrated and later expropriated. The production and sales of gases were greatly impaired in the early post-war period by shortages of cylinders and carbide, transport problems, and cutbacks in the electric power supply in 1946 and 1947. Sporadic restrictions on power usage continued even into 1948, but sales began to increase – in part as a 'result of the ongoing expansion of our cylinder supply and the adequacy of our carbide supply'.[46] In the wake of the currency reform, sales of oxygen and acetylene rose in a 'remarkable manner'.[47]

Resumption of industrial gas production faced special problems in Berlin. Despite the uncertain circumstances in the city under the four occupying powers, Linde decided to re-enter its previously largest sales market for oxygen, albeit on a small scale for the time being.[48] In February 1947, Linde requested permission simultaneously from the Neukölln District Office and from the US headquarters in Berlin-Steglitz to acquire 'from military surplus a mobile Heylandt oxygen plant with 80 cbm hourly output' and set it up in Britz to produce 'urgently needed oxygen for industrial uses as well as for separation cutting and welding work'.[49] Even in 1949, the Berlin situation seemed so uncertain to Executive Board members in Munich that Rudolf Wucherer baulked at accepting credit from the American Marshall Plan to rebuild Linde's representative office with a 'worthy exhibition room for small refrigerators ... [and an] office and workshop'. He wrote that he considered 'credit that, when all is said and done, comes from America, to be far from innocuous. If things go awry in Berlin, we will probably have to pay for it anyhow.'[50] Only in 1952 did Linde regain a firm foothold in the (West) Berlin market with the acquisition of another company's production facilities in Neukölln.[51]

Figure 4.1 Rebuilding post-war Berlin: the provisional plant in Berlin in 1948 (above) and the new facilities in 1950

Division C, Maschinenfabrik Sürth, had suffered especially severe bombing damage compared to the other divisions. Its remaining machine tools and devices were repaired mostly in the division's own workshops, and reactivated. During the clean-up and repair work phase, the transitory Reichmark period was a blessing, in a way, for most West German industrial enterprises, because it facilitated the continued employment of the clean-up workers with 'cheap' money. An immediate currency reform in 1945–46, comparable to the introduction of the Deutschmark in the former East German states in 1990, would probably have forced the dismissal of many employees at devastated companies that needed to be cleaned up. For the employees, who were unsettled by the discontinuity in politics and society, the 'anchor' of steady employment was important and attractive, even with payment in Reichsmarks.[52] In any case, the reconstruction of damaged buildings and facilities made such good progress that by 1948 it was once again possible in Sürth 'to satisfy the strong demand for small refrigeration units and refrigerators'.[53]

The Güldner plant in Aschaffenburg was a different proposition. To be sure, the plant had been hard hit and the production facilities for low-speed diesel motors almost completely destroyed. But the relocated plant in Gottmadingen on the 'Fahr' factory premises was able to resume production of engine spare parts and tow tractors, mostly with wood gas motors, in the second half of 1945,[54] despite delays due to the 'removal of valuable factory equipment by the occupation force'.[55] By 1948, about 1,000 units had been produced before operations were completely moved back to Aschaffenburg.[56] In early 1946, repair work in Aschaffenburg had progressed far enough that the production of spare parts could resume; starting in mid-1946, the foundry operated again, too.[57] At the orders of the American occupation authorities, Güldner's manufacturing programme remained restricted to low-power diesel engines and tractors; the production of large diesels belonged to the past once and for all.[58]

Due to the conversion of many agricultural machine manufacturers to armaments during the war, there was substantial pent-up demand in many areas during the early post-war period. This was the case for refrigeration units, field tractors, and diesel engines, too. Industrial goods were therefore subject to quotas in the early post-war period. Of course, quotas also applied to vendor parts (tyres, ball bearings, and so on). In short, the initial reconstruction occurred in a seller's market. Production could satisfy 'demand only with great difficulty'.[59] Until at least the mid-1950s, Sürth, Kostheim, and Aschaffenburg enjoyed this seller's market, in which companies merely distributed their goods rather than having to market them. This experience made it difficult after the end of the 1950s for Divisions A (refrigeration units) and D (agricultural tractors and engines) to adjust to a buyer's market with fierce competition, to improve advertising and marketing, to develop new markets and products, and to optimize production.

Moreover, the early years of reconstruction and an economy of scarcity were marked by the situational exploitation of local opportunities – of material allocations, barter transactions, or wartime relocations. The growth of individual divisions was accordingly uneven. The development of Linde's divisions went much like the remnants of the airplane maker Heinkel in Stuttgart – which temporarily produced pots after the war before moving on to motor-scooters and small cars and finally returning to aeroplane development in the mid-1950s. The adoption of new lines of production, such as the wood gasifier, had nothing to do with strategy; it was merely a reaction to short-term needs. The voices calling for more strategy and long-term concepts were still rare, but from the mid-1950s they increased significantly, sometimes dissociating themselves from the spontaneous (re)actions of the early post-war period.

Denazification and reparations

Not a single member of Linde's Executive Board had belonged to the NSDAP during the Third Reich. Richard Linde unequivocally belonged to the opponents of the National Socialist regime. He remarked on 7 January 1945 in Höllriegelskreuth in a speech celebrating the 50-year service anniversary of his brother Friedrich Linde that everyone present knew that

> our honouree was anything but a supporter of the NSDAP and its goals; that as managing director he resisted, as much as possible, attempts to infiltrate our company with party influence; that he put not the slightest value on blue- and white-collar employees belonging to the party – quite the opposite.

He further emphasized that 'there probably were not many managing directors in Bavaria' who, like Friedrich Linde, had not even belonged to the 'German Labour Front' [Deutsche Arbeitsfront, or DAF]'.[60] Richard Linde also attributed to this stance the fact 'that till now our company has been relatively unmolested, at least by the occupation authorities – apart from [my brother] himself'.[61] Specifically, there were denazification proceedings pending against Friedrich Linde on account of his title of 'military economy leader' [Wehrwirtschaftsführer]. The proceedings dragged out over several years.[62] But after a series of exculpatory statements, a ruling was rendered on 28 January 1948 that classified Friedrich Linde as 'exonerated', and thus as not needing to expect any punishment or sanctions for his actions as managing director.[63] However, Friedrich Linde had already resigned as CEO and chairman of the Executive Board as of 1 June 1946 and moved to the Supervisory Board, where he took over as chairman.[64] It is not entirely clear what role the still unresolved denazification proceedings played in this. At the celebration of Friedrich Linde's 50-year service anniversary in January 1946, his brother had emphasized in his speech that 'at the urgent request of the gentlemen on the Supervisory Board' Friedrich would not step down.

However, there most certainly were staunch National Socialists at Linde, too (see Chapter 3), who had to be let go after the war or could work only in a limited capacity. For example, at the Ellira Division for electric hand welding apparatuses, 'the division's work [in 1946] was greatly obstructed by the fact that all of its heads had coincidentally been party members, who on the basis of law no. 8 had to resign or were only allowed to be employed with subordinate

Figure 4.2   Friedrich Linde in private with his second wife Ilona *ca.* 1960

work'.[65] Similar problems emerged in other plants of the company,[66] but they soon became less important as the denazification requirements were rapidly liberalized.[67]

In October 1946 Philipp Borchardt, who was forced to flee during the Third Reich due to his 'Jewish ancestry', received 'the news from Berlin that my permit to return to my homeland has been approved'. He had fought for this since the end of the war but had been stymied by Allied regulations. In frustration he had written to Richard Linde in March 1946: 'I still do not know if there will be enough space for me in my homeland, but this much I know for sure: that there is absolutely no space for me elsewhere, and thus my path is mapped out, and I will follow none other, come what may.'[68] On 17 December 1946 Borchardt returned to Höllriegelskreuth for the first time in eight years. Meanwhile, Richard Linde had written to the chairman of the Supervisory Board, Otto Meyer of MAN, and suggested that Borchardt should replace him (Linde) on the Executive Board as soon as possible. Borchardt noted in his diary: 'I agreed to this for an interim period.' But the Supervisory Board rejected this suggestion; it deemed Richard Linde indispensable, and so he stayed on the Executive Board until he retired on 30 June 1950.[69] Borchardt returned to his pre-war position as chief technology officer with power of representation for negotiations and signing contracts abroad. Borchardt's return was a great stroke of luck for the Linde Company. His work was particularly important for exports to Anglo-Saxon countries.

In contrast, another former executive's return to his old domain in Germany proved difficult and encountered resistance from the Linde Company. In a letter of 14 January 1947, Erich Marx, then living in London, reasserted his claim to a stake in Matra. In doing so, he presupposed 'a very generous attitude on the part of the Linde Company'.[70] But the company reacted with surprise and chagrin, because especially in the view of Friedrich Linde, the sale had been made properly. The Supervisory Board and Executive Board grappled with Erich Marx's claims on several occasions.[71] Friedrich Linde, however, refused 'to get involved in a dispute with him'.[72] In protracted correspondence, the company offered one year later, on 9 February 1948, to return to Marx his former nominal share of 100,000 RM. But after the last increase in share capital to 5 million RM at the start of 1948, this would have resulted in only a 2 per cent share for Marx, instead of the 50 per cent stake he demanded. Marx called this offer 'completely inadequate, neither satisfying legal requirements nor doing justice to our moral obligations, and downright ludicrous'.[73] Marx's main demand was the restoration of his 50 per cent of voting rights; in his view, the 'dispute over capital and property' should be postponed. For a considerable time, Linde was uncertain, in legal terms, as to how the Allied regulations on restitution and reparations should be applied. The first American reparations law, the so-called Law 59 on the restitution of property stolen in the 'Aryanization of the economy', was enacted only on 10 November 1947.[74] And on 26 April 1949, the council of the states in the American zone passed the first compensation law that set a uniform standard for those states. It laid out the fundamental definition of what constituted persecution, who was regarded as a victim, which damages (including intangible ones) were to be compensated, and what form compensation would take.[75] Linde referred principally to the growth that had occurred at the company since Marx's departure and the impossibility of restoring the company to its state as of 1935. But in the course of negotiations, Linde changed its position significantly and at one point offered Marx 26 per cent of the voting rights (a qualified minority).[76] The dispute, in which Marx proved to be a tough negotiator, lasted until the start of the 1950s, when Marx finally achieved a settlement that compensated him with about 500,000 DM. This sum was 50 times greater than Linde's original offer.[77] In retrospect, it is unclear why Linde baulked at granting Marx management authority again. After all, in the initial years of the collaboration he had been managing director. Marx thereupon remained in England, though at times he had been interested in returning to Germany and ready to do so. In other companies, too, exiled businessmen of Jewish descent were unsuccessful in resuming their business activities in Germany after the war. They had to settle for financial compensation. The Freundlich family, mentioned in Chapter 2, is one example of this. The restrictive attitude in the direct post-war period toward businessmen 'of Jewish descent' who were willing to return was a sad chapter and a further great loss to the German economy and culture.[78]

Property and commitments in Eastern Europe

For quite some time, the situation of subsidiaries and holdings in the Soviet occupation zone remained uncertain in both operational and legal terms. The heavily damaged crystal ice factory and cold storage in Dresden were able to resume operations in 1945, thanks to the 'self-sacrificing efforts of the workforce'.[79] By 1948, the cold store in Dresden was even turning a slight profit. In Leipzig, the crystal ice factory and cold storage were almost undamaged and were immediately able to operate again, until dismantlement halted work in the new section of the factory and the cold store was rented out as an uncooled warehouse.[80] Under these conditions, it comes as no surprise that the cold stores in Leipzig remained unprofitable. It soon became apparent, however, that these facilities would be lost altogether. With the founding of the GDR in 1949, the cold stores in Dresden and Leipzig were declared state-owned enterprises 'and thus divested of our control', according to the 1949 annual report.[81] In the following year, 1950, the state industrial agency also expropriated the cold store in Magdeburg.

In Berlin, Linde's most important sales region for oxygen up to the end of the war, the Soviet Union also stripped industrial sites in the western part of the city. Berlin was initially conquered by Soviet troops in 1945, who seized the opportunity to rapidly dismantle modern production facilities, since jurisdiction under the occupation remained unclear until the Potsdam Conference. Under these conditions, the technical equipment from the Berlin-Borsigwalde oxygen plant, as well as the facilities in Berlin-Britz on the Heylandt property and the gas cylinders, 'were loaded off to Russia'.[82]

In the medium term, Linde lost all of its facilities in the Soviet zone and later GDR: four oxygen plants, two acetylene plants, and the cold stores in Dresden and Leipzig.[83] Even before final legal clarification, Linde wrote off these properties east of the Elbe River (in the Soviet zone) and also stopped including them in the company report. In isolated instances there were repercussions and reciprocal claims across the Iron Curtain even decades later. In 1955, for example, Linde unexpectedly faced a reimbursement claim from the Stalinwerke (Stalinovy Závody, Národni Podnik) in Czechoslovakia, which as a state-owned successor company attempted to collect on the Linde Company's obligations vis-à-vis the Sudetendeutsche Treibstoffwerke AG in Maltheuern by Brüx. By then, Article 10 of the London Debt Agreement of 27 February 1953 had confirmed the legal validity of these sorts of commitments. Nonetheless, Linde successfully refused to pay its debt of 26,212.82 Reichsmarks, since the nationalized company lacked the capacity to sue. Although the Czech company had taken possession of its predecessor's real property and production facilities, it did not have the right to act in the name of and on behalf of the legitimate owner. If payment was made to an 'unauthorized party', argued Linde's attorneys, there would be a risk that 'the true authorized party would make a claim to be paid again'.[84] After the fall of the Iron Curtain and the Berlin Wall, Linde regained a large part of its former property, at least within the territory of the former GDR.

Loss of Linde's foreign holdings

During the war, Linde lost all of its foreign holdings, even those in neutral countries like Spain. It was for Linde therefore acceptable to lose 'two factories that were built with our own funds in Alsace and Lorraine'.[85] In contrast, the Linde Company saw itself in the role of an 'innocent victim' in 1945 with the threatened loss of other foreign holdings. In addition to Linde-Riedinger in Austria, which was involved in refrigeration, Linde mainly lost companies that produced industrial gases. Linde assumed in 1945 that its shares of Hydroxygen GmbH in Vienna, Hydroxygen AG in Prague, the Sauerstoff- und Wasserstoffwerke AG in Lucerne, Abello Oxigeno Linde SA in Barcelona, the Dansk Ilt- og Brintfabrik AS in Copenhagen, Ungarischer Krypton GmbH in Budapest, and the Sauerstoff- Wasserstoffwerks GmbH in Lambach would be 'lost due to the outcome of the war'.[86] Its holdings in Eastern European countries and in Austria were indeed nationalized. Yet in the following years, Linde succeeded, for the most part, in rebuilding subsidiaries or holdings in the West. This took more than 20 years. Internationally, however, it had fallen far behind its large American competitors and L'Air Liquide. Linde's traditional strength in the East was gone, and its role in the West was greatly weakened. For many years, Linde was strong only in Germany in the gas area. Only with the acquisition of the Swedish AGA Group in 1999 has Linde undisputedly rejoined the ranks of the world's largest gas producers.

The break-up of IG Farben and the end of the VSW

Since 1925, Linde had sold gases in Germany through the Vereinigte Sauerstoffwerke (VSW) in Berlin. It was half owned by the Chemische Fabrik Griesheim-Elektron, which belonged to IG Farben. The VSW was a sales cartel or syndicate that set the 'prices and sales of compressed gases and acetylene'.[87] Linde had pushed through the 50/50 distribution in tough negotiations with the powerful chemical industry, and was very satisfied with its work prior to the war.[88]

After the war, IG Farben was regarded by the Allied occupying powers as an economic organization that had been close to the National Socialist state, directly involved in the exploitation and economic subjugation of the countries Germany occupied during the war, and thereby assumed a market-dominating position.[89] The break-up of IG Farben ranked among the deepest interventions by the Allied occupying powers into the structure of large German industry after 1945.[90] However, the individual occupying forces interpreted the provisions of the Allied Control Council law quite differently.[91] While the Soviet Union initially operated the large IG Farben factories in Leuna, Schkopau, Bitterfeld, and Wolfen as the Sowjetische Aktiengesellschaften (SAG) and later turned the factories over to the GDR's central economic agency (usually after comprehensive dismantlement), the French occupation force put the Ludwigshafen plant under French management and leased other factories in Rheinfelden and Rottweil to corporations controlled by French joint-stock companies. The British control officer named trustees for the IG Farben plants in Leverkusen, Dormagen, Uerdingen, and Elberfeld; the co-operation between these factories was left untouched. The American occupation authority, in contrast, promptly began breaking up IG Farben in its zone. The agency founded to this end, the 'IG Farben Control Office', placing trustees in the individual factories; later, the German federal states were also involved in their appointment. Each of the successor companies was to be guided toward complete economic independence.[92]

Since the Vereinigte Sauerstoffwerke maintained sales branches in all four occupation zones, the occupying powers' different strategies in the break-up of IG Farben had a direct impact on Linde's gas trade. In the American and French zones, the VSW branch offices came under control of the military government. At the same time, it was decreed that the factories' sales would no longer be permitted under the auspices of the VSW but would have to be transferred to IG Farben. Even trade with the VSW headquarters in Berlin was prohibited. The plants Linde owned in the American and French zones (Höllriegelskreuth, Augsburg, Illertissen, Nuremberg, Rheinfelden) were also barred from balancing accounts with the Berlin headquarters; they now had to prepare accounts independently, although they were nonindependent and not registered as corporations. 'But it is to be anticipated', Linde wrote to its Berlin lawyer in January 1947, 'that in the foreseeable future sales for our factories in the Bavarian US zone will be passed to us'. In the Soviet zone, the VSW branch offices were nationalized or sequestrated. 'We are making every effort to get them released again', Linde informed its lawyer, and added: 'The works council of our factories in Magdeburg decided a while ago that sales would no longer take place under the auspices of VSW but through our company. We gave our consent to this.'[93] This series of events highlights the political and economic uncertainty in 1946–47. But the complications would increase even further. On 29 March 1946 the VSW headquarters was put under Allied trusteeship in accordance with Control Council Law 52. The appointed trustee, Mr Arnold, an authorized agent of the VSW, now attempted to relaunch VSW throughout Germany or at least to uphold the appearance of this. But Linde was interested in ending its involvement with IG Farben as embodied in VSW, and it had thus begun selling gases itself in the British zone, with the consent of the British military government. Arnold now protested against this, although he had absolutely no authority in the British zone, and in a letter to Linde of 31 May 1946 he 'even threatened to have the upper management of our company arrested'.[94] In a bitter dispute, Linde finally pushed through the separation of sales into IG Farben and Linde sales outlets, retroactive to 1 January 1946. From this point on, Linde sold the gases it produced solely under its own name through its gas sales headquarters in Höllriegelskreuth.

The IG Farben dispute had brought home to Linde the new, pejorative connotation of the terms 'Konzern' ('group' or 'conglomerate') and 'Kartell' ('cartel'). In order to avoid further conflicts, Richard Linde proposed to his brother Friedrich (who was on the Supervisory Board) in August 1946 that the accounts receivable and payable of Linde's subsidiaries should no longer be reported as the accounts receivable and payable of the 'Konzernunternehmen'. Richard argued 'that the Linde Company was not a conglomerate' and accordingly the proper term for its subsidiaries was the more neutral sounding 'Tochtergesellschaften' [literally, 'daughter companies']. In this way, one could 'avoid the ominous word "Konzern" and 'still do justice to the legal disclosure requirements'.[95] Thus times change.

The economic miracle, 1950–66

The long 1950s, 1950–66

After export restrictions had been gradually lifted, the West German economy oriented itself primarily toward exports. Starting in 1950, the 'Korea boom' accelerated the growth of foreign trade.[96] Ludwig Erhard, the Minister of Economic Affairs, promoted the undervaluation of the D-Mark during the 1950s in order to gain export advantages for West German industry in the 1950s and 1960s. This export orientation, which resulted in the early post-war period from the weakness of the domestic market, thus became permanent, and persisted even when the domestic market later grew explosively[97] The 'economic miracle' in West Germany that began in the early 1950s caused productivity and real wages to rise rapidly in tandem, so that the buying power of the populace grew at a tempo that had been completely unimaginable in 1949. The dynamic of consumer needs increasingly determined the demands on industrial production, and the seller's market of the reconstruction phase with its insatiable hunger for goods turned into a buyer's market. In the consumer culture, ownership of electric household appliances – such as a radio, refrigerator, washing machine, and finally a television set – came in waves and gradually became taken for granted. For industry, this created a need to cater more to consumers' demands. In the refrigeration industry, too, the trend intensified from commercial to household refrigeration, from the cold store to the refrigerator. Linde was thus convinced that it absolutely had to lead the household refrigerator market despite its traditional strengths in manufacturing commercial refrigeration systems.

Starting in the mid-1950s, rising wages and ancillary wage costs increased the pressure to become more efficient. A 15-year phase of full employment began in 1957 when unemployment fell below 4 per cent for the first time. The labour shortage became 'more than ever before in German history an impulse toward mechanization' in industry.[98] Technical innovation was increasingly highly valued by businesses and by society at large.[99] Up to the mid-1950s, the economy had grown mostly by extending its size; now, technological advances took on a decisive role in sustaining growth.[100] The West German industrial structure underwent a transformation, as well. The southern and southwestern states, Bavaria and Baden-Württemberg, profited most from the shift away from the old heavy industries toward light industry, and since the 1960s they have distinguished themselves as strongholds of new technologies. After the construction of the trans-alpine natural gas pipeline, Bavaria became a centre for petrol chemistry. Agriculture, too, underwent major changes starting in the late 1950s. The mechanization of agricultural work with tractors, combines, and milking machines has required increasingly large capital expenditures, which in turn has produced an economic imperative to create larger farms.[101] The establishment of a European single market in 1957 with the European Economic Community (EEC) increased the competitive pressure on farmers.[102] Extensive national and European subsidies and support buying were utilized to at least cushion the impact of structural change on agriculture. Especially from the 1960s onward, appreciation has increased for technologies that reduce the wage labour that goes into a product. Labour-intensive manufacturing, such as textile production, was transferred into low-wage countries at an accelerated pace. At the same time, the recruitment of workers from Southern and Southeastern Europe was stepped up; the portion of foreign workers in West German agriculture and industry rose from 2 to 10 per cent during the 1960s.

Linde's growth with the economic miracle

Apart from foreign trade, which at first suffered greatly under legal restrictions, Linde was lucky in the early post-war period – with the location of its production sites, the small number of plants dismantled, and the minor obstructions to operations. But its management also made clever decisions in the clean-up and resumption of production, in absorbing employees from the east, and in offering products needed for reconstruction. Now Linde profited from the excellent macroeconomic growth of the 1950s, borne up with the rest of the economy. Indeed, Linde's sales and new orders rose even faster than the gross national product. Overall, Linde's development during the 1950s can be described as a continuous upward trend; the persistent export barriers and the rapidly rising wage costs (the wage index climbed faster than productivity) did not appreciably mar this picture. Foreign contacts that had been interrupted by the war were gradually resumed, often on the basis of old personal connections. A typical example of this was the initiation of renewed co-operation with the Spanish oxygen producer Abello, whose operations Linde later took over. This collaboration had been started by Carl von Linde.

As early as the second half of the 1950s, structural problems began to loom. They were not initially recognized as such during the boom, as long as the strong economy and Linde's good reputation meant that its products practically sold themselves. The company's leadership thus came under no pressure to define long-term objectives, and indeed did not do so, though it expressed increasing unease about the company's rapid growth. Sales grew 22 per cent in 1950 compared with 1949, an increase to which 'all divisions of the company contributed'.[103] In the following three years, as well, Linde grew disproportionately quickly. To be sure, its annual reports noted rising production costs and pressure on prices. But even by international comparison, production costs were not yet a problem – not least due to the incipient rationalization of manufacturing.[104] Germany was not yet a 'high-wage country'.[105] In line with the upward economic trend, Linde was able to pay higher dividends each year:

1951: 5 per cent, 1952: 6 per cent, 1953 and 1954: 9 per cent, 1955 and 1956: 10 per cent, 1957: 12 per cent, 1958: 15 per cent, 1959: 17 per cent, and 1960 even 18 per cent![106] Given these numbers, it is surprising that the Executive Board took a thoroughly critical view of the strong economic growth. It perceived the approach of increasing problems in connection with this growth, such as in purchasing semi-manufactured goods. Linde reacted by maintaining ever larger supply inventories as a hedge against missing deadlines for its own deliveries.[107] Despite efforts to increase efficiency, wage and salary expenses continued to rise disproportionately at Linde, as throughout Germany. Labour costs climbed by 16 per cent in 1955 in comparison to the previous year, and by another 14 per cent in 1956.[108] In order to cover investment needs and adjust the authorized capital to the growing business volume, it was decided at an extraordinary general meeting on 25 May 1956 to increase share capital from 17,133,000 DM to 51,399,000 DM. In addition to the 51,000,000 DM in bearer shares, there were another 399,000 DM in personal shares with 20-fold voting rights, intended to guarantee independence and held predominantly by family members.[109] Annual general meetings in the 1950s were thus also meetings of the families that sustained the company. Sitting at the head table were not just Lindes but also members of related families – the Michaelis, Seiler, and Ranke families – as well as descendents of the Supervisory Board members from the founding years. With Rudolf Wucherer as CEO and Friedrich Linde as chairman of the Supervisory Board, the speeches stayed within the family, too, until 1955, when the 80-year-old Rudolf Wucherer succeeded his brother-in-law Fritz as chairman of the Supervisory Board.

The Executive Board was reluctant to issue a long-term prognosis on the company's growth in 1956 'given uncertainty about the extent to which wages and material prices can be stabilized, which burdens lawmakers will yet impose on business for social and political reasons, and which direction the highly-politicized world economy will take'.[110] The satisfactorily high proportion of exports in unfilled orders – about 40 per cent overall, ranging between 7 and 85 per cent depending on the division – actually provoked concern on the Executive Board: 'In view of the current economic trend abroad and the progressive cost increases in our operations, our dependence on exports represents a growing risk.'[111] The creation of a single European market, too, was initially regarded with scepticism by the Executive Board, which feared the end of Linde's regional monopoly.[112] On the one hand, this sceptical view of the economic miracle shows how independently the Executive Board formed its judgements; on the other, this documents a desire for smooth sailing, due in part to its members' advanced age.

The year 1958 marked the zenith of the economic miracle. In retrospect it can be seen that the still impressive boom in the Western world and the Federal Republic of Germany now began to slow slightly. Of course, the growth rate varied considerably by economic sector. At Linde, with its widely diversified production programme, this led to different economic trends, depending on the division. Two groups formed at Linde for the next 15 years: the factories that earned stable profits, and those that consistently posted losses.

Within the company, this development would bring about major reforms. As a result, the unprofitable engine construction business, in particular, was upgraded and transformed into the Materials Handling Division, the profits of which were increasing slowly as of 1971. A further result was the departure of the Linde family from upper management. According to their detractors, the Lindes were unable to get a grasp on the problems, particularly in the unsuccessful divisions. But there had also already been organizational reforms from the mid-1950s onward, albeit on a smaller, gentler, and more harmonious scale. In 1954, the divisions were renamed with more transparent monikers. Division A in Wiesbaden became the 'Commercial Refrigeration Division'; the previous Division B was now called the 'Gas Liquefaction Division, Höllriegelskreuth'; Division C, Maschinenfabrik Sürth, lost first the 'C' and then the 'Division', and was now simply called 'Maschinenfabrik Sürth'. A new Household Refrigerator Division was created in Kostheim in 1958, which in 1960 was renamed the 'Household Refrigerator Factory Mainz-Kostheim'. Division D, Güldner Engine Works (Güldnermotorenwerke), underwent the slightest changes; after 1954, it was called the Güldner Engine Works Division, Aschaffenburg. In 1958 Linde's pension fund and its four benevolent societies were combined to form the Pension Fund for the Linde Company (Unterstützungseinrichtungs-GmbH der Gesellschaft für Linde's Eismaschinen). One year later, the Organization Division was newly created at company headquarters. With this, the head office in Wiesbaden gradually became more important for the structure and activities of the divisions.

Changing of the guard on the Executive Board

By today's standards, the Executive Board and Supervisory Board in the 1950s and 1960s were rather advanced in years. But in the 1950s and 1960s, active participation remained possible even in old age. In many areas of the West German economy, politics, and society, there was a strong continuity in personnel from 1945 until the mid- or late 1960s. The chancellor, Konrad Adenauer, was only one example of this. The reason for this continuity can be found not just in the reform-averse, encrusted society of the 1950s – for which Adenauer's 1957 campaign slogan was paradigmatic: 'no experiments!' – but also in the loss of the war generation. Thus, relatively elderly individuals held or assumed leadership positions in the late 1940s, which they then continued to hold for many years – very successfully, by the way. The current political debates on a productive life in old age and on raising the retirement age could profit from detailed studies of the 1950s.

There were simply no young family members at Linde in the late 1940s who could have been entrusted with leading the company. But there was a consensus among company executives that the familial tradition should be preserved. Thus, the composition of the Executive Board and the Supervisory Board changed only gradually. At the start of the 1950s Richard Linde moved from the Executive Board, to which he had belonged since 1928, to the Supervisory Board. At the Executive Board meeting of 15 March 1950, Richard Linde had suggested as his successor his close colleague, the engineer Walter Ruckdeschel, who had worked for the company since 1925; the Executive Board, in turn, had recommended that Richard Linde be named to the Supervisory Board.[113] Under different circumstances, his two elder sons would certainly have moved into positions of responsibility then, at the latest.[114] Rudolf Wucherer, the son-in-law of the company founder, who had served as CEO since 1945, retired from the Executive Board on 30 September 1954 and joined the Supervisory Board in July 1955.[115] Hugo Ombeck, who resided in Wiesbaden, became the new CEO. But it was foreseeable that Ombeck would hold the reins only for a transitional period, and not just due to his age. At the time that Rudolf Wucherer stepped down, his son Johannes joined the Executive Board, initially as an acting member.[116] In 1957 he became a regular member (for Plant Engineering) and in 1961 CEO. Christian Megerlin (Mainz) and Otto Wagner (Wiesbaden) were also admitted in 1955 as acting members; they were appointed as regular members of the Executive Board as of 1 January 1960. Within a few years, the focus of the Executive Board members had thus shifted geographically from Munich back toward Wiesbaden, and from the family toward outsiders. This shift was an expression of the void left by the death of Richard's elder sons, and an attempt to get a grip on the increasingly obvious problems in refrigeration technology and engine construction.

A countertrend appeared in 1961 when Hermann Linde joined the Executive Board on behalf of the Gas Liquefaction Division and took over management of Plant Engineering. The grandson of Carl von Linde and the son of Richard, he held a doctorate in physics and had worked for the company since 1948.[117] Only now did the son of the company founder, 'Dr phil. Dr-Ing. E.h. Friedrich Linde, the retired long-time general manager who served the company's growth so commendably' resign from the Supervisory Board 'due to old age'.[118] He died in 1965.

Employee retention, social benefits, and workplace atmosphere

The Linde Company was no longer an engineering office in the 1950s, but had become a widely ramified conglomerate. Since the 1930s, the majority of its employees had worked in production. The company's character varied considerably from one site to another. At Division A in Wiesbaden, the style was still that of a large international engineering office. This atmosphere was even more palpable in Höllriegelskreuth. By contrast, Sürth, Aschaffenburg, and Kostheim, as well as the production areas in Höllriegelskreuth and Schalchen, had the character of manufacturing operations.

For upper management, an important goal was to strengthen cohesion in the geographically dispersed company and to retain its employees over the long term. Especially during the phase of full employment, keeping good employees was a matter of survival for Linde. The resources and efforts committed to this end were substantial: a wage compensation system that rewarded seniority, social fringe benefits, a company newsletter, employee training programmes, holiday homes, housing programmes, and much more. The company newsletter that began publication in 1936, 'Werkszeitschrift Linde', did not survive the war, but it was revived in 1950. In its first post-war year, it appeared in 'German script' and in a format almost identical to the first nine volumes published during the Third Reich. This seems to have met with criticism, for it was redesigned the very next year. In the early 1950s, Linde founded a company training centre in Höllriegelskreuth, with evening courses open to all employees. It offered individual lectures, film presentations, and certificates of apprenticeship in various work areas of the factory. Education was even more popular in the form of entertainment. Linde opened up its head office in Höllriegelskreuth as a rehearsal hall for an orchestra. Many of its musicians worked at Linde.[119] The technical library was expanded. In view of the lack of municipal and state services, these were quite attractive fringe benefits. But of course material fringe benefits were even more important. Linde had thus decided at the beginning of the 1950s to invest in housing construction for its employees, despite scarce resources.[120] There were two reasons for this: for one thing, housing shortages were tremendous in cities such as Aschaffenburg, Cologne, and Munich due to wartime destruction. Housing was thus truly a central question in hiring and retaining employees. For another, the state supported company housing construction programmes and company-affiliated housing associations, thus enabling the company to do something good for its employees with partial financing. The capital tied up in housing construction enhanced the company's stability in two respects; it not only helped retain employees, it also indirectly expanded the company's equity. From the beginning, Linde invested in the construction of both apartment buildings and single-family homes, as well as in home mortgages, because home ownership was very effective in retaining executives, too. The programme started in 1951, when Linde invested about 420,000 DM in interest-free employer loans for Wohnungsbau GmbH Linde, a non-profit housing association it founded in Wiesbaden in 1951, to build a total of 67 apartments. In 1953, 18 apartments in Sürth, 17 in Mainz, and 32 in Aschaffenburg were ready for occupancy.[121] By the early 1960s, when the dramatic housing shortage ended, an additional 532 apartments and 101 single-family homes had been built.[122] Many of these single-family homes (and some apartments, too) were later sold to employees. To this day, the company is prepared to support the purchase of a house or apartment with loans on an individual basis.

Also in the early 1950s, Linde decided to offer its employees a company health insurance fund (Betriebskrankenkasse). In a ballot on 19 November 1951, the vast majority of employees who were required to enrol in compulsory insurance voted in favour of it. Since its founding in October 1952, a single company health insurance fund, the BKK Linde, has served all of the company's branches and factories.[123] Additional social benefits at Linde AG were the Linde Foundation (Linde-Stiftung), scholarships, anniversary gifts, bonuses, social services for families, holiday pay, and not least the company retirement pension plan, which was also an instrument for long-term employee retention. Carl von Linde's prestigious old home on the Prinz-Ludwigs-Höhe near Munich served from the 1950s onward as a guest house; it was also available to Linde employees who had business in Höllriegelskreuth. The two company-owned holiday homes, one in Schalchen and the other at the Spitzingsee, had a special function in cultivating a sense of community. The recreation home in Schalchen was reserved 'primarily for our blue-collar workers',[124] while blue- and white-collar workers could holiday in summer and winter at the Spitzingsee, a lake in the Bavarian Alps.[125] The very different treatment of female and male employees during the 1950s was reflected in an Executive Board decision in 1952 that male employees' wives and children would also be allowed to use the recreation homes, but that husbands of female employees (with the exception of disabled veterans) would be categorically barred from admission. For decades, the two holiday homes in the Bavarian Alps and their foothills strengthened a sense of belonging to the Linde 'company family'. Only in the 1970s did their usage decline noticeably because people could afford (and preferred) to stay elsewhere, and the facilities were sold off.

In retrospect it is clear that efforts to provide social fringe benefits for employees were especially pronounced during the 1950s. State-guaranteed benefits were still meagre. But when they were expanded – by, for example, reducing the working week and improving the state social welfare system – Linde's company reports consistently took a critical stance, stating that the company would have rather taken these steps voluntarily and independently.[126] The economic miracle provided adequate scope for such action. Linde was regarded as an attractive employer from the perspective of social benefits, too. Due to its attractiveness – in terms of the work itself, its workplace atmosphere, and its medium-term benefits – Linde could even afford to offer quite modest starting salaries for engineers.

Concerns about commercial refrigeration

Since the beginning of the 1950s, Division A had consistently been one of the 'problem children' at Linde. For one thing, refrigeration technology took off later than other sectors in the early 1950s; for another, it suffered from intensified competitive conditions in the late 1950s. Annual reports noted that commercial refrigeration systems 'held sales steady' in 1950; 'new orders matched last year's level' in 1951; 'growing customer restraint and an intensification of competition were ascertained in the domestic and foreign markets' in 1952, with the previous year's sales holding steady and earnings improving slightly.[127] In subsequent years, too, business development remained quite static.[128] In 1953, the previous year's domestic sales were 'not quite achieved'; in return, the foreign market improved significantly. In the following year, 1954, foreign business already accounted for nearly 75 per cent of commercial refrigeration machine sales. Until the end of the 1950s, the Commercial Refrigeration Division traditionally had other companies build the majority of the refrigeration units and other system components it needed, and it scarcely used the Linde Group's own production capacities. But in the early 1950s, the Executive Board had finally convinced itself to move the Commercial Refrigeration Division from Wiesbaden to Sürth and to combine the two units organizationally. Expansion measures in Sürth paved the way for the final transfer of the division in 1960.[129] With the consolidation of commercial and domestic refrigeration came the hope that co-operation between planning and production would improve as well. In the late autumn of 1960, the Commercial Refrigeration Division moved to Sürth – after almost 82 uninterrupted years of being headquartered in Wiesbaden – and was renamed the 'Industrial Refrigeration System Division'.[130] But this merger failed to solve the problems in commercial refrigeration. The market remained small and production expensive, and the potential leap into new areas of large plant construction was actually hampered by the organizational commitment to the refrigeration area.

A new boom for the cold stores in the Cold War

As in previous decades, the utilization and profitability of the cold stores during the 1950s depended primarily on governmental stockpiling policy for agricultural products, because the large retail chains built their own cold stores to meet their needs. In the early 1950s, governmental agricultural stockpiles were still small, despite the Cold War. There simply was not enough money for this function. Business was accordingly sluggish. Linde's holdings in the Gesellschaft für Markt- und Kühlhallen, Hamburg, and the Blockeisfabrik Köln von Gottfr. Linde GmbH, Cologne, remained unprofitable during these years. Occupancy of Linde's own cold stores in Munich and Nuremberg was at least 'satisfactory'. Linde naturally considered West German stockpiling to be 'insufficient' and worked to convince governmental agencies to increase stockpiles for national security reasons. This effort bore fruit. With increasing financial scope in its budget, the Federal Republic devoted more resources to this national security problem starting in 1956. From then on, Linde's company reports expressed low-key praise for state policy: 'The West German stockpiling agencies have finally accommodated public needs with appropriate stockpiling.' From this point on, the cold stores in Munich and Nuremberg were 'quite busy and yielded a reasonable return'.[131] In 1962, a quarter of a century after Linde had last built a cold store, it was worthwhile to build a new one in Regensburg; this was followed in 1965 by another large, modern cold store in the north end of Munich. In addition to steady government stockpiling, another state-supported market for the cold stores opened up in the early 1960s: the increasing storage of agricultural surpluses under the auspices of EEC price guarantees. For several decades, the storage of mountains of meat and butter and oceans of milk would be a predictable load factor for the cold stores.

The boom in gases and plant engineering

The Gas Liquefaction Division (Gases and Plant Engineering) grew very successfully during the 1950s. Plant Engineering, as described, was already working at full capacity in 1950, primarily due to foreign orders. Linde therefore decided to keep its branch factory in Schalchen, Upper Bavaria, which was established during the war when operations were shifted to the countryside; it moved out of its rented quarters and into its own factory halls in Schalchen.[132] At the same time, the Höllriegelskreuth factory buildings were expanded 'in view of the large number of export orders on hand'. The first large order – or more accurately, the 'largest order that the Linde Company had ever received' – came from South America in 1952.[133] It was for a large oxygen plant with 50,000 m³/h. The main problem in its construction was the difficulty in obtaining semi-manufactured goods, especially sheet steel and steel pipes, due to the steel shortage resulting from the Korean War. During the late 1950s and early 1960s, copper was replaced by aluminum as the most important material in plant construction. Linde had already begun to use aluminum during the war due to the limited availability of copper and continued to do so afterwards. Georg Plötz, the director of engineering and later Executive Board member (as of 1970) was an early advocate of aluminum. It took, however, a number of years to solve thermal and manufacturing problems with the material. Concerns on the part of the most valued professionals in manufacturing at Höllriegelskreuth and Schalchen, that is the coppersmiths themselves, were voiced slowly.

Linde profited in the 1950s not only from an expansion of the oxygen market but above all from a constant stream of new applications for gas separation. The chemical industry once again became its most important customer in the 1950s, but this time for the production of plastics instead of synthetic gasoline.[134]

For Linde, plant orders often meant a thrust into uncharted technological territory, and thus a high risk. The large plant business required, then as now, concrete commitments to the client at the signing of the contract. It thus called for considerable boldness and mutual trust between client and contractor, as well as among those responsible for development, construction, and assembly on the contractor's side. The close relationships among those responsible in Höllriegelskreuth, including personal friendships, and the trustful relationships between Linde and large German chemical companies were the indispensable basis for the growth of this technologically fascinating and highly lucrative business. In this segment, Linde competed very little on the basis of price, but rather on technological competence, reliability, and promised efficiency.

The 1950s and 1960s were also an era of fascination with the peaceful uses of nuclear energy. A wide variety of technological and industrial fields that took pride in themselves – ranging from shipbuilding to medical technology to thermodynamics – sought a connection to nuclear engineering. Linde, too, was proudly involved in this future-oriented field from the mid-1950s onward. In 1955, it built a factory for separating heavy hydrogen (deuterium) by rectifying liquid hydrogen for nuclear power plants.[135] To avoid missing the boat, Linde even invested 100,000 DM in the financing company for the Karlsruhe Nuclear Research Centre (Kernforschungszentrum Karlsruhe). But the construction of plants to produce deuterium was not destined for a great future, for light-water reactors would prevail internationally. Linde dropped out of the Karlsruhe centre just a few years later. But the excursion into nuclear energy had positive side effects, as was so often the case in Linde's development work. With its deuterium production plant, the company achieved –252 degrees Celsius for the first time in an industrial process. In subsequent years, cryogenics would develop into one of Linde's domains, which was at first not very profitable, but has been increasingly so since the 1970s.

Back to Höllriegelskreuth and Schalchen, which in the second half of the 1950s were running at full capacity and demanded a lot of overtime despite ongoing expansion of the workforce.[136] As is typical in the large plant construction business, short-term gaps appeared time and again, as in the absolute boom year of 1958.[137] A basic trend in plant construction was the demand for ever larger plants and for complete 'turnkey' solutions. The engineers and factories in Höllriegelskreuth and Schalchen were not prepared for either of these. It became apparent that the market for traditional planning and production methods for gas separation plants would dry up over the medium term. The division had to decide whether it should take the leap into offering complete large plants, with all the attendant risks.

The production and sale of industrial gases also grew very satisfactorily during the 1950s. Throughout the decade, gas sales climbed continuously and necessitated a progressive expansion of capacity.[138] The Gas Division, led by Walther Ruckdeschel since 1950, successfully aimed at securing the domestic market and re-establishing itself internationally. Because German companies were, until the late 1950s, in part prohibited from founding foreign companies, Linde's acquisition of shares was done via the Swiss Corporation for Refrigeration in Lucerne, which Linde had founded precisely for this purpose. Ruckdeschel was able to found over 20 associated companies into the 1970s, thus ensuring that his division would once again turn the highest profits in the corporation.[139] From Ruckdeschel's perspective, Plant Engineering functioned to help the Gas Division. Consequently, requests to Plant Engineering were sent first to Ruckdeschel in the 1950s, who wanted to determine whether the construction of a plant would benefit gases. It was not until the 1960s that this practice was stopped. To guarantee its regional dominance, Linde bought up excess oxygen in Germany that was a by-product of chemical processes.[140] In addition to cylinder sales, which dominated the market overall, Linde also had an increasing number of large customers who bought oxygen in liquid form and fed it on-site into their factories via a distribution network with the help of a cold gasifier system.[141] The steel industry developed into a major buyer of liquid oxygen as it increasingly went over to the LD process for raw steel production. Linde built new gas plants at the end of the 1950s in Krefeld, Witten, Regensburg, and Meiderich. Noble gases increasingly joined the classic products, acetylene and oxygen.

Refrigeration business

Starting in the 1950s, the refrigerator became an increasingly common household appliance in Germany. Was Linde well advised to enter this market and compete for its share of customers? Linde's small refrigeration unit consisted of two factories in Sürth and Kostheim. During the 1950s, Sürth had two different departments: the AM department dealt with general machine construction, which was, for the most part, the manufacture of machines for air and gas splitting, while the KM department constructed refrigeration units and systems.[142] In the KM department, refrigeration units and devices were also produced for the factory in Kostheim, which distributed refrigerators, freezers and deep-freezers, and commercial refrigeration systems. In general, small refrigeration units increased and took a decisive upturn mid-decade, when stiff market competition forced Linde to rationalize.[143] In general, the serial production of refrigerators in Kostheim during the 1950s ought to have developed into one of the most important providers in the field. It took some time, however, before Linde was able to adjust to this mass market's special sales and production conditions.[144] During the 1950s and early 1960s, the refrigerator changed – like many other durable consumer goods – from an exclusive luxury good to an indispensable element of the modern household. In 1951, the German parliament was still discussing whether a luxury tax should be imposed on refrigerators and other products not absolutely necessary for living. Only ten years later at the beginning of the 1960s, 54 per cent of West German households already owned a refrigerator.[145] In the US, by contrast, this process had taken place before the Second World War. Successful sales at the end of the 1950s brought more planning security for refrigeration firms. Until this point, refrigerators had been seasonal articles whose sales fluctuated accordingly. It had only been worth advertising in early summer or before Christmas. Even in 1956, about 248,000 refrigeration units were produced during the second quarter, while about 149,000 were produced during the fourth quarter.[146]

The refrigerator's success did not, however, only depend on increasing prosperity and sinking refrigerator prices.[147] Pre-war houses had been built with storage basements, larders, and pantries, which had been used to store foodstuffs over long periods. These storage options were missing in new houses built after the war; council housing, therefore, had no such spaces, making refrigerators a necessary item. 'At the same time, it also became a visible sign of the beginning of a better life', and was also a 'safety box of the new affluence'.[148] Most of the refrigerators of the 1950s only had a small freezer; storing frozen foods over longer periods was not practical. Thus, communal freezers were built in many housing complexes during the period of reconstruction. Two such freezer systems existed in 1950, and ten years later there were 9,300. Thus began the popularization of frozen foods in the German Federal Republic. Nevertheless, the freezer in one's own apartment was what made frozen foods really popular.

In 1956, Linde set up a completely new factory for the expanded serial production of domestic refrigerators in Kostheim. In the following years, steps were taken to rationalize its administration. Distribution of ready-to-use products – basically, domestic refrigerators and ice chests – was moved from Sürth to Kostheim in 1957. In 1959, refrigerators for export were also moved to Kostheim, and in 1960, the distribution department for small commercial refrigeration units was combined with the distribution of domestic refrigerators due to 'special rationalization measures'.[149] These measures freed space in Sürth to move the large-scale refrigeration units department from Wiesbaden there during the same year.

Solid ground for field tractors

Department D (Güldner) had a successful start in the post-war period. Its most important product, the field tractor, developed during the late 1930s with a wood-fuelled carburettor and subsequently improved, was in high demand. It was a seller's market, in which the manufacturer could sell his products without much effort. In 1951, the factory's capacity had to be increased.[150] Moreover, Güldner's market share of tractor sales in Germany at 5.5 per cent (1950) rose slightly the following year in addition to its export business.[151] Nevertheless, sales stagnated in 1952, clashing with the cyclical growth cycle then happening during the economic wonder in Germany. The short post-war boom in agricultural machinery was nearing its end. In the 1953 business report, Department D attributed its first drop in sales from the previous year to 'a general restraint that has set in domestically and overseas in the tractor market'.[152] From this point on, tractor production was a problem child for the firm until production was cut at the end of the 1960s. The market was too small for the large number of manufacturers in Germany. Demand fluctuated dramatically both cyclically and seasonally.[153]

Güldner also suffered from self-created structural deficits. The business sector had no concept of long-term development beyond field tractors. The transition from seller's to buyer's market was appreciated late, and incurred losses caused debts to accumulate. Still, marketing remained insufficient; those at Güldner continued to believe that one could simply depend on its high quality work to secure a good reputation and a core clientele.[154] Güldner was considered particularly conservative in the branch; something that one agro-technology newspaper highly valued when it remarked in 1954 that the 'Güldner-Motor-Works displayed a noteworthy restraint by continuing to manufacture and improve models already in use despite the hasty production of new models witnessed during the past years.'[155] In 1954 it was nevertheless decided in Aschaffenburg to follow the trends of the time and, for the first time, to produce air-cooled fitted-diesel motors and tractors. In 1955, associates at Güldner thought that there was still a future in agricultural machinery and took over the appliance carrier, 'Multitrak' from the firm Karl Ritscher GmbH, a universal machine with extendable frames for various agricultural purposes.[156]

A first try at freeing themselves from their dependence on the inconsistent and low-growth agricultural machine sector was the 'Güldner Hydrocar', a platform vehicle (capable of carrying up to 4 tons) with a variable hydrostatic gearbox.[157] It was a true technical breakthrough, a revolution that had not actually been planned as one. Güldner had built the vehicle in order to test its newly developed, variable transmission. Sales remained predictably small at the beginning. Nonetheless, this variable transmission provided Güldner with the technical basis to build up this sector over the next decades as the most important area of sales for Linde.

The division in Aschaffenburg was built within the city limits and offered no space for further expansion during the mid-1950s. In the business, one was convinced that motors, tractors, and hydrovehicle construction did not stand a chance without modernized assembly line production. Thus, the Executive Board decided to operate a new vehicle factory in Aschaffenburg-Nilkheim, which was only a few kilometres from the motor factory. Starting in 1956, vehicle parts were produced, and tractors as well as hydrovehicles were assembled on this factory's assembly line.[158] It was a brave and far-sighted decision, which the Executive Board implemented even as sales during the summer had become so poor that part-time work had to be introduced – and all of this during a booming economic situation marked by full employment and even a shortage of workers![159] In 1957, part-time working was no longer enough and dismissals could not be avoided. Better sales in the motor sector could not compensate for the fall in tractor sales. In order to 'mitigate the dependence on agriculture', work was started on the 'development of new products', as the business sector cryptically called it without actually naming what these new products would be. During the next year, the future became clearer: in 1958, Güldner Aschaffenburg took over the hydraulic power unit from the firm Gusswerk Paul Saalmann & Söhne OHG in Velbert. Thus, the foundation was laid for the creation of the later field of floor conveyors and hydraulics. In the same year, Güldner began the serial production of hydraulic power units and floor conveyors after reorganizing its production. These efforts did not at first, however, lead to an improvement of the economic situation; at the end of the year, temporary part-time work had to be reintroduced.[160]

At the end of the 1950s, Güldner was not any worse off than other tractor manufacturers. Indeed, there was a general structural crisis, leading to agriculture's loss of significance in the national economy and the breaking-up of the market. The market share of Germany's ten leading companies together was smaller than that of the three leading American and English companies. In 1958, at the latest, a wave of takeovers, joint ventures, and mergers started in Germany to help German firms stand up to the British and French tractor industries.[161] An attempt to reach a 'rationalization agreement on a large scale' faltered. Attempts included negotiations with MAN with which Güldner was connected through the parent company.[162] Despite these setbacks, Güldner was able in 1958 to declare a far-reaching co-operation with the machine factory Fahr, where Güldner had been housed during the war. This was announced on the occasion of its 100,000th diesel motor.[163] All of the tractors were produced on the same construction basis and called the 'Europe Line' as part of a common marketing programme.[164] Linde's model was the large American and English manufacturers who had more standardized parts in all of their tractors and offered a wide range of air-cooled and water-cooled motors.

Starting in 1958, Güldner had clearly changed direction, marked by new tractor types, the strategic co-operation with Fahr, the newly developed Hydrocar-models, and the additional purchase of hydraulic technology. Storm clouds, however, remained on the horizon. After just two years, co-operation with Fahr fell apart. In addition, competition became stiffer in agricultural machinery.

Figure 4.3   Cost reductions through rationalization of production at Güldner, 1955. The working process of the old milling machine (left picture) and the new one (right picture): the time per piece for three work pieces is reduced by over 50 per cent. The saving for 1,000 work pieces amounts to 13,000 minutes or 1,500 Marks.

Strategies and worries about the future, 1961–66

Cyclical dip or economic crisis?

In economic history, the exact year marking the end of the rapid upward trend of the West German post-war economy, the 'economic wonder', remains contested. Into the 1960s, the Federal Republic belonged to the world's strongest national economies, but overall economic development started to slow down at the beginning of the 1960s. After record growth rates were reached in 1964, a board of experts commissioned by the German parliament warned of the threat of a sudden economic downturn, which in fact occurred in 1966.[165] Growth rates fell dramatically, high prices hurt workers, and unemployment rose to over 500,000 by the end of the year.[166] The steadily increasing state handouts of the previous years could no longer be covered and resulted in a deficit of 4 billion DM at the national level. From our present perspective, it is difficult to evaluate this as a symptom of a serious crisis; nonetheless, the economic collapse in 1966 marked the end of the post-war boom.[167] The higher than average growth rate in the Federal Republic at an international level had to slow down some time, but neither the government nor the economy had reckoned that it would occur at this point. State expenditures, investment, and wage increases had been planned according to permanent growth. The state responded to the economic plateau with a Keynesian-style solution – the raising of debts. This was new. And it helped very little. The state began slowly to realize that during the late 1960s, it would be much more difficult than it had been during the post-war period of reconstruction to direct the economy towards full employment in a highly-industrialized economy that was closely connected to the world economy.[168]

Structural and leadership problems

At the beginning of the 1960s, Linde started to feel the sharp edge of competition in ever more markets. In refrigeration technology and tractor and motor construction, the company had allowed itself to rely on its good name and on the belief that its clients would pay higher prices for Linde products. Radical changes in the structure of the company including the dissolution of certain factories were not yet on the horizon for most of the employees, not even in the 'threatened' sectors. In the first place, one believed in a diverse range of product, which was supposed to make better use of available production capacities without resorting to old, unprofitable production lines. Naturally, rationalization policies from the 1950s were implemented, but always in small areas and in relation to existing production lines. Serious economic problems arose in the domestic refrigerator sector (Mainz-Kostheim), in general machine construction (Sürth), and in Güldner (Aschaffenburg), but market demands also affected successfully operating plant engineering in a completely different way. Industrial clients wanted steadily larger systems in ever shorter intervals. Thus, the financial risks of mid-size plant engineering rose for Linde since, on the one hand, materials and labour costs had to be financed in advance, and, on the other, the customary production process oriented around system engineers' experience ran up against its limits. It became necessary to provide comprehensive estimates, calculations, and documentation for a profit margin that was shrinking due to sharper competition. In this sense, the overall economic result for Linde was that a positive expansion of plant engineering could not be separated from increasing risks.

In order to prepare for the necessary investments, the capital stock was renewed during the general meeting on 20 July 1961 and raised by 13,734,000 DM to 65 million DM. Due to higher stock prices, Linde ceased worrying about possible takeovers and ended the protective measure of registered shares, which had by this time become an old-fashioned policy. They were exchanged at a rate of 3:2 into bearer shares.[169] Since the Linde family had held the majority of registered shares, their influence thereby sank dramatically in 1961 despite the nominal revaluation of their shares. In this year, Linde had no large shareholder with over 5 per cent of the capital among its 7,760 shareholders. In terms of figures, the largest group of shareholders according to a survey of the shareholders in 1963 was housewives and widows. They accounted for 31 per cent of shareholders and 20.5 per cent of capital shares. They were followed by salaried and wage-workers, who accounted for 20 per cent of shareholders and 11 per cent of capital shares. According to the survey, private individuals (agricultural and forest engineers, salaried and wage-workers, housewives and widows, retirees and pensioners) accounted for 61 per cent of shareholders and 38.5 per cent of the capital. The

Linde Society's capital stock was also widely spread – 'its largest part seemed to be in firm hands; Linde shares were highly valued by small-scale savers'.[170] Linde shares appeared to be minimally speculative, have high earning power, and be secure, like a 'blue chip'. The dispersion of shareholdings was treated by the company as a strength.

At Linde, feelings of crisis came in waves. One reason for this was the good overall economic situation, on the one hand, which resulted in ever 'new high points' in production, sales, and orders,[171] and, on the other, the 'bad profit situation especially in Aschaffenburg and Kostheim', which the Executive Board knew about and which had the tendency to climax over the years.[172] One thus wavered between optimism and worry. In order to understand the measures taken by the Executive Board to solve the structural problems, we need to explain the business's information- and decision-making structures. The Executive Board was, in fact, formally accountable for all of the divisions' economic results. Nevertheless, the individual board members were neither obliged nor able to oversee or direct the numerous sectors either technically or financially due to the different ways that the individual sectors were organized and the resulting decentralized structure. A comprehensive analysis of the economic activity of one sector by a board responsible for another sector completely contradicted the customary practices of the business leadership at Linde. The leadership was organized so that complete trust was put in the competence of the current manager and the belief that this person would be in a position to improve his department or sector alone.[173]

The most important reason for creating this decentralized structure lay in how the factories had been conceived. Starting in 1902, members of the Linde family had founded the first sectors (gas liquefaction and splitting, and then, gas factories) in Munich, which they directed from there – far from the company's headquarters in Wiesbaden. It was unimaginable that the strategies for these sectors should come from Wiesbaden. Sürth and Aschaffenburg, in contrast, were additional purchases from large manufacturing factories with a long history of independent business culture. The spatially disparate structure of the factories alone would not have automatically led to this decentralized business and administration culture. But, since the different factories were also separated institutionally and by field, the spatial separation further emphasized the mental and cultural differences between the factories in a negative way. Due to this separation at three levels, the sense of belonging and common identity was weak between the factories. A common strategy could only be developed with difficulty.[174]

The Supervisory Board postponed discussion on the difficult situation in the individual divisions for many years. When the board member Simon reported on the significant worsening of the company's profit situation during the board meeting on 9 December 1965, it was feared that there would be further turbulence in the Supervisory Board: 'The disclosure will cause a great shock to Dr Oetken and Dr Brandi, and the Executive Board will have to prepare for

Figure 4.4   The upper ranks in the early Sixties: front row, left to right:
Dr Johann Simon, Prof. Dr Kurt Nesselmann, Dr Otto Meyer, Walther
Ruckdeschel, Dr Friedrich August Oetken, Dr Hermann Brandi,
Dr Johannes Wucherer.

unpleasant questions.'[175] Besides structural problems, there were clearly also
problems of communication.

The new evaluation of the firm's situation, the realization that Linde was now
in a difficult and perhaps dangerous situation, as well as new worries about an
unfriendly takeover led to clear decisions and steps in the divisions over the next
few years but not to drastic changes in the leadership structure itself. The years
leading up to 1969 are characterized as a period of insecurity, a confrontation
over the correct way into the future. During this period, it was unclear which
individuals and which concepts would win out; several attempts were made,
and it was rare that the Executive Board was able to muster unity behind the
necessary measures.

The wish to rejuvenate the company by going public led to a cosmetic
name change in 1965. The somewhat long-winded title 'Society for Linde's
Ice Machines Limited Company' was turned into the short and concise Linde
AG. A year earlier the branch offices in Mainz-Kostheim and Sürth had been
integrated into the 'Refrigeration Division'. They shared a common marketing
and distribution department; the company's production lines[176] were now called
'Fields'. In 1965 there was again a new redivision of divisions within the entire
company, this time oriented – apart from the cold stores – on the geographic
distribution of the production sites; they were now separated into divisions
in Munich, Sürth, Ascheffenburg Güldner, and the cold stores. These changes
were, however, more rhetorical in nature, hiding the original four divisions
behind the new location names. In addition, the new geographic titles actually

emphasized the disparate structure. Wiesbaden had lost out as it was no longer the headquarters of a division, allowing, however, the central administration to expand. Was this enough to deal with the structural problems?

The Executive Board and the Supervisory Board felt unsure and suspected that the company lacked strategy. Their worried look into the future accompanied by a new concern with structural problems was not unique in Germany, indeed, it was rather typical for this period. All over Germany, worries about the end of the economic boom were growing. The technological and organizational head-start in the United States had been much greater during the 1950s, but it was now that German businesses were beginning to complain. Wake-up calls and warnings of various sorts echoed through the business landscape, particularly at the end of the 1960s.[177] Organization and planning appeared underdeveloped, businesses too small (economies of scale), and research and development neglected. The technological gap with the US increased year after year[178] for firms that did not take quick and decisive action to set up their positions internationally, bolster research, and improve their organizational and strategic positions. Planning and organization departments started appearing everywhere, not only in companies but also in associations and state institutions. This atmosphere also took hold of the Executive Board at Linde. Board deputy Hans Meinhardt took advantage of this mood to improve Linde's organization using an American model. In literature from this period, we also find a cool, partially ironic treatment of the fearful view across the Atlantic, such as the characterization in certain technological journals of those issuing warnings as 'Gapists'. But, at Linde, this cool attitude was viewed as out of step. The majority of board members, apart from Hermann Linde, assessed the future with concern. At this time, the readiness to undertake radical steps towards reform in the business increased. Another necessary condition for this mood change was the business leadership's age range. From the early to mid-1960s, there was still a group of very elderly men with leadership responsibilities.[179] This group had until this point secured the continuity of leadership ideas in the company. By the mid-1960s, this continuity appeared old-fashioned, and the younger generation had the chance to call for new ideas, a process similar to that which one now regularly finds in political parties. The generational break encouraged the elderly and the young to accept change, since it seemed that the traditional principles that had functioned under Carl von Linde no longer sufficed in adapting a tradition-rich business to a global market.

Hurdles in Sürth

Despite this changing mood, Linde dragged around various unsolved structural problems in several divisions for many years. The clearest case was in the Industrial Refrigeration department. The department was created in 1960 when the Commercial Refrigeration System department in Wiesbaden was merged with the machine factory in Sürth.[180] The divisions also had their individual suppliers, such as the foundry in Hennef/Sieg that belonged to Sürth.[181] Sürth had substantially expanded due to the merger of the refrigeration technology programme.[182] No less comprehensive and varied was Sürth's second main area of production, general machine construction.[183] The working group had access to internationally recognized knowledge such as the production of dry-run compressors with lubricant-free compression. Nonetheless, Sürth had two main problems after 1964: it produced too many different products in small amounts at high, unmarketable prices. The prices that Mainz-Kostheim paid internally to Sürth for semi-hermitic compressors were higher than market prices.[184] In 1962 the Executive Board decided that Sürth must be competitive with market prices in internal settlements. While a meagre profit of about 1 million DM was made in 1962, there was a loss of about 2.2 million DM in 1963 because of this policy change. In the refrigeration equipment business as well, costs were decidedly higher than those of its competitors. For the Executive Board, the reasons lay in organization and were, thus, still solvable.[185] The alternative position understood the problem as unsolvable and favoured selling off a portion of the refrigeration business.

The next organizational reform in 1964 was the uniting of Sürth and Mainz-Kostheim into one department. When the reform was suggested, it provoked the first big conflict over structural reforms. The parties working on the corresponding plans were the Executive Board, factory managements in Sürth and Kostheim, and the central administration and organizational department in Wiesbaden.[186] The possible alternatives included selling Kostheim or merging it with Sürth. At the beginning of 1964, the marketing director in Kostheim, Klein, used strong words when he called for a merger that would lead to a powerful unit:[187]

> In an expansionary market [referring to refrigeration technology], red figures are explained by the competitive weakness of one's own firm. As long as we have not exhausted all possibilities to establish our competitiveness, a retreat from the business is the wrong decision.[188]

Klein's conclusions on the causes of refrigeration losses clearly diverged from those of his colleagues: 'out-of-date construction, poor quality, uncompetitive prices, unmarketable sales policies, and heterogeneous leadership', and furthermore, many Linde products were no longer up to date with modern developments and construction methods. High expenditures on guarantees, qualitatively uncompetitive products, and generally overpriced products made Linde uncompetitive. The general conditions were also an omen for the future since there was 'neither in Kostheim nor in Sürth long-term, systematic product planning with a competition-oriented programme', and the current marketing and distribution apparatus did not act as 'an instrument for the modern competitive market system' because important functions such as marketing and product research' were 'not applied'. Consequently, in order to 'remove the red figures from the refrigeration sector at Linde', Klein saw the establishment of Sürth and Kostheim as 'a single economic unit "Linde-Refrigeration" integrated with a shared leadership body for development, production, finance, and sales' as the only possibility.[189] In 1964, the Executive Board became convinced that the integration of the two divisions under a common administration was the best solution; Kostheim was reduced to a subsidiary of Sürth. The headquarters of these two divisions was at Sürth, where they shared marketing, research, and development departments.

The problems were not, however, fundamentally redressed; the situation remained difficult since no one in the Executive Board wanted to be responsible for the 'Sürth problem child'. The board member responsible for Sürth, Simon, stated during the board meeting on 19 November 1965 that he was no longer prepared to continue being responsible for the division in Sürth and suggested that 'a factory management committee composed of one technician and one economist be created immediately'.[190] This radical step in the business leadership documents refrigeration technology's problematic situation at Linde during the mid-1960s. At the same time, however, it marks a milestone on its way to creating a new model for the Executive Board, in which the board no longer directed one division but instead oversaw the appointed management.

The sale of the 'white goods'

The founding of the new division in Sürth did not solve the problems of the refrigeration sector because most of the strategic problems, such as the future of the 'white goods', for example domestic refrigerators and ice chests, remained open even after the restructuring in 1964. In 1960, production was at full capacity and accompanied by higher sales.[191] Improved models, in particular the new duo thermal refrigerator that had separate refrigerator and freezer compartments, which met the 'the wish for ample storage possibilities for frozen foods', were built. Despite this success, household appliance sales were already stagnating by the next year, 'especially during the seasonal sales standstill'.[192] By the beginning of the 1960s, post-war hunger had been largely satisfied, and the mail-order business was looking for businesses in Southern and Southeastern Europe with lower costs.[193] Prices were sinking while wage and material costs were rising. Sales stagnated, and in many cases, thousands of finished appliances remained in storage, which tied up capital and delayed production. This sales crisis had dramatic consequences for domestic refrigerator manufacturers in Germany. Smaller firms using skilled manual labour and partial-industrial production did not have enough capital to invest in the necessary large-scale serial production. Even large and well-known businesses such as 'Ate' in Frankfurt gave up production around 1960; and Opel transferred its 'Frigidaire' production to a factory in Paris. At any rate, the 'clearing-up of the market' had positive consequences for the surviving businesses – they could now produce larger numbers of units. 'After a few years, the increasing serial production led, however, to a concentration of a small number of manufacturers, which were in the position of implementing production methods with the fastest flow of materials in new factories.'[194] Not only did sales and production conditions change, but the refrigerators themselves as well. The trend went from a large standing appliance to a small built-in refrigerator, whose size was adapted to standard kitchen units. Technology became more compact so that instead of an entire compartment for the motor, a small empty space in the cupboard casing sufficed. Space was also saved with new insulation materials; at the beginning of the 1960s, firms at the forefront of artificial materials in the Federal Republic, including AEG and Alaska, Escher-Wyss and Linde, BBC, and Teves, replaced polystyrene with polyurethane as the insulator in refrigeration construction. By the mid-1960s, the table model had become the West German housing standard.[195] In the 1964 business report, it was hoped that frozen foods expansion would 'stimulate the sales possibilities of our domestic refrigerators, also in municipal institutions'.[196] This hope was not actually fulfilled. The year's new refrigerator models received a 'good response', but the domestic refrigerator business was already 'declining' in 1965.[197] In July 1965, the Executive Board was, thus, once again concerned with the future of the refrigeration sector.

One of the plans prepared by both the Sürth sales director Klein and the organization department and then approved by the Executive Board foresaw Linde offering a complete programme of 'white goods', including refrigerators, washing machines, dishwashers, and stovetops for the medium term[198] as either a licenser or in co-operation with other manufacturers.[199] The sales manager in the refrigeration department at Sürth, Karlhanns Polonius,[200] had in the meantime also come out in favour of this form of expansion.[201] Other German firms including Bosch, Siemens, AEG, and Bauknecht also harboured great expectations from this turn to a larger product programme, which, according to the American model, led to brand recognition and the greater likelihood that customers would stick to one brand for all their domestic appliance purchases. The name was intended to act as a distinctive marker for the product. Although this plan remained a marginal vision, it did lead to the selling of a complete spectrum of domestic appliances in the refrigeration sector. Because the majority of board members were convinced that they should either make a complete offering or relinquish the market, they came out in favour of a joint venture in 1966. A majority in the Executive Board preferred the Italian company Zanussi; only Hermann Linde and Walther Ruckdeschel were in favour of an agreement with Siemens.[202] Negotiations with Zanussi, however, fell apart over discussions on how to create a common production and sales organization,[203] because it quickly became clear that Zanussi, above all else, wanted to use the Linde name and the common company to market its own products more effectively in the Federal Republic.[204] The Executive Board subsequently feared that Zanussi would quickly dry out manufacturing in Kostheim.[205] Meanwhile, Siemens joined Bosch to create a common 'white goods' programme, which is today's successful Bosch-Siemens Household Appliance GmbH. Linde's later negotiations with AEG proved to be just as successful. AEG made clear that it was prepared to buy Linde's refrigeration business.[206] In 1967, domestic refrigerators and refrigeration units were then branched off into an independent business. At the beginning, Linde maintained 25 per cent ownership of the newly founded Linde Household Appliance GmbH., while AEG held 75 per cent. In the factory in Mainz-Kostheim, the parts of the factory engaged in household appliance manufacture were also sold to this new company.[207]

The refrigeration sector profited greatly from this sale. First, money became available for innovations. Second, Linde could now concentrate on commercial refrigeration. The mass-produced commercial refrigeration units, especially the new 'Europe-Program' (Europa-Programm) for self-service shops,[208] sold well. Linde became the market leader in this area over subsequent decades and continues to be one to the present day due to its strategic purchases. But, with the sale of Linde's single serial production for private consumers, the growing company slowly lost its public name recognition.[209] Even if refrigerators had not helped Linde's capital goods production or marketing, they had provided a good advertisement of Linde's excellent quality.[210]

The end of tractor and diesel engine production

We also find substantial organizational reforms taking place at Güldner in the 1960s. During the 1950s, agricultural machines were already facing a difficult sales market. This was due to the significant seasonal shifts and the general decline of agriculture's contribution to the gross national product. International competition from such heavy-hitters as Massey-Ferguson, International Harvester and John Deere, whose large-scale production resulted in lower prices, also became a factor. Güldner's joint venture with Fahr in 1959 in the hope of lowering costs was dissolved just two years later in 1961 after a competing firm, Deutz, had taken over 25 per cent of Fahr's shares; Deutz produced its own tractors and was not interested in further co-operation with Güldner. The Europe Line of tractors was discontinued.[211] Despite the setback, Güldner was able to maintain its market position 'with its reliable technology and its readiness to fulfil the wishes of its clientele'. Its market share increased, in fact, but the factory still suffered worrisome losses. Güldner hid its debts by pushing them onto subsidiaries, which were then transferred to foreign manufacturers and distorted competition.[212] But it was common internal knowledge that: 'Like all German tractors, Güldner tractors are too heavy and too expensive (loading capacity 51 to 73 kg per HP compared to Ferguson at 38 to 59 kg per HP).'[213] By the end of the joint venture with Fahr, ideas for factory expansion were concentrated on other areas.[214]

One area of growth was the manufacture of hydrostatic transmissions, which was transferred to and expanded in the Nilkheim Factory in 1961.[215] In 1964, the lively demand for hydrostable transmissions made further expansion necessary; the R&D department was also expanded.[216] During 1963 and 1964, the management (Pöhlein and Meinhardt) at the Aschaffenburg factory wanted to continue producing tractor and motor parts until the growth sector in hydraulics and floor conveyors was in the position to manufacture at full capacity after the necessary expansion.[217] The Executive Board at Linde made the following decision at its meeting on 24 April 1965: 'The further development of hydraulics and fork lifts requires the full capacity of tractors and motors.'[218] A possible terminus for motor and tractor production was set for the years 1970–71. Sales and results between 1963 and 1965 seemed to confirm this idea: in 1965, the number of newly registered tractors was again greater than the number of tractors on the market; Güldner's market share lay at 6 per cent. In 1966, however, sales in the German market dramatically collapsed.[219] Although Güldner was able to hold onto its market share, this development led to a renewed worsening of the accounts. The successes in forklift sales and hydraulics were now endangered by the plight of tractor and motor production.

Subsequently, management modified its business policy and planned the phasing-out of the motor and tractor sector for 1969–70 with the simultaneous forced development of the growth sectors. The question of a joint venture in the tractor sector was examined; negotiations with the firms Klöckner-Humboldt-Deutz and Renault did not, however, lead to satisfactory results because neither of the potential partners was prepared to take the risk of becoming the majority partner in a newly formed Güldner Tractor Society. As a result, Linde examined the possibilities for continuing tractor production using motors from Deutz, MWM, or Daimler-Benz instead of its own. In order to remain competitive, however, large additional investment was necessary, so that instead of becoming immediately profitable, further losses were expected. Under these circumstances, the most practical option was to close down tractor production at Güldner. On 29 January 1969, the managers at Aschaffenburg, Pöhlein, and Meinhardt thus confidentially suggested to CEO Wucherer and his deputy Dr Simon that the motor and tractor sector be dissolved and requested that the Executive Board decide immediately to stop producing Güldner tractors and retreat from the tractor business.[220] The Board of Management followed his suggestion and on 21 March 1969, publicly announced to general surprise that the Linde Company could no longer 'maintain its motor and tractor business despite its 5.5 per cent domestic market share (3,254 tractors) in 1968 that put it in 7th place'.[221] Klöckner-Humboldt-Deutz took over Güldner's diesel motor and tractor customer service and replacement parts service. At the time, it was the second largest manufacturer of air-cooled diesel motors and the largest German tractor manufacturer. Overall, Güldner had until this point built approximately 300,000 diesel motors and over 100,000 field tractors. By making this decision, Linde forced itself to undertake a significant reform programme in its factories.

Conclusion

The 1950s and '60s witnessed an economic boom without precedent in the German economy. In the industrial gas and process technology sectors, the Linde Corporation, as we now refer to the company, grew at a rate faster than the booming national economy until the beginning of the 1960s. All of its other fields also grew substantially. This growth hid, however, three structural problems at Linde: the unprofitably wide range of refrigeration products, the lack of a main product at Güldner, and the inability of the firm to reach decisions at the central level.

During the mid-1960s, just under 50 per cent of Linde's sales were made in the Munich Division, just under 35 per cent in Sürth, 15 per cent in Aschaffenburg, and about 1 per cent in the cold stores.[222] The crisis-ridden climax of the administrative situation in the northern business sectors in Sürth and Aschaffenburg resulted in the dissolution of the refrigerator section (and creation of a refrigeration technology factory in Sürth) and the shift in production at Güldner from motors and field tractors to hydraulics and floor conveyors. Overall, it was a persuasive concept. With the sale of its refrigerator sector, Linde bid adieu to the problematic mass market. Important steps in commercial refrigeration were taken to increase productivity. In Aschaffenburg, a new factory was purchased. This area remained, however, stuck in a precarious administrative situation.

How assignments would be delegated and the structure of the company's leadership organized remained unclear. How should work in the Executive Board be allocated? How large and strong should Group Headquarters be? Where should decisions about investment be made – by the divisions or the Executive Board? Did the individual board members represent the divisions at the executive level, or did the Executive Board control the divisions? Heavy battles ensued during the 1970s over how to solve these organizational problems. This confrontation is described in the next chapter.

At the same time, somewhat exaggerated worries about the future and the allegedly superior US competition, where industry was perceived as more international, efficient, and better organized and financed, grew during the 1960s. At Linde, thoughts about the company's organizational deficits became increasingly troublesome, a development that was common among firms in Germany at the time. Contemporaries in many German companies and commentators on the national economy were having similar thoughts. After years of self-assuredness as to the company's strength, the board of directors, the Supervisory Board, and the second-tier leadership began to question whether good technology alone was enough to win the future and whether the company needed more competence in planning and organization in order to resecure and build upon the great successes of the past. The team surrounding Hans Meinhardt in Wiesbaden and Aschaffenburg answered this important question with a clear yes. Without a comprehensive systematization of organizing and planning, the firm had no real chances of survival, even if the general business management data did not look bad. Standing in opposition to this group was the technological leadership surrounding Hermann Linde, who was responsible for plant construction at Höllriegelskreuth. These groups regarded the possibilities of organizing and planning sceptically and, at the same time, were more optimistic about Linde's future without restructuring.

This conflict, which was becoming more serious, was not, however, the old disagreement between engineers and businessmen about who was running the company. In the 'classic' disagreement between the technology and business departments, engineers argued that the 'cold calculating businessmen' wanted to destroy their good technological ideas and visions, while the business side denounced the engineers' unrealistic and economically dangerous visions.[223] This pattern was used in the disagreements at Linde, but did not play the decisive role. Moreover, businessmen and engineers were on both sides of the argument even if more of the Wiesbaden centralists supporting Hans Meinhardt came from the business sector and more of the Höllriegelskreuth inventors supporting Hermann Linde came from the physics-technology sector. It was actually more about sympathies and antipathies towards certain ways of interacting and about strategies. These preferences and reservations were expressions of ways of life and fundamental beliefs and had, thus, a very personal and emotional aspect.

5
The 1970s: From Engineering- to Management-Driven Business

Linde in the 1970s

The 1970s saw the end of the post-war boom to the German national economy. While the unemployment woes and low economic growth rate in Germany seems less gloomy from today's perspective,[1] the oil crises of 1973–74 and 1979 nonetheless had a major psychological impact on the German population. The Linde AG, as was often the case in its 100-year history, expanded in a counter-cyclical manner during the Seventies, experiencing significant growth in comparison to the previous decade.

More salient than this growth however, was the structural shift that occurred within the Linde Company itself during the 1970s. Linde underwent a profound change, moving as it did from a family-run (that is, an engineering- and scientist-driven) business to a management-led corporate group. To be sure, company leaders during the first 100 years were, and considered themselves to be, both managers and engineers or scientists, yet the proportional weight and perspectives granted to each shifted remarkably. The shift in power from Hermann Linde as Executive Board spokesman to Hans Meinhardt as CEO of the Executive Board was paradigmatic of this change. Initiated in the 1960s with the attempt to create a strong Group Headquarters, the company's internal restructuring continued by centralizing decision making within a new Executive Board. A centralized administration and Executive Board was to represent and realize the visible interests of the business as a whole rather than those of individual working groups. This chapter takes a close look at the internal struggles over this reform.

The Seventies was a decade of social upheaval and structural change in Germany, the elements of which were already present in the 1960s.[2] This included improved access to higher education, but also an increase in individual mobility and holidays taken abroad and – because of an economically induced immigration – the involuntary and slow transformation towards a multicultural society, the 'sexual revolution', the women's movement, a shift in gender roles, an increasing interest in politics, and the democratization of society.[3] The student protests of 1967–68 were a visible expression of this social change, as was the shift to a social-liberal coalition in the government in 1969 and the move towards détente with Eastern Europe. In addition to this, the Seventies were marked by the experience with leftist political terrorism. Social manners transformed, the modes of social interaction grew increasingly casual and past constraints began to fade as the baby boomers started their university education. We will explore first the economic, political, and social upheaval in the years following 1968 and its impact on the Linde Company as manifested in the internal restructuring and change in leadership at the very top.

Détente and business in Eastern Europe

The construction of the Berlin Wall in August 1961 and the Cuban missile crisis in 1962 mark turning points in the Cold War between two politically antagonistic blocks. During the years following these crises, the Federal Republic of Germany increasingly sought rapprochement with the Eastern Bloc. Foreign minister Gerhard Schröder (CDU) attempted in the early 1960s to engage Eastern Bloc states in a dialogue, while avoiding the USSR and excluding the German Democratic Republic (GDR). Ultimately, this assertion to the right of exclusive representation by the Federal Republic of Germany (FRG) failed.[4] It was not until after 1969, when the social liberal coalition came into power and recognized the existence of the GDR that treaties with individual Eastern Bloc states could be made and in rapid succession: 1970 in Warsaw, 1972 the Basic Treaty with the GDR, and 1973 with Czechoslovakia. All of these treaties centred upon the exchange of ambassadors, the recognition of borders, a renunciation of the use of force, the relaxation of travel across borders (particularly with the GDR), cultural exchange, and the enhancement of trade relations. The FRG then became the most important trading partner in the West for all Eastern Bloc states. By the mid-Seventies, trade between many Eastern Bloc states and the FRG surged ahead of that with the GDR.[5]

Linde also fared well under the Federal Republic's policy of Ostpolitik. The 1970s are characterized by the wealth of business conducted with the Soviet Union, the GDR, Hungary, Poland, Rumania, Czechoslovakia, and – although under different circumstances – with China.[6] With a strong presence at Eastern European trade fairs, the Linde AG also sought to build up a network with individuals in the Eastern Bloc.[7] For Linde, customers from behind the Iron Curtain were more reliable and punctual in terms of making their payments on time. West German industry enjoyed particularly good relations with its Soviet trading partners. The Soviet Union bought turnkey-ready systems from Linde, with oxygen systems for steel manufacturing and ethylene plants being particularly popular. In contrast to this more traditional business arrangement, trade with other Eastern Bloc states, such as Hungary and the GDR, was often conducted on a barter basis and characterized by co-operation in development and erection. These initial co-operative efforts proved instrumental later to Linde's ease in founding subsidiaries in Eastern Europe after 1990. Central to its successful business relations with Eastern Europe was the company's good name both in business and science, which had been established before 1945 and to which many in the East could still refer. Indeed, Carl Linde's name was to be found in a number of Russian physics textbooks. Trust in offers made by the Linde Company was established relatively quickly throughout Eastern Europe.

One example in particular illustrates the specific character of Linde's business with Eastern Europe. In the early 1970s, the Russian Minister of Chemistry Kostandov attended the opening of a Linde ethyl plant in Hungary, as its products were to be delivered across the Hungarian–Soviet border to be used in a Russian plastics manufacturing facility. The Linde delegation knew that at that time a similar plant was being planned for the Soviet Union in the Urals. Hermann Linde, who was at that time the board member responsible for plant engineering, later remembered:

> And so there we sat, not exactly at the same table, when suddenly Kostandov raises his glass to toast us and says: 'So – the next time we'll celebrate in the Urals!' And that was of course a clear indication that we could, shall we say, be rather flexible in setting a price for the project, which was determined six months later.

Two years later at the ACHEMA, a trade fair for the chemicals industry at Frankfurt am Main, Kostandov referred smilingly to his statement in Hungary as 'the biggest mistake of my career'.[8] As a Linde speciality, ethyl plants were crucial as well to the improvement in economic relations with the GDR.[9] The first of such plants in the GDR was built by Linde in Böhlen. In great part due to political support, Linde continued to receive a healthy flow of orders through 1990, which ran in stark contrast to the increasing erosion witnessed by the GDR's social system and industry.[10]

The energy crisis, a new environmental awareness, environmental and power engineering

It was the so-called oil crisis of 1973–74 which illustrated to Western industrial nations for the first time in over two decades, that low energy prices were anything but stable and definitely susceptible to significant increases.[11] From October to December 1973, the price of oil rose from 3 to nearly 12 dollars a barrel.[12] On 19 November 1973, German federal chancellor Willy Brandt imposed a nationwide ban on driving for the following four Sundays. Public discourse itself was affected by the oil price crisis: terms such as energy-saving, alternative energies, thermal solar facilities, heat pumps, or combined heat and power cycles made their break into the public's vernacular as domestic coal and coal-fired power stations were reconsidered as viable energy sources. Environmental historians now consider the high consumption of resources of the 1950s to be a historical anomaly.[13] In the 1970s the conservation of resources once again played a major role in politics, society, and the economy, as it had before the 1950s, albeit for different reasons. Indeed, it was the environmental impact of resource consumption and the mid-term limited availability rather than cost price which fuelled the drive to save resources and protect the environment. As of 1973, oil became the illustrative example in the Western world for discussions over the shortage of raw materials and the limits of growth. Warnings of industrial damage to the environment had been increasing in the years before the oil crisis. Rachel Carson's provocative book, *Silent Spring* had already appeared in 1962 and in 1972, as his report to the Club of Rome, Dennis Meadows published his bleak prognosis *The Limits to Growth*.[14] Reactions to the energy crisis and environmental movement were particularly significant in the Federal Republic of Germany.[15] The second report to the Club of Rome in 1974 was co-edited by the German systems engineer Eduard Pestel.[16] In 1975, demonstrators from the growing anti-nuclear movement stopped the construction of a nuclear-power plant by occupying its construction site in Wyhl near Kaiserstuhl. In 1980, nature conservationists, activists from the student protest days and the peace movement, third-world advocates, and energy and environmental protectionists combined to form the Green Party. The Greens have successfully acted as proponents of political reform, the restructuring of industries to conform to ecological principles, and have been represented at state-level since 1983 and at Federal level since 1998.[17] Particularly in Germany, there were many dramatic opinions predicting everything from the exhaustion of oil resources by the turn of the century to massive river floods to the widespread destruction of forests. At the same time, the efforts made in Germany in power and environmental engineering – or more generally the decoupling of economic growth from resource consumption – were unrivalled. Having consistently faced ecological challenges, German industry often is and continues to be, within this market segment, a world leader.

Total energy prices increased by 44 per cent at Linde from 1972 to 1975. This comparatively low increase was directly related to the fact that the lion's share of energy used by Linde was electricity, the price of which increased less than that of fuel oil.[18] Since then, Linde has systematically improved the technology for heat recovery and strengthened its market position particularly in the areas of refrigeration and installation systems. In 1974, Linde introduced an air conditioner with an elaborate heat recovery system to the market.[19] Production lines for what became and continue to be the world's best spiral heat exchangers were built in Höllriegelskreuth. Generally speaking, industry instituted energy conservation earlier and more quickly than the private household.[20]

The oil price crisis prompted a reconsideration of future energy demands. Power engineering became a popular new course of study at universities, making the development of energy-saving technologies (again, as in the 1920s) a focus within engineering. A forerunner to these developments, Linde profited from them. One such development included the superconductive power cable for the transport of large amounts of energy. Since its inception in 1968, Linde, along with AEG Telefunken, AEG Kabel, and Kabelmetall had a share in the Cryogenic Cable Consortium, 50 per cent of the costs of which were paid for by the then-existing Federal Ministry for Research and Development.[21] However, this technology proved not to be cost-effective in energy terms. Nonetheless, by promoting the application of superconductivity, this technology ushered in a number of 'side effects' from computer tomography to electronic synchrotron, with Linde contributing through cryogenics.

Similarly, environmental engineering was also important for Linde. The plant engineers at Höllriegelskreuth in particular profited from the growing market in environmental engineering and expanded their range of products. Linde thus made developments in sewage treatment and bought patents from the American company Union Carbide. In 1977, Linde completed the largest 'Lindox' plant for sewage processing in Platting, Bavaria. Then in 1982 Linde handed over the – at the time – world's largest PSA hydrogen purification plant to Union Rheinische Braunkohlen Kraftstoff Corporation in Wesseling. Despite the increasing controversy over nuclear energy, Linde remained loyal to its development, particularly that of the high temperature reactor. This was a specifically German contribution to nuclear power engineering, the development of which, similarly to that of the fast breeder reactor, was discontinued.[22] In 1971 Linde received the until then largest ever contract in the field, the construction of a helium loop for a 300 megawatt thorium high-temperature reactor including the gas purifiers and helium storage for the reactor at the nuclear power plant in Uentrop-Schmehausen.[23] These contracts were followed by others, including those for the nuclear power plant Mühlheim-Kärlich and the nuclear research centre at Karlsruhe.[24]

Rivers of milk and mountains of butter

A European Union agricultural policy which led to increasing subsidies for agricultural products proved favourable to growth for the Linde Company. The EU guaranty of purchasing goods at fixed support prices meant that the cold stores division experienced an increased demand for cooling space. Meat, milk, and butter products in particular were stored in Linde facilities. Over the years, the market for cold storage changed as eating habits increasingly gravitated towards lighter foods such as fruits and vegetables as well as frozen foods. As the demand for stored goods such as ice cream and portioned frozen foods grew, the market base for cold stores became increasingly independent of subsidies.[25] Linde expanded its cold stores in Munich, Kassel, Markgröningen, and Garching. Until 1980, the cold stores enjoyed an above-average occupancy rate. A successful division, cold stores eventually felt the pressure of competition from the cold stores of large food chains themselves.

The debate over equal employee participation on the board

In the 1970s the Federal Republic of Germany witnessed a profound strengthening of co-determination for employee representation in the advisory boards of publicly traded companies as Parity Co-determination was one of the Social-Liberal Coalition's central political goals. Since the early post-war days, German employees had enjoyed the right of participation more than those in many other Western nations. Enacted in 1952, the Betriebsverfassungsgesetz (Works Council Constitution Act) assumed a social partnership between employees and employers. This act lay the building blocks of trust in co-operation between employee representatives and employers, thereby ensuring peaceful relations – which were unique in the West at the time – in German industry for decades. Works councils then became essentially independent of the unions. However, they possessed a true right of co-determination only in matters of social and personal consequence.[26] Nonetheless, employee representatives were suddenly present at Supervisory Board meetings.[27] At Linde there were two such representatives – meaning, in fact, they had the right to information and consultation rather than true co-determination. Paul Derkum from Sürth and Karl Kreß from Aschaffenburg were the two representing workers' and employees' interests until 1961; their successors were Herbert Neuendorff (Wiesbaden) and Fritz Schaller (Höllriegelskreuth).[28]

During the 1966–67 economic crisis and the social-liberal coalition, unions pushed for an extension of co-determination at the company level. Changes made to the Works Council Constitution Act in 1972 gave works councils on the one hand new rights of co-determination in issues of working hours, wages, and the introduction of new technologies, yet on the other stopped short of providing an effective right of co-determination at the strategic level of Supervisory Board.[29] During those years, the company felt as if it were protecting an essential component of democratic freedom with its own freedom. Hans-Martin Schleyer, as president of both the Confederation of German Employers' Association and the Association of German Industries, was a paradigmatic figure of this attitude. Relations amongst social partners became tense as the co-determination law passed in 1976 gave workers' representatives a near parity co-determination in Supervisory Boards. Suddenly, employers and employees respectively comprised nearly 50 per cent of the Supervisory Board. However, when it came to counting votes, the Supervisory Board chairman, representing the employers, enjoyed twice the voting power. One of the employees' seats on the Supervisory Board was reserved for senior staff employees, who often voted with the employer. Nevertheless, the umbrella organization for employers' associations expressed its desire to challenge this law at the German Constitutional Court, prompting a conciliatory response from many companies concerned about public opinion. However, the Linde Company with Hans Meinhardt at its helm, joined in this legal battle, risking – or rather accepting – the prospect of internal corporate conflict. The atmosphere at the first meetings with an expanded Supervisory Board was anything but warm and relaxed. The number of Executive Board meetings with an 'expanded' board – that is Executive and Supervisory Board representatives at the expense of employee representatives – increased. Hans Meinhardt and Linde's Executive Board contested in particular the inclusion of non-company union representatives on the Supervisory Board. When the German Federal Constitutional Court rejected the case in November 1978, Linde's Executive Board accepted the status quo and business continued under less tense conditions.[30] The debate over co-determination on the Supervisory Board has been the only incidence of politicized confrontation between employers and workers in the history of the Linde Company to date. It was a battle that did not suit the image held by many company employees of Linde as a large engineering company – itself an anachronistic image which no longer fitted the international corporate group that Linde had become by the mid-1970s with the majority of its employees working in production and organized in unions. The numbers employed in production sank over the next few decades. In recent years, as the group has increasingly globalized and since the AGA takeover in 2000, which has left only one-third of Linde's employees in Germany, issues of co-determination have shifted. Currently, within Europe, harmonization is the key issue, whereas internationally, the reigning issues are concerned with Central Works Councils and minimum standards for wages, working hours, and co-determination.[31]

Developing the divisions

A boom in Höllriegelskreuth

Both plant construction and industrial gases boomed during the 1970s. The aformentioned conditions – trade with Eastern Europe and ecologically efficient technologies – helped cultivate this growth. In the 1960s, Plant Engineering made headway in the field of commercial plant engineering and was able to enjoy the fruits of its efforts in the face of increasing risk from contract to contract. The department of Plant Engineering had in Herman Linde, who was its director from 1961 and board spokesman from 1972, an enthusiastic manager capable of evoking enthusiasm for the field and – as a representative of the Linde name – a powerful force for the acquisition of contracts up until 1976. However, as spokesman of the board, Linde had not been focused enough on a strategy for the entire company – that is, on his position in the company – which led ultimately to him losing it.[32]

The contract boom centred primarily upon ethylene and olefin plants, along with oxygen and nitrogen plants – in which chemical processes were used to produce various preliminary products for the manufacture of synthetic materials using oil. At the ethylene plants the mineral oil Naphtha was split into ethylene and other products in crack ovens, and then separated and fractionated via low-temperature distillation. Other source materials such as natural gas, butane, and propane were later used at ethylene plants.

Linde's work in the field of low-temperature engineering in particular was highly successful until well into the 1960s. Department B (as it was then called) planned and built a majority of the new air and gas separation plants in over 60 countries around the globe. Once a Linde customer, most clients remained loyal to the company, which had a reputation for being client-oriented and highly competent in matters of order processing. Nonetheless, during the 1960s, Linde faced tougher international competition as American companies began to offer complete, single-source, turnkey-ready chemical plants, which were easier to deal with and more attractive to buyers. For the first time in the business, risks were no longer divided – which was to the considerable advantage of investors. Services, costs, and appointments were guaranteed and the prime contractor bore the weight of responsibility.[33] This development was a threat to the Linde

Company because low-temperature installations, other than those used for air separation, were in most applications only one step in the majority of chemical processes in which they were used. If customers had not been prepared to factor out low-temperature engineering from a prime contractor's responsibility and hand this over to a specialist, Linde would have disappeared from the chemicals industry.

In the case of ethylene plants, this is exactly what threatened to happen. Despite excellent references for the ethylene plants' 'cooling section' – that is, the gas separation section – Linde received fewer and fewer contracts in this growing market. At the beginning of the 1960s Linde was faced with the decision to do as the Americans had done – that is to take on all the risks and offer complete ethyl plants with both cooling and heating sections for a set price and to offer complete responsibility for the plants or to give this field up. This meant not only a great jump in the level of financial responsibility taken by the company, but also a considerable expansion of its own core competence in commercial plant engineering. Linde took this chance and succeeded. As was the case with coal hydrogenation, it was the long-established trust between Linde and numerous German chemical groups which made this jump easier. In 1965, Veba Chemie (Scholven) approached Linde with an attractive offer. Together with Linde, Veba wanted to buy the components, machines, and piping materials, and then build and erect an ethylene plant themselves. This made it possible for Linde to take over the engineering of a complete ethylene plant without having to bear sole financial responsibility for the complex in its entirety. After a controversial discussion, the division decided to go for this nevertheless risky project and succeeded. The Scholven project led to further contracts for Linde from BASF, Shell and others for turnkey-ready plants at set prices.[34] This was followed by further major projects in France, Spain, and Holland. From 1969 to 1972, nearly 50 per cent of the contracts for the construction of petrochemical plants in Europe were landed by Linde,[35] and there were a number of such contracts overseas.[36]

There were significant fluctuations in the relative importance of individual countries to the Linde Company throughout the 1970s. In 1976, two contracts for commercial plants in South Africa made up 48 per cent of the contract volume, whereas in 1980 Australia provided 26 per cent thereof. The share of contracts from the USA sway between 2 and 12 per cent.[37] In the mid-term however, there were regional hotspots. In addition to Western and Eastern Europe, these included the Middle East (for example Iran), South America and increasingly the Far East.

The few major contracts that determined the success of the department were of a high-risk nature and demanded great courage as the performance and construction timeline guarantee could have seriously jeopardized the company's inventory, particularly in the 1960s and 1970s. Most of the important contracts ran the risk of failure at crucial points. Under these conditions of performance guaranty, the sellers – usually the company management – and developers had to rely upon one another.[38]

In the following years, Linde invested resolutely in expanding its competence in the heating devices of chemical plant engineering and in 1975 acquired 35 per cent of the Selas-Kirchner GmbH, a German subsidiary of the American industrial furnace manufacturer Selas Corporation America with headquarters in Hamburg. Linde thus intensified its work with Selas, which was working on the planning, construction, and maintenance of industrial furnaces for olefin and steam-reform plants in the chemicals industry and refineries.[39] This was an important step for Linde towards creating its own process ovens. Up until then, process ovens had been additional appliances that had to be bought from another company, which impaired the Linde AG's entrance to the chemicals and petrochemical industries. Over time, Linde increased its shares, eventually acquiring 100 per cent of Selas-Kirchner.[40]

In 1972, Linde made another strategic acquisition in the gas separator sector. The Linde AG agreed to take over the division of low-temperature engineering at Messer-Griesheim, a subsidiary of the Farbwerke Hoechst AG and also a Linde competitor in the gas industry. At the same time, the Linde AG handed over its department of welding engineering to Messer-Griesheim, the market leader in Germany for welding.[41] This exchange allowed Linde to ensure and enlarge its position as market leader in air and gas separation plants. Along with the division of welding engineering (Elira), Messer-Griesheim also took over new technologies for shielded and submerged arc welding. Both companies transferred all commercial trademarks and production facilities for the divisions they gave up to the other company. This included Linde's welding powder factory in Hennef. Linde also leased a part of its Lohhof Works to Messer-Griesheim for the long term. At the same time, Linde and Messer-Griesheim jontly founded the Likos AG for a part of their international gas sales and to be represented with production and sales companies in Belgium, France, the Netherlands, and South Africa, at least for the mid-term.[42]

In 1972, the Linde AG's Executive Board decided on another important internal change and sub-divided the Munich Division into a Division of Plant Engineering, and a Division of Industrial Gases.[43] The goal was to achieve greater independence in the gas industry, which stood in the shadows of plant engineering.[44] For plant engineering, an independent gas division made for a good customer and test operation. However, it posed a problem as well: the Division of Plant Engineering supplied plants to competitors in the gas industry who could respond with peevishness to Linde's dual role as supplier and competitor. Internationally, Linde's gas division lapsed for a number of decades and remained a market leader only in Germany and in a few individual European countries. It was, however, the most reliable source of revenue for the corporation.

The division of responsibility led as well to new projects in industrial gases. Linde acquired 50 per cent of the shares of a well-known German acetylene manufacturer, Industriegas GmbH & Co. KG, Cologne to improve their market position. In 1974, Linde decided to boost its activities in the Brazilian and Australian gas markets. The simplest way to do so was to once again engage in co-operation with major clients for gas separators. Linde built a large industrial gas separator not far from Rio de Janeiro that could pipe not only oxygen to a steel mill, but other liquid products to other factories as well.[45] The co-operation and synergy profits in Höllriegelskreuth clearly outweighed the disadvantages of a vertical integration of plant engineering and gas sales. The advantages of a common division first became apparent in the year 2000 when the divisions were reunited.

Refrigeration is de- and re-centralized

The Executive Board's agenda in 1972 was to divide and thus decrease the might of the powerful divisions. As they had done with the Munich Division, the Executive Board sub-divided the Sürth Division in 1972 into two new divisions, 'Industrial Refrigeration' and 'Cooling and Installation Systems', which had been combined in 1964.[46] However, in the long run, this regrouping failed to solve problems in refrigeration. By 1979, the Executive Board decided to reunite the two divisions. The separation in 1972 was meant to deal with the then-existing discrepancy between the profitable air-conditioning technology market and Linde's internal organizational problems. At the time, the Executive Board expected the refrigeration market to see 25 per cent annual growth until 1980. The two new Sürth divisions sold their products separately to consumers in industry and trade. Another reason to separate the divisions was the attempt to sell the turbomachines, which failed in 1979, so that the division could reunite. The new Division of Industrial Refrigeration sold refrigeration and air-conditioning systems for industrial use as well as piston compressors and turbo machines. The Division of Refrigeration and Installation Systems focused on domestic or household refrigerators and retail equipment as well as refrigeration and air-conditioning machines in the durable consumer goods trade.

Linde took a similar route in commercial refrigeration to that it had pursued in plant engineering: Linde offered complete refrigeration systems and equipment for the retail grocery industry. In January 1971, Linde bought the Variant GmbH in Bad Hersfeld, manufacturer for retail equipment construction.[47] Customer service moved to Sürth, while manufacturing and construction remained in Kostheim. There were a few remaining manufacturing sections of Hausgeräte GmbH in Mainz-Kostheim that were sold to AEG. There were two major fires at the Mainz-Kostheim factory in the same year. Restructuring had created an atmosphere of fear amongst the staff, which led in the short term to warning strikes as rumours spread that production was to be given up.[48] Yet Linde aimed to experience growth rather than shrinkage in the commercial refrigeration sector, acquiring in 1976 new manufacturing capacities with the Tyler Refrigeration International GmbH in Schwelm/Westfalen.[49]

At the beginning of 1979, the divisions of Refrigeration and Air-conditioning Installation systems as well as Piston and Turbo Machines were combined anew to form the Division of Refrigeration and Installation Engineering. Once again, as was the case in 1972, the idea was to strengthen sales. However, this time sales were to be strengthened not via division, but through bundling. At the same time, however, Linde undermined this argument by closing foreign representations in industrial refrigeration. While in the short term this move saved costs, in the long run it pushed Industrial Refrigeration to the periphery. The central idea was to shrink the Division of Industrial Refrigeration, which enjoyed the longest tradition within the Linde Group and a highly lucrative maintenance, repair and renewal business, thanks to the old systems of the past. However, compared to commercial refrigeration, the market was smaller and less dynamic, even though spectacularly immense contracts continued to come in sporadically, such as the contract for three natural gas-lit absorption refrigeration plants for the Italian Unilever subsidiaries in 1973,[50] or for marine refrigeration plants at Polish and Brazilian shipyards.[51] The international commercial refrigeration business savvily took advantage of the contacts in industrial refrigeration built up over the years in places such as the Middle and Far East, South Africa, and Brazil.[52] Although there were technological innovations at Linde during these years, they were little used. Linde had been a technological leader since 1968 in both low-pressure and high-pressure snowmaking machines for winter sports, which it even exported to Japan in 1972–73. Linde stopped their production in the 1970s for 'technical and economic' reasons,[53] some years before the product was beginning to be put to use internationally in ski tourism.

Figure 5.1   Franz Josef Strauss admires the Linde snow cannon in Höllriegelskreuth, 1972

Aschaffenburg succeeds at structural transformation

The decision to expand Linde's tractor business and make a quick, consistent entry into forklift manufacturing at the end of the 1960s was in contrast one of the success stories that furthered the rise of Hans Meinhardt to eventually become CEO. The move into the growth sector of material handling was well prepared by the marketing departments in Aschaffenburg and Wiesbaden, yet its success had much to do with existing favourable conditions.

First, competition was sparse in the forklift branch, with few large companies to worry about.[54] Linde could use its size to its advantage and grow through takeovers. Linde enjoyed one important advantage over its competitors, as it had Güldner's long-standing retail network at its disposal.[55] International market leaders did not regard Linde as a serious competitor in forklift manufacturing at the start of the Seventies. This allowed Linde to stage a surprise coup of sorts. Keeping its secrets strictly internal, Linde entered the market as a powerful rival seemingly overnight.

Linde also offered a truly new innovation with its hydrostatic transmission. The Linde forklifts were efficient, with an infinitely variable clutch that functioned in reverse as well. Although forklifts with hydrostatic transmissions had existed since 1960, it was with Linde in 1970 that this technology experienced a true breakthrough.[56] In the 1970s, Linde presented new technological developments each year at the Hannover trade fair – such as the first electric forklift in 1971 for use in large closed warehouses.[57]

In contrast to plant engineering, the business segment of material handling could only grow through the purchase of competing companies. Important acquisitions included Still in Hamburg (1973), Baker in the USA (1977), Fenwick in France (1984), Lansing and Saxby in England (1989), Wagner Fördertechnik in Germany (1986–88) and Fiat OM in Italy in the 1990s. In addition, Linde began in 1993 the joint venture Linde-Xiamen in China and at the end of the century acquired shares in the Japanese forklift manufacturer Komatsu. Linde thus became the world market leader in forklifts.

Let us return to the early 1970s. The restructuring and development of the Material Handling Division was financed by profits from Höllriegelskreuth. It was a major and worthy investment. The first important acquisition came in 1973 as Linde took over all business segments of the mid-sized SE vehicle works established by Hans Still in Hamburg in 1920. Before the war, Still was a major manufacturer of electric motors and had already built his first forklift in 1949. In the 1960s, Still merged into Still Esslingen (SE) and built both electrical and combustion motor forklifts. Still is, to this day, the only large North German subsidiary within the Linde group, which continues in both style and habit to be a Southern German company. Whereas Aschaffenburg enjoys certain advantages both geo-economically and within the corporate group, there continues even today a sportsmanlike competition between Hamburg and Aschaffenburg, a true result of a multiple-brand strategy.[58] This internal competition was planned by the group management in Wiesbaden to incur division growth,[59] a strategy relied upon over and over again.

The next important acquisition came in 1984 with the addition of the largest French forklift manufacturer, Fenwick Manutention. The buyout was drawn out for months, despite French government support. Eventually, Linde was able to buy Fenwick for a very reasonable price. The buyout immediately strengthened

Linde's position in France. More important than the export market was the fact that the acquisition helped secure the European market against imports from the Far East.[60] Linde's employee numbers at the end of the 1980s were double those of 1975. Growth within the Linde AG since the mid-1970s was primarily due to growth in the Material Handling Division. Starting in the mid-1980s, Material Handling became the largest division within Linde, while Plant Engineering delivered 10–15 per cent of total growth within the group. In terms of profits, however, the situation in Höllriegelskreuth was significantly better than those in Hamburg and Aschaffenburg.

The path to a strong group headquarters

In the 1950s, Linde Group Headquarters was preoccupied with accounting, company health insurance, the housing association, and other such issues.[61] Since the 1960 transfer of the department of commercial refrigeration systems, Wiesbaden had waned in importance. A former manager described the relationship between Wiesbaden and the various divisions in the 1960s:

> In the past, when somebody from Group Headquarters would come to the division, he would be treated like an envoy of the Vatican, who, despite his position, had no say. He was treated well, invited to dinner, told what had been improved or altered (which was already known) ... it was all nothing more than a flash in the pan.[62]

This was a situation to be changed in the future.[63] Hans Meinhardt and his reformers pushing for reorganization were determined to extend Group Headquarters at Wiesbaden. Preparatory work for strategic decisions made by the Executive Board was to be done by Group Headquarters rather than the divisions. The development of centralized staff units was a difficult, 15-year process that began in the 1950s and ended in the 1970s.[64] Supporting staff units within divisions were established to complement the central staff units and placed under their auspices. In this way, the central staff units were able to implement their decisions at division level and maintain a minimum of authority over the divisions. Group Headquarters therefore retained its right to have a say in decisions regarding personnel made by the staff units of individual divisions and insisted on the regular exchange of personnel between the divisions and Group Headquarters. This second pyramid with the Group Headquarters at Wiesbaden on top and each division's administration below it, was subject to the Executive Board but maintained a right of direct access to the divisions.

Marketing and advertising departments

Marketing and advertising at Linde were carried out from the very start both intuitively and successfully. There was, however, no specific staff for such purposes. Before 1960, marketing was essentially unknown as a concept and the importance of advertising was underestimated. Individual clients had been showered with attention, but the market itself had yet to be thoroughly and broadly addressed.[65] Starting in the early 1960s, a separate advertising department was created at the Wiesbaden Group Headquarters, with complementary departments in the various divisions.[66] Advertising, which had been conducted through brochures, went unnoticed until then.[67] In 1965, at the opening of an internal exhibit 'Advertisement Then and Now', which was intended to convince Linde employees of the value of advertising, then-CEO Johannes Wucherer expressed praise for 'the reserve and elegance' of older Linde advertisements in comparison to the 'often unintended bizarre products of so-called advertising experts of today'.[68]

The creation of advertising departments was not critical to the power structure between the divisions and Group Headquarters. This however, was different for the marketing departments established in 1962, as they had an impact on product and corporate strategy. Presumably due to practical reasons, advertising was integrated within marketing. As of 1965, the areas of market research, advertising, sales promotion, market policy, press relations, and information became the responsibility of the marketing department in Wiesbaden.[69] However, the division heads had little time for the notion of marketing and spurned the Wiesbaden attempt to exercise influence in this area.[70] Therefore, temporary exceptions were made to the scope of the central marketing department's territory, as laid out in a plan started in 1965 to be completed by 1966. The Department of Market Research in Cologne-Sürth was to stay there for one year before being transferred to Group Headquarters. Exceptions were made as well for Plant Engineering and Industrial Gases at Höllriegelskreuth, where, due to 'the complex subject matter of the product and the rather large distance between Wiesbaden and Munich [...] market research should be left in Munich for the near future'. Yet from Group Headquarters' point of view, this was a temporary situation. The marketing department aimed for the mid-term to 'establish a cooperative system in which to gradually gain both insight and influence'.[71] Two years later, the independence of the Höllriegelskreuth staff unit was removed in the 'restructuring of company organization'.[72] The marketing department however, was not to be limited to gaining mere insight into the corporation's internal happenings. It was to systematize press relations and the activities of associations[73] and began purposefully analysing the organization structures of other large German companies. The goal was to gather data, learn from other examples, and use this information to argue for the expansion of Linde's own competencies.[74] The newly established influence of the marketing department was seen first in the preliminary work of market analyses for household appliance sales (Mainz-Kostheim) as well as the shift to hydraulics and material handling engineering at Aschaffenburg.

Organization departments

Documentation and organization were unable to keep up with the rapid growth of divisions such as Plant Engineering in the 1960s.[75] In the mid-1960s, no thorough records of Linde-wide construction data existed for Plant Engineering.

Thus, when an important client asked to have the 'Linde Standard' sent to him, he received a shorthand note from the then-department head of the engineering department in Höllriegelskreuth, Georg Plötz,[76] that this was not possible per post because of the immense amount of information.[77] He was, however, to receive this during his next visit. Under great pressure to finish quickly, Höllriegelskreuth comprised a Linde Standard from all available documents.[78] The same was true for a streamlined and regular documentation of the corporate leadership structure. While – in contrast to the company's self-image – thoroughly detailed organization plans had existed since the 1920s,[79] there was nonetheless no single office or team with primary responsibility for this objective. The organization of production lines was more developed than this.[80]

From 1959–62 Hans Meinhardt pushed to establish independent organization departments, each with its own management and staff, first at Güldner Aschaffenburg, then, as of 1960, at Headquarters and the other divisions at Aschaffenburg, Mainz-Kostheim, Munich, and Cologne.[81] Each department was granted a broad index of authority.[82] Headquarters in Wiesbaden, however, retained its right to dictate directives to the division-level organization departments and take drastic measures. The Wiesbaden Organization Department thus became a powerful entity within the corporation, which even the Executive Board could only minimally affect.[83] Certainly, there were differences in the way authority was wielded from division to division. Höllriegelskreuth ignored the expertise of Group Headquarters longer than others.[84] Yet the trend was clear: Headquarters' numerous regulations made their impact and led to a centralization of strategic corporate decision making.

The planning department

The creation of an independent strategic planning department in 1964 furthered the strengthening of Wiesbaden Group Headquarters. Johann Simon, an Executive Board member in charge of Group Headquarters, was a proponent of and his colleague Hans Meinhardt the decisive motor behind this development.[85] It was the right time to institute this department, as faith in the possibilities offered by systematic, scientific planning was particularly high in the 1960s in both the West and East. Planning departments were shooting up like mushrooms all over the place – not only in the industrial world but also in administration and politics.[86] In the East, where the planning bug had already bitten, this included the introduction of centralized planning at all levels from production to central planning commissions with four-year, one-year, three-month, one-month and weekly plans.[87] In the West this included policy planning staffs at state agencies, reform planning for federal transportation routes and the introduction of planning sciences as courses of academic study. On both sides of the Iron Curtain, the possibilities offered by cybernetics in maths and technology were fused with a widespread faith in rationality and planning. The complexity of tasks was to be mastered through complexity planning, systems technology, and cybernetics; decisions were to be rationalized and modernized.

The main planning department was not only to record already existing plans, but also to create a new planning system as a guideline for future planning, which was to be approved by the Executive Board. Once this planning system was introduced, the main planning department, together with individual planning staffs, was to review the 'realism and logic' of each division's planning. In addition to this, the planning department was to take over co-ordination within the divisions. Furthermore, each division's planning department was to be combined, streamlined, and brought under the auspices of central planning at the Wiesbaden Group Headquarters. In the end, it was the planning department which analysed and commented on the general plan, ultimately translating it into concrete terms.[88] The planning department appears in an organization plan for the first time in 1970,[89] and in 1971 it appears under 'corporate planning'.[90] The planning department's documentation increased Headquarters' influence on individual divisions.[91]

From Hermann Linde to Hans Meinhardt

The starting point

To understand the switch in power from Hermann Linde to Hans Meinhardt, which spelled a fundamental shift in the structure and alignment of the company, we must examine the transition that occurred in the Executive Board's duties during the 1970s. In an effort to demonstrate competency in the field, Hermann Linde had suggested numerous organizational reforms in the early 1970s. There was a vociferous debate over structural issues. At the heart of the debate lie concerns about the manner of corporate leadership and its duties. For Hermann Linde, an applied physicist for technology, the management of a technology corporation was to exhibit first and foremost technological leadership, as he had witnessed with his father. Because of the complexity of the technology, he was in favour of decentralized management structures, situational leadership and diversified product strategies. In contrast to business strategies such as the strategic buyout of another company, individual engineering projects and product lines were given higher priority. Hans Meinhardt argued instead for a centralized, strategically inclined management, preferring economic over technological data. Meinhardt's vision for the entire corporation included divisions with equal power, the courage to become a market leader in relevant markets, and to achieve balance amongst major shareholders. His guiding tools were organizational reform, personnel decisions, sell-offs, and acquisitions. Both sides took pains to work effectively together until 1975 and keep their differences professional, which were known throughout the corporation and reflected in a number of details.[92] Let us look more closely at the various positions regarding the Executive Board's duties which arose out of these differences.

At the beginning of the 1960s, the management structure at Linde was decentralized. Each division had at least one representative on the Executive Board, which rarely met more than once a month, at times once every six weeks.

This structure allowed for powerful and powerless board members. The Höllriegelskreuth representatives, emboldened as they were with higher turnover and greater profits than the remaining divisions, enjoyed greater power. Division autonomy was justified by the large differences in engineering orientation. They were each given complete control over their own product from planning to delivery. Höllriegelskreuth therefore had its own training, research, development, and testing department as well as its own engineering department.[93]

Early reformers: Johannes Wucherer and Johann Simon

In 1961 Johannes Wucherer replaced Hugo Ombeck as CEO, which put a member of the Linde family once again at the helm of the corporation after a seven-year hiatus.[94] Ombeck was a typical CEO of the 1950s – a self-confident man of quick decisions and a difficult person who was feared by those around him. In contrast, Johannes Wucherer was well-liked, amiable, and careful. Nonetheless, it was Wucherer who in 1961 suggested changes to the Executive Board's structure and duties. Ombeck, as a transitional figure, and being neither a member of the Linde family nor the Munich Group Headquarters, did not address this issue, suggesting instead reforms at the level of staff units and manufacturing planning in the divisions. Both Hans Meinhardt (at least during his first years at Linde) and Group Headquarters benefited from these reforms.

The establishment and strengthening of (centralized) staff units continued under Wucherer's leadership. Johann Simon, who had left the Hoechst dye works in 1947 to join Linde at Headquarters, also encouraged this development.[95] As deputy member of the Executive Board starting in 1954, Simon was Ombeck's right arm in the office. In 1961 he moved up to deputy chair of the Executive Board, which gave Group Headquarters and issues of organization and administration more weight within the corporation. He was less enamoured by the prospect of taking responsibility for a division. Whereas Wucherer saw himself in the tradition of a technical corporate manager who accepts the division of labour in duties, Simon in contrast acted more like a stern accountant.[96]

Wucherer, as a member of the Linde family with one foot in Munich and the other in issues of technology development, was able to suggest fundamental structural reform more easily than Ombeck.

A look back to the 1970s reveals that already by the 1950s the corporate leadership at Munich and Wiesbaden had begun to fray. Alongside the Executive Boards there were ambitious second and third levels of management at both. Next to the Linde family itself, there was at Plant Engineering in Munich a circle of development-oriented engineers and physicists with common musical interests who associated with each other regularly.[97] This included the two Baldus brothers, the three Hailer brothers,[98] and – from the family – Gert Linde (1920–) as deputy sales director in Höllriegelskreuth as well as Carl Ranke's (Ranke was one of Carl von Linde's sons-in-law and sat on the Supervisory Board until 1950) son Hans Ranke (1897–1957) and his son Friedrich (1924–) and nephew Gerhard (1929–), who rose to become department heads in the 1960s.[99]

The mood was rather different at Industrial Gases, which had been managed by Walther Ruckdeschel since 1950 and who was replaced in 1970 by Georg Plötz. This department was formally tied to Plant Engineering until 1972. Thanks to Ruckdeschel's persistence in expansion, Industrial Gases bypassed Plant Engineering in turnover in the 1960s. In 1950, Richard Linde had suggested Ruckdeschel instead of his nephew Johannes Wucherer as his successor, indicating the opportunities for advancement for non-family members even in Munich. Wucherer, however, took over parts of Plant Engineering (Development) a few years later. After Hermann Linde had been appointed to the Executive Board and head of Plant Engineering, Ruckdeschel gave up production (Schalchen) and focused on the growing sector of industrial gases.[100] Ruckdeschel had a fair amount of say on the board, bringing Hermann Brandi and Hermann Holzrichter on to the Supervisory Board.[101] Both Ruckdeschel and Plötz came out of Plant Engineering, which highlights the strategic position of this department within the corporation. Both initiated important developments there. After 1970, Plötz was thus in support of Wiesbaden's chosen path.

Executive Board members Karl Beichert, Hugo Ombeck, and Johann Simon all came from the Wiesbaden old-boys' network; so too did Meinhardt, Klein, Polunius, and others.[102] In 1961 Wucherer suggested giving the Executive Board a more effective role. Prompted by suggestions from Wiesbaden and startled 'by the poor returns particularly at Aschaffenburg and Kostheim', he concluded early on in his tenure that it was indeed possible 'to produce a satisfactory balance sheet for this year – but further setbacks can't be taken'.[103] He located one of the roots of the problem in 'the physical separation of the branches ... [making] it impossible for the corporation's Executive Board to decide quickly on important business matters'.[104] He wanted to 'keep the collegial system [of top management]', but insisted as CEO, on 'basic co-determination for all important decisions regarding branch operations, for example business and procedures that entail pronounced technological and economic risks'.[105] Expanded central staff units were to provide the support for this strengthened position. Wucherer's suggestion was agreed upon within the Executive Board, leading to increased staff unit positions. However, the suggestion ultimately went unrealized because it was Simon and not Wucherer who held the ties to Wiesbaden.

Simon came up with a realizable solution two years later. In 1964, he returned from a two-month educational sabbatical of sorts with Meinhardt in the USA armed with American models of Executive Board activities. Simon proposed that the CEO and his deputy (the model referred to two CEOs) – himself and Wucherer – form an 'Executive Organ' within the Executive Board, as was the practice in many American companies at the time.[106] This Executive Organ, supported by central staff units, was to be capable of quick action in the short term. It was to keep sight of the corporation's overall interest as the Executive Board represented division interests.[107]

The proposal hinted in part at a degree of diminished power in the Executive Board and restrained divisional autonomy for the mid-term, despite their expressions of trust.[108] Because of this, Munich Executive Board members

Ruckdeschel and Linde stood in opposition to this model and expected a clear demarcation of division competencies from those of Group Headquarters. Hermann Linde pushed for an expansion of the Executive Organ to include an Executive Board member from a respective division in decisions to be made regarding the relevant division. Ruckdeschel and Linde together suggested that Group Headquarters be transferred to Munich – making a rather clear challenge to Wiesbaden.[109] However, they either could not or did not wish to push this through. There were cosmetic changes made, such as alternating Executive Board meetings between Munich and Wiesbaden in 1965.[110] Yet it was not until the beginning of 1967, after long, arduous debates in which Oetken, Brandi, and Holzrichter on the Supervisory Board argued in favour of Simon's and Wucherer's proposal,[111] that the new concept was agreed upon. It was feasible because Wucherer and Simon formed a unified front, calling the change 'necessary immediately'.[112]

The concept was, of course, a compromise on numerous points. In the weeks previous to the resolution, the issue at hand was less the Executive Organ's competencies than those of Group Headquarters. This was the true motor driving reform. The resolution paper 'Corporate Organization Restructuring' describes in detail the duties and competencies of the central staff units for development, marketing, the legal and insurance department, organization, planning, finance, and social/personnel issues. The Executive Organ's competencies were also listed: determining business policy, to represent the corporation publicly, preparing and realization of Executive Board resolutions, controlling measures passed, cultivating ties with the Supervisory Board, maintaining discussions and consultations with division management and other duties including negotiating with the Central Works Council or reporting on major contracts.[113]

This resolution posed a problem insofar as it rendered the competencies of other Executive Board members unclear. Instead of providing clear definitions, this issue was passed on to Hans Meinhardt. He was to negotiate Headquarters' responsibilities with individual board members.

Hans Meinhardt's corporate concept

Hans Meinhardt, an educated economist, started at Linde in 1955. Following internal corporate training at various divisions, he took on the job of revising Group Headquarters. The Wiesbaden Executive Board members Ombeck and Simon challenged the young man from the start. It was said that Ombeck had already proposed Meinhardt become a member of the Executive Board in the early 1960s.

In 1959 Meinhardt left Wiesbaden to restructure serial production at the Güldner division in Aschaffenburg and to keep Ombeck updated. Starting in 1961, he took over the business administration and became a member of division management in 1963 with his duties including administration, sales, and later production as well. He held onto this function even after he returned to Wiesbaden. Up until then, Aschaffenburg had been somewhat of an appendage of the Linde Group. A problem for years, Aschaffenburg was important neither in terms of turnover nor identity for the corporation. Over decades, Meinhardt built it up into a powerful division – at times the most powerful in the group. Essentially shifting the Linde AG's character, this development was costly in terms of investment and therefore not free of risk for the engineering company – but it was one of its most important strategic moves. Meinhardt's preference for expanding the Material Handling Division presumably had to do with the fact that this division required improvements to structural and process organization in serial production, manufacturing efficiency, and automatic assembly rather than individual technical innovations.

At first glance, the expansion of Material Handling contradicted another strategic goal: corporate focus on fewer products and putting an end to 'mixed-products' at Linde.[114] Meinhardt was, however, for the expansion of those areas he felt could become highly profitable – gases, plant engineering, refrigeration and material handling. He accepted the company's technological spread, because the advantages outweighed the disadvantages. As an economist, Meinhardt could recognize complementary relationships between the business segments with different business and investment cycles. This fascinated him because it allowed for a theory of interwovenness between Höllriegelskreuth and Aschaffenburg. He repeatedly strove – through the Eighties and Nineties – towards a corporate model of equilibrium and balance.[115] Peter von Zahn's 1984 film on Linde uses the hanging mobile as a symbol of the mechanics of the divisions' connectedness.[116] Meinhardt's model could be criticized for not establishing the necessity of Aschaffenburg's expansion. Both Technical University Munich divisions (Industrial Gases and Plant Engineering) had differing business and investment cycles. Similar complementary cash-flow and investment cycles existed between Industrial Gases and Material Handling as well.

The different cycles not only provided stability throughout the entire group; they made it less vulnerable to things such as hostile takeovers. This was certainly important, as a successful business with a narrowly defined family of products was easier to take over than a company with widely varying divisions. Meinhardt could safely assume that there was no other company in the world offering the same combination of industrial gases and logistics as Linde. The security of independence continues today to affirm this strategy.

Later, as CEO, Meinhardt spoke of the principle of divisionalism and placed a high value on division independence in issues specific to their product. His motto was: decentralize as much as possible and centralize as much as is needed. Divisions were to be responsible for the operational side of business. Decisions made about specific divisions were based on the market size of their respective sectors and their potential for becoming market leader.[117] However, due to limitations with the market itself, the Linde AG has not been able to put this principle of equal business segments into practice in the past few decades. Refrigeration has continued to remain weaker in profits and smaller than the Munich departments.

Meinhardt returned to Wiesbaden by the early 1960s while holding onto his role in Aschaffenburg. He focused on strengthening the internal organization of Group Headquarters and establishing strong staff units with the right to information and to take drastic action at division level. These rights were to be backed up with the support of the Executive Board. Meinhardt had the opportunity to familiarize himself with American management methods during his first educational visit to the USA with Johann Simon in 1964 and again during a months-long stay at the diversified armaments manufacturer Raytheon in 1968. Enthusiastic about American management models, he used them to argue for his reforms.[118]

In Meinhardt's opinion, Linde the technology corporation possessed much more potential than engineering knowledge and good products. The modernization of its organization, the strategic combination of divisions from an industrial and not technical perspective, were the unseen treasures of the company, their potential being the blind spot of the company until his arrival. Meinhardt was less concerned about research or client satisfaction, as Linde was already strong in both areas. He focused instead on the group's structure and developed his vision of a powerful Group Headquarters early on – a headquarters with the reins of financial control in its hands but one that lent a fair amount of slack on decisions in technology. In contrast to the engineers with their technological vision, Meinhardt both felt and was indeed superior when it came to strategic thinking. Even Munich's successful Plant Engineering appeared narrow-minded, technologically stagnant and out of tune with the times in the 1970s. In the Eighties, as the global economic climate turned once again towards technology – producing a flurry of technology centres, inventors' trade fairs, and technology prizes – the strategic plan at Linde was already in place.

One of Meinhardt's major virtues was his stamina, or ability to be patient and therefore effective in the long term.[119] He turned down an appointment as a member of the Executive Board in the 1960s, being satisfied with his roles as assistant to the CEO, protocol director at board meetings and as a member of the Aschaffenburg Management Board.[120] Meinhardt envisioned a small Executive Board. Executive Board members were to manage the corporation as a team, each member taking on a central duty – finance, controlling, personnel, operations, production – and supervising a division without dictating engineering or technology issues. In exchange, each division was to have its own business management. In certain ways, this concept resembled that of a Holding Board.

As long as Meinhardt was not CEO, he could continue to work on increasing Group Headquarters' power.[121] Group Headquarters introduced job descriptions at all management levels.[122] Continued education training in the fields of employee leadership and organization increased. As in many other German companies at the time, the compasses were directed at the 'Harzburger Model'. Reinhard Höhn, the founder of the Harzburger Leadership Academy (Germany`s most important management school in the 1950s) and former SS General, promoted a concept of leadership as the delegation of responsibility and conferral of individual duties.[123] To intensify co-operation between Group Headquarters and the divisions, new positions were to be filled that would encourage the systematic exchange of personnel between Wiesbaden and the divisions.[124] This spawned a systemized personnel development strategy that planned for potential leadership careers within the corporation.[125]

Group Headquarters' systematic expansion of its responsibilities resulted in increased levels of formalization within the corporation. Organization plans were thus regularly produced and updated.[126] The resistance to such a policy in some areas, particularly Plant Engineering, was to be expected. Group Headquarters was often seen as 'an unnecessary headache'.[127] Everyone had grown accustomed to the (up until then) relatively weak status of Group Headquarters and was suddenly taken aback by its numerous competencies and the large number of high-level positions that began to appear in Wiesbaden.[128]

It is worth noting how weak the resistance to these changes was and how successful Group Headquarters was in achieving its goals. There were a number of reasons for this: first of all, Group Headquarters pursued expansion at Aschaffenburg and, albeit with some reductions, the refrigeration division using profits particularly from the gas sector. Little criticism was thus heard from Aschaffenburg and Sürth. In Munich, Georg Plötz replaced Walther Ruckdeschel as director of gases in 1970. Because he did not get along well with his colleague Hermann Linde at Plant Engineering, there was rarely a unified voice out of Munich. Furthermore, the profound impact of the structural division and division of labour between Group Headquarters, the Executive Board and Plant Engineering that had been introduced was not acknowledged in the first few years. While the new executive structure may have formally done away with division autonomy, it had not been cemented in top management – particularly not in the case of Plant Engineering. Being more interested in details, Hermann Linde considered the situation less from the position of an Executive Board member and more from that of his pet division.

Another important reason for weak resistance was related to the relative weakness of Hermann Linde himself in his discussions with Hans Meinhardt and the Executive Board as a whole. Having completely underestimated the importance of Group Headquarters, Hermann Linde did not take the conflict over corporate structure seriously enough, which meant he had, outside of his division, only a few trusted partners. Linde placed a great deal of trust in the strengths of the division he himself led and found it difficult to realize the changes pursued by a powerful Group Headquarters. What appeared to him a morally dubious pretence to power, were in fact the expectations of an expanded administrative management organ. He also failed to acknowledge the strategic importance of smaller conflicts, such as those over the position of Executive Board spokesman or the question of direct access to the Supervisory Board and acted punctually, but not strategically.

Implementing Executive Board reform in 1970

During the 1960s, the average number of Executive Board members was six, one member for each larger field – two from Munich, one each from Sürth, Kostheim, Aschaffenburg, and Group Headquarters. The number fluctuated somewhat, due to the overlap of successor members who would usually spend one or two years of induction as deputy members. There were also accidents. Kurt Nesselmann, for example – a professor emeritus of refrigeration engineering at the Technical University of Karlsruhe, became an Executive Board member due to a misunderstanding. Nesselmann, who had worked for Linde in the 1950s, was approached regarding a Supervisory Board mandate. Taking this as a concrete offer, he spoke publicly of his board membership. Because no seat was open, the Supervisory Board chairman offered him an Executive Board post instead, which Nesselmann took for two years.

In 1970, Ruckdeschel, the Executive Board member from Munich responsible for gases, retired and two new board members were brought in: Hans Meinhardt (who was to replace Simon in 1971) and Georg Georg Plötz for Ruckdeschel. By 1970, Hermann Linde found himself increasingly isolated in the Executive Board. In the next few years, Pöhlein (Aschaffenburg) and Wucherer (CEO) retired due to both age and ill health reasons.[129] This generational shift was advantageous to changes being made in board organization. As Meinhardt stated in 1969: 'The decision to re-organize the Executive Board is of great importance. There is no better time than now, as nearly the entire Board will be retiring in the next few years.'[130] The new small Executive Board was to be comprised of members who resided in the same area. Up until then, the majority of members lived near their respective divisions and came together only for the purpose of meetings.

The most important and soon-to-be only opponent on the Executive Board to the organizational restructuring of Linde in the 1970s was Hermann Linde himself. In 1972, he was to replace his recently retired cousin Johannes Wucherer as CEO, while at the same time Hans Meinhardt was to succeed his former boss Johann Simon as Executive Board member for Group Headquarters. Under these new personnel conditions, the concept of the Executive Organ could not be realistically implemented, as Meinhardt and Linde had different business concepts. Both the Supervisory and Executive Boards decided to do away with the position of CEO and replace it with the weaker concept of an Executive Board spokesman, despite the fact that Meinhardt's 1969 proposal for Linde's leadership clearly included the position of CEO in a small Executive Board. All of this occurred while Hermann Linde was at Harvard Business School in Boston.[131] At the shareholders' meetings, Linde and Meinhardt appeared in tandem: Linde spoke to the engineering and Meinhardt to the business issues.

As the concept of the Executive Organ receded, Meinhardt reduced the term to himself, becoming the representative voice for Group Headquarters on the Executive Board. Communication with the Supervisory Board was to be conducted via the Group Headquarters board member, rather than the spokesman.[132]

As of 1970, the new Executive Board concept included exact descriptions of each member's duties. This became a never-ending issue, because the areas of responsibility were always being changed due to restructuring at division level.[133]

This included defining the Executive Board's duties with regard to the divisions under the concept of 'supervision'. Furthermore, 'special management directives for supervision' were drawn up for all board members to follow. The Executive Board members were to know their respective business sections thoroughly and represent them on the board. These new concepts were accompanied with concrete descriptions of how a board member was to develop contacts within his area, which financial decision competencies he could delegate or when he should confer with the Executive Board.

Consequently, a new administrative layer was introduced beneath the Executive Board, the Divisional Board of the divisions. In contrast to Executive Board members, they were responsible for their business sectors only.[134] They were to handle 'general management directives',[135] and develop their business sector according to the plans, regulations, and directions stipulated by the Executive Board. In terms of hierarchy, Divisional Board members were at the same level as staff unit superiors at Group Headquarters. They all had the rank of director.[136] This reform also enhanced the role of Group Headquarters. The greatly increased number of Executive Board meetings in the first half of the 1970s was a manifestation of the energetic debates between Meihardt and Linde. Never before had the board been so occupied with itself or issues concerning its duties.

Arenas of controversy

Nearly every strategic issue debated within the Executive Board from 1970 to 1976 reflects at some level competing concepts of corporate organization. Hans Meinhardt brought up the majority of issues, continuously producing papers on organizational reform from Group Headquarters, which was under his leadership. He persisted on most issues, as they were his ideas that he enthusiastically and meticulously put to paper, whereas Hermann Linde responded haltingly, at times bemusingly, but essentially with resistance to questions of organizational reform. Let us look at a few examples.

Organizational reforms

The implementation of ambitious reforms within the corporation was a constant issue for the Executive Board. On a few occasions Hermann Linde himself took up this issue and initiated organizational reform proposals, perhaps as a means of demonstrating his prowess to Meinhardt in this area. Inspired by a recently completed seminar, Linde proposed in 1972 a new decision-making process along the lines of the 'decision-tree' principle. Decisions were to be reduced to pro and contra questions and all participating levels of hierarchy were to be set. A decision was thus acknowledged when more than 50 per cent at each decision-making level were in favour. However, this proposal appeared rather spontaneous and not well-thought-out. It was tested, but not implemented.[137] In general, Hermann Linde failed to effectively carry out any initiative in organization structure reform.

Figure 5.2   The third generation on the Linde AG Board of Management: Dr Hermann Linde as spokesman for the Board of Management, 1972

Expanding Aschaffenburg

Hermann Linde followed the expansion of Aschaffenburg, which had found success in forklifts and hydraulics, from a critical distance. He saw himself isolated, as even his colleague Georg Plötz supported investment in material handling. Linde regarded with scepticism the arguments for fast expansion (good conditions, competitors' underestimation of Linde, faster delivery than the competitors, new technology, and a strong market position) and the reports of success at Aschaffenburg.[138] Linde pointed to a strong increase in orders received in the last years of his term.[139] He was bothered by the fact that this remained unnoticed and that more was invested in other areas. Yet he provided little in the way of thorough alternatives. Linde went on the offensive late in the game, formulating in 1976 radical ideas – albeit orally and theoretical – to sell Aschaffenburg and even Sürth, then use the money to acquire a large plant engineering facility as a means of maintaining the corporation's image as an international leader in engineering. However, by the 1970s ideas such as these had no hope. Instead, it was through strategic acquisitions over the next two decades that Aschaffenburg became first Europe's and then the world's largest manufacturer of industrial trucks. The profits of Linde's largest division from 1985 up until the acquisition of Sweden's AGA remained, however, often below those of the Munich departments.

Invest in Munich?

The department of development in Plant Engineering produced in 1973 a tanker system to make the transport of liquefied natural gas safer.[140] This was conceived as a bottle-like gas tank reservoir that could be manufactured independently of a shipyard and be quickly installed for retrofitted ships. Intended to compete against the widely used ball-shaped tank, researchers were focused on entering the promising market of natural gas, which began to boom because of the growing demand for energy supplies. Hermann Linde argued heavily in favour of developing natural gas tanks, yet the Supervisory Board voted against the project.[141]

Relocate Group Headquarters to Munich?

The magnitude and reach of Group Headquarter's importance increased so much during the late 1960s that it outgrew its location on Hildastraße in Wiesbaden. A new, larger, and grander building was to be built. Thus, the debate over the best location for Group Headquarters arose once again. Should they stay and build in Wiesbaden or move to Munich/Höllriegelskreuth, where nearly a half of Linde's employees, the two profitable divisions and the Linde family as well as then-CEO Wucherer were? At the shareholders' meeting in 1970, the future location for Group Headquarters and the Executive Board was discussed openly.

Hermann Linde argued in favour of Munich. He pointed to important client contacts in plant engineering, arguing that the presence of an Executive Board member is necessary for this kind of business and not so in manufacturing, where the physical separation of Executive Board and production is easier to handle.[142] Meinhardt and the rest of the Executive Board argued in favour of Wiesbaden.[143] Meinhardt was concerned about his concept of a neutral Executive Board floating above the divisions being damaged in the event of 'a personal union between administrative headquarters and a business sector' in Munich, which could raise fears of discrimination amongst other divisions. Furthermore, the other half of Linde's employees in extended Group Headquarters departments worked in Cologne and Aschaffenburg.[144] For both sides, it was the capacity for influence which lay at the core of their arguments. For Hans Meinhardt, it was precisely the point of access to Group Headquarters, which he had restructured, that lay at the foundation of his strategic concept. He sought to strengthen his argument for keeping Wiesbaden as Group Hedquarters and presented compelling facts: here in Wiesbaden, he argued, where the corporation had maintained its headquarters for 92 years and where he had risen through the ranks, he could negotiate with the city and the state of Hessen to acquire an 8,000 sq. metre site for a reasonable price that could be expanded upon and was close to the autobahn. While Group Headquarters, as an establishment without production facilities, paid relatively little in terms of commercial taxes, it did pay income, profit, and property taxes that in 1970 totalled 15.2 million DM. This option rendered Group Headquarters non-dependent upon Munich financially and could finance the lot purchase, construction, and move virtually alone.[145] In light of these cost-effective conditions, a decision was made in favour of Wiesbaden instead of Munich, where four of the five Executive Board members resided. The prohibitive cost of relocating employees was one argument against moving to Munich. In 1974 Group Headquarters moved to its new location on Abraham Lincoln Street.

In 1976, shortly before his retirement, Hermann Linde tried once more to bring Hans Meinhardt personally, and thus for the mid-term Group Headquarters, to Munich by suggesting he take over as director of the Industrial Gas Division. However, at this point, Hermann Linde was much too isolated and weak within the corporation to have any success with this idea. Hans Meinhardt politely refused. To this day, the debate over the proper location for Headquarters continues to arise, despite further expansion in Wiesbaden.

Hermann Linde resigns

In March 1975, Hermann Linde's contract as Executive Board spokesman was extended by five years to 1980. Nonetheless, in October 1975, the Supervisory Board chair Hermann Holzrichter approached Linde, proposing he switch to the Supervisory Board in 1977 at the shareholders' meeting, replacing his cousin Johannes Wucherer. Holzrichter wanted to find an elegant and peaceful solution to the struggles on the Executive Board. He therefore spoke in terms not referring to problems in the Executive Board, but rather in terms of the need of a good successor to Johannes Wucherer on the Supervisory Board. He also pointed out the need to keep a member of the Divisional Board at Höllriegelskreuth, Joachim Müller, by offering him a seat on the Executive Board. For the 58-year-old Linde, this offer seemed unacceptable, as it meant his withdrawal from operations. In his view, the argument seemed disingenuous. Hermann Linde rejected the offer to switch to the Supervisory Board, announcing instead in August 1976, his surprising decision to leave the Linde AG on 30 September. He wanted to spare himself and others a long transitional period. A spontaneous cry of protest arose from employees when they learned of his resignation at a works meeting in Höllriegelskreuth on August 31. The Höllriegelskreuth works council sent a resolution to the Supervisory Board in favour of Linde's continued work on the Executive Board. It was signed by over 85 per cent of the employees.[146] The German magazine *Spiegel* published an article about Linde: 'Horror in the Executive Board'.[147] Yet this kind of attention merely expedited the dissolution of Linde's contract, as the conflict had become public.[148]

Hermann Linde left the corporation under dissent and turned his attention to the sciences – albeit not as willingly as his grandfather had – becoming an honorary professor at the Technical University Munich and publishing, together with the famous thermodynamics specialist Helmuth Hausen, a standard edition on cryogenics.[149] Hermann Linde's resignation was the first public scandal involving personnel at the Linde AG. Much more than a mere personnel scandal, his resignation spelled the end of an era, in which technical feasibility and plant engineering in Munich stood at the forefront of the company's image. For the business-minded manager Meinhardt, the path had finally cleared for him to take command of the Executive Board and carry out his strategic goals.

Restructuring comes to an end

New shareholder structure

Having suffered many blows to its independence as a stock corporation in the 1960s and 1970s, the Linde AG became increasingly concerned with securing this independence. In the press, Linde was often regarded as a candidate for a takeover, in large part due to its economic success. Three years before he took the reins as CEO, Hans Meinhardt influenced a decision at the shareholders' meeting in 1973 aimed at keeping Linde a publicly held corporation with diversified stock by limiting to 10 per cent the voting rights of shareholders holding (either directly or indirectly) more than that of the capital stock. Despite turbulent rumblings at the shareholders' meeting, the proposal passed with 20 per cent dissenting.[150] As Linde's largest investor at the time, the insurance company Allianz AG stood out in support of the amendment to the articles of incorporation. After 1976, Meinhardt continued with determination to acquire major shareholders for Linde. In addition to Allianz insurance, he also brought both Commerzbank and Deutsche Bank on board, each holding 10–12 per cent of Linde's shares and helping to stabilize the aspiring group. In 1980, Meinhardt invited the CEO of Allianz, Wolfgang Schieren, to join Linde's Supervisory Board. Shortly thereafter, Schieren became chairman, a position he held for nearly 15 years. In Schieren, Meinhardt had a trusted partner and ally on the Supervisory Board who ensured continuity of support for Meinhardt's business strategy.

Cold stores and turbine engineering are sold off

Despite its glowing success in the 1970s, the cold stores division within Linde was comparatively small and no longer fitted the strategic balance model.[151] As signs of a weakening cold stores market began to appear in the early Eighties, mostly due to changes in the European Community's agricultural subsidies and an increase in self-owned cold stores by large chains, Linde decided to sell the division.[152] In 1981, due to irrational operations, Linde shut down its Nuremberg cold store, which had been founded in 1910 as one of Germany's first. In 1984 it consolidated the cold stores together into a legally independent company and transferred its shares to the Markt- und Kühlhallen AG (MuK) in Hamburg, in which Linde had shares since its founding.[153] Corporate strategy did not allow for continued investment in relatively insignificant sectors.

In 1984, Linde also sold its turbomachine business in the Refrigeration and Installation Engineering Division, which manufactured gas-piston compressors, turbo compressors, expansion turbines, and compressed air tools. At the time, this sector employed 760 people and turnover was 100 million DM. The division and its employees went to the Swedish machine engineering and manufacturing global giant Atlas Copco AG.[154] To prevent raising concerns amongst other employees, Linde insisted that the main divisions in Cologne were not to be affected by this plan.[155] This sector was to be sold as part of the larger strategy of product concentration.

Negotiations on the acquisition of the French forklift manufacturer Fenwick were already completed by 1984. Hans Meinhardt announced a tentative end to restructuring on the heels of the Fenwick and two other significant purchases.[156] From his point of view, important goals had been reached: management organization had been tightened; Group Headquarters had the ability to intervene on important, fundamental decisions; and if one combined Plant Engineering and Gas, there were three relatively equal divisions. Certainly, Refrigeration had shrunk not only in size but in power and its turnover had grown only minimally since 1981, but both profits and share price illustrated a robust increase in the mid-Eighties.

Conclusion

Both the structure and face of the Linde AG changed significantly in the 1970s. Economically speaking, the group had overcome the weak phase of the mid-Sixties by 1970, tripling both turnover and profits and increasing employee numbers by 50 per cent by 1980. These achievements are particularly impressive considering the gloomy global economic climate, the internal struggles at Linde over its thorough restructuring, and divided leadership. Despite all the strife, the years preceding 1976 were a golden era of sorts for plant engineering. Industrial gases were also highly profitable. The number of products manufactured at Sürth decreased significantly and parts of the mechanical engineering and manufacturing divisions were sold off, while a new field of forklift production was established in Aschaffenburg that proved successful on the market.

The 1970s were distinguished by a move away from a model that had characterized the development of the Linde AG for decades. Hans Meinhardt was the force behind these changes, which put an end to the influence exercised by the founder's family. The most significant representative of the traditional Linde corporate model as a research-oriented engineering company was, until 1976, Hermann Linde.

Hans Meinhardt implemented and realized the model of a centralized business with a small, but powerful (Holding) Executive Board and an expanded Group Headquarters. Management meant above all providing strategic vision, economic control, and business leadership. Research and development issues were tangential issues for top management. Hermann Linde stood in opposition to this, advocating instead a decentralized structure, in which individual divisions would enjoy wide-reaching autonomy over decisions of central importance to them and the Executive Board would function more as a helping, supervisory hand that posed important questions. Keeping his eyes on his strategic goals, Hans Meinhardt persistently pursued them over a long period of time. This included the vision of fewer but equally significant divisions which stabilized each other in their diverse spending and boom cycles, centralizing decisions regarding investment, becoming market leader in all divisions, and achieving independence from major shareholders (that is, establishing a balance between a number of shareholders).

The strategic acquisition of other businesses and a change in the landscape of Linde's portfolio were central to the group's development in Meinhardt's opinion. Hermann Linde, however, placed more faith in the group's own research and development as a source for new products and manufacturing opportunities.

The pressure to change this model of an engineering business came in the 1970s not from Munich, but rather from Group Headquarters and indirectly from the manufacturers in Aschaffenburg, Kostheim, and Sürth. They suffered more than the rest under the traditional model – a model under which Munich prospered. For those sectors involving heavy manufacturing however, a new model was necessary. The long-term need for change in the overall structure of the corporation was not taken into consideration during the acquisitions of the first half of the twentieth century. The research and development style of a large engineering company has persisted in Munich even today, years after Hermann Linde's resignation.

The strategic model of business divisions nearly equal in power, with mutually stabilizing differing business cycles, was particularly attractive for balance within a diversified business. However, in reality, a balance of this sort has proved difficult to realize. Certain areas, such as forklifts and refrigeration, were smaller and weaker than others. To create an image of balance, Meinhardt compiled the divisions differently over the years and spoke varyingly of three divisions (Plant Engineering, Industrial Gases, Vehicle and Mechanical Engineering),[157] later of four (Plant Engineering, Gases, Refrigeration, Material Handling) and by the end of the century of three (Gas and Engineering, Material Handling, Refrigeration).

Despite its long duration, Linde's structural reform was a great business success. Meinhardt took a floundering division such as Material Handling and turned it into a global market leader. In addition to repairing deficits in organization, marketing, and strategy, the group won its independence, continued to earn profits and set attainable growth goals. A more engineering-driven business would probably have appeared peculiar and anachronistic in the 1970s. This type of shift was seen also in other science-based large groups, particularly in the chemicals industry. This was to change once again in the 1990s, when the independent inventor enjoyed renewed respect and was entrusted with the ability to make Germany an economic powerhouse once again. The engineering or technology expert with a vision is now considered to be a wise choice for business leadership.

Before the first decisive debates had begun, Hermann Linde had already written out in 1969 his 'Principles of Corporate Management' which he used in his discussions with the Munich – not the Wiesbaden – leadership. He spoke of team spirit, trust, close co-operation, and research and development as fundamental to economic success. From a business ethics standpoint, he insisted that the goods produced should 'at least do more help than damage' to humanity. This idea reflected the engineer's image as a 'Master Builder for a Better World' as once declared by the Association of German Engineers.

After the completion of restructuring in the mid-Eighties, Hans Meinhardt wrote down the principles of his corporate management in a document entitled 'The Linde AG's Business and Management Principles and Group Headquarters' Duties and Status' which he then presented to top management at Group Headquarters. He laid out a clear corporate philosophy for employees: 'When corporate management demands creativity, commitment and responsibility of its employees, it is essential that they state their principles and goals as business policy so that employees can achieve their potential with a goal in mind. A situation in which losing sight of the goal spells double the work is to be avoided.'[158] At the same time, Meinhardt considered the necessary austerity directed towards himself and others in the corporation as a secret of the Linde AG's success. Hard-nosed in his approach even towards other management leaders, he implemented rationalization measures and corporate decisions with rigour.[159] The economic gains of his strategic corporate decisions succeeded not only in stabilizing the trust of his employees, but also in legitimizing the debates of the 1970s in hindsight.

6

Normal Operations for the First Time: From Reorganization to the AGA-Purchase, 1984–99

General economic development

The late 1970s seemed to promise the continuation of a stable economic upswing. The economic balance was still positive in 1979. In fact, with a growth of 4.5 per cent in the gross national product, Germany was the international leader. However, in 1979 oil prices increased again, causing worldwide inflation and throwing Germany into a three-year deficit for the first time in decades.[1] While deficits are normal in the United States, they caused a great deal of insecurity in German politics and in the German public sphere. Developing countries, however, faced even graver problems. As a result of the recession, the price of raw goods – with the exception of oil – sank, and deprived many third-world countries of their incomes. At the same time, industrialized countries restricted the importation of inexpensive industrially manufactured goods from developing countries, and, in addition, an unexpected inflation of the US dollar sharpened the global debt problem. At the beginning of the summer of 1982, the Falkland crisis and the related inability of Argentina to make debt payments were the forerunners of an international debt crisis. Mexico's declaration of a moratorium on its interest and debt payments, which was followed by other Latin American countries, posed a menacing threat to the international finance system.[2] The private bank industry was overwhelmed by the recession, the turbulent foreign exchange market, and especially the debt crisis. The banks were strained by the inverse credit structure: short-term interest rates were higher than long-term interest rates. Commerzbank, for example, faced a severe crisis in the early 1980s.

With the recession of 1981–82, the economic polices of the Social-Liberal coalition under Helmut Schmidt, which had been successful for years, were pushed to their limits. The demand-oriented Keynesian policy, with its sharp increase in public spending, seemed to reveal itself as a delusion. The major economic recovery programmes left few permanent traces, and the economy fell into recession again. The number of business failures increased rapidly, and, as a result, unemployment increased. The deficit of the public budget increased:

in 1980 the public debt surpassed the amount of the overall budget for the first time, to the tune of 225 million DM; 22 per cent of the state budget had to be used for debt payments.[3] Particularly troubling was the unemployment figure – two million – which was shocking at that time. Full employment, which had been taken to be the normal condition, became increasingly unachievable. Particularly in areas lacking infrastructure and in some traditionally industrial areas, unemployment grew to around 20 per cent. In addition to the economic slowdown, there had been crises in particular industries since the middle of the 1970s. For example, mining, steel, shipbuilding, and textiles underwent major structural changes, during which the industries did not disappear from Germany but rather continued to exist through an increase in efficiency and international networking and a sharp decrease in personnel.[4] Public structural policy, however, was too often merely a subsidization policy, which slowed the decline but was unable to stop it. The large overcapacity of coal and steel since 1974 caused an unprecedented subsidy race, which cost European taxpayers well over 100 billion dollars by 1983.

Although Nordrhein-Westfalen and other traditionally industrial areas were able to make the transition into new trades in the medium term, the economic collapse at the beginning of the 1980s greatly destabilized the political atmosphere and led to the end of the Social-Liberal era.[5] Circumstances in domestic and foreign policy also contributed to the economic sense of catastrophe, leading to the downfall of the Schmidt government. The lack of unity of the SPD in defence policy, nuclear energy policy and in questions of weapons export hindered the ruling party. In addition, the ruling party seemed at a loss to handle the social changes of the times. Increased ecological awareness and the anti-nuclear energy movement allowed the rise of a new ecologically-oriented party (the Green Party), which would change the political landscape of the Federal Republic in the near future. The FDP did not want to follow the guidelines emphasized by Chancellor Schmidt. On 1 October 1982 it voted with the CDU/CSU faction in a constructive vote of no confidence in Helmut Schmidt, and elected Helmut Kohl as the sixth chancellor of the Federal Republic of Germany.

The Christian-Liberal coalition transformed the economic policy from one oriented towards supply into one oriented towards demand. It also privatized public enterprises and created tax breaks for businesses. These policies, however, came at the cost of considerable reductions in social services and an increase in the value-added tax. The export economy became the motor of an economic boom, aided by an increase in the value of the American dollar and a decrease in the price of raw goods.

The liberal-conservative government focused especially on innovation and modern technology. Environmental protection technologies played a greater role in research and development funding, indirectly promoted by the oppositional Green Party, which had been in the parliament since 1983. Examples include the applied research in the reduction and optimization of energy use as well as the technical reduction of sound and toxin emissions. Continuing research of atomic energy as an emissions-poor form of energy production was initially part of the programme. But the trust in the safety and environmental soundness of atomic energy, which up to this point was only doubted by a small percentage of the population of the Federal Republic, decreased dramatically after the Chernobyl accident of 1986. As a result, the nuclear research institutes switched over to environmental and other research. Another incident of worldwide significance – which was not as widely known in Germany as the Chernobyl accident – was an accident in a Union Carbide chemical factory in Indian Bhopal in 1984, which caused thousands of fatalities.[6] Union Carbide, one of the largest chemical firms in the world, never recovered from the negative publicity and economic burdens it suffered as a result of this tragedy. It was taken over by Dow Chemicals 15 years later. Union Carbide had owned the rights to the Linde trademark in the USA since 1917, and Linde was able to recover these rights in 1999 in connection with the sale of Union Carbide to Dow.

Back to 1983. The price of oil sank again as of 1983, due to the fact that the conflict-ridden OPEC countries could not come to an agreement about maximum capacities, and their percentage in the world market sank. The International Monetary Fund (IMF) took charge of the debt conversion negotiations with the overly indebted Latin American countries and disbursed emergency credits. In return, the debtor countries had to commit to a strict economic recovery plan, an opening of markets and liberal economic reforms under the supervision of the IMF. Through this process, banks and debtor countries gained time and leeway. The stability of the world economic system was thus ensured for more than a decade.

After its turbulent beginnings, the 1980s developed into an economically successful decade. Interrupted only by short downturn phases, the value of the German blue chip Judex DAX quadrupled from about 500 points in 1982 to 2,000 points in 1995. Politically there were also signs of relief. The gradual withdrawal of the Soviet Union from Afghanistan and the slowing of the arms race between the United States and the Soviet Union contributed to a gradual thawing of the frosty climate of the Cold War. The Federal Republic granted several billion Marks of credit to the GDR, in order to aid Germans there as well as to promote the export economy of the FRG. Linde profited from this as well.[7] The 1986 election of Michael Gorbachev as general secretary of the Communist Party was the beginning of an era of reform in the Soviet Union, which ultimately led to the fall of state socialism in Eastern Europe. As of 1986, a broad citizens' movement began to form in the GDR, which faced strong repression by the state. At the beginning of September 1989 the important 'Monday demonstrations' in Leipzig began, during which the populace called for changes in the GDR. During this period, Hungary opened its border to Austria for GDR citizens; this was the beginning of a mass flight of GDR citizens to the border. On 18 October, Erich Honecker, the state and party leader of the GDR, stepped down; his successor was Erich Krenz. The attempts of the state to win back its citizens through relatively minor concessions came too late. The division of Germany into East and West came to an end with the fall of the wall between

East and West Berlin on 9 November 1989, 44 years after the end of the Second World War and 28 years after the construction of the wall.

As part of the monetary, economic, and social union, most of the laws of the Federal Republic were taken over by the GDR parliament on 1 July1990. The complete unification of the FRG with the reconstituted states of the former DDR was completed on 3 October 1990. With the addition of the new states to the Federal Republic, an unusual boom period began, which continued past the mid-1990s. However, the reunification had radically different consequences in the old and the new states. In the old states, the gross national product increased in real terms to 4.5 per cent, the largest economic growth in the FRG since 1976. The rate of increase in the economic sectors that participated foremost in meeting the backlog demand of the former GDR was in some cases significantly higher. These sectors included retail trade, consumer-related branches of industry and machine building. In contrast, the general economic performance of the new states decreased sharply. The transition from the state-controlled large-scale enterprises of a planned economy to the requirements of a market-based economy led to a failure of the economic structures of the new states. For the most part, the large combine businesses did not survive; in practice, the new states underwent a rapid deindustrialization. The 'blossoming landscapes' that Chancellor Helmut Kohl promised the East Germans during the elections turned out to be primarily branches of industry upon which weeds grew. Because the new states now also had to calculate their goods and services in 'hard currency', the demand from Eastern Europe, which had been the basis of the majority of East German exports before reunification, almost disappeared. In addition, the new citizens of the Federal Republic favoured 'Western products' and for a long time shunned products produced in Eastern Germany. These factors led to collapses in production, and especially in employment. For the population of the new states, this usually entailed a loss of jobs; only some of the employed were later able to find comparable employment. The state trust attempted to create economically viable businesses by separating the combines into smaller units. Often the new owners, most of whom came from the old states, found it impossible to make these businesses economically successful, despite government assistance. Newly formed businesses and the continuation of smaller business under new management were the most successful. Structural changes that for the FRG began in the 1960s and developed over many years in a relatively socially acceptable form, occurred within a few years for the new states in the early 1990s.

However, through widespread transfers of services it was possible to maintain the long-neglected communication and transportation infrastructure in the new states. Housing was renovated and modernized in many areas. The boom in the building trade led to economic growth in the reunified Germany, which, with an average of 2.2 per cent, surpassed other EU countries (2.0 per cent).[8]

In 1992 there was also an economic and monetary union at the European level. In the Maastricht treaties, the 12 EC countries decided that they wanted to have broad economic freedoms among themselves and that they wanted to try to achieve or maintain certain stability and convergence criteria in their economies until the introduction of a common currency in 2001. The goal was to create a unified trading area in Europe that would be able to compete with the trade giant USA and the up-and-coming trade nations of the Far East. Nationally and internationally the 1990s was a decade of consolidation. In addition to pharmaceutical companies, banks, and insurance companies, Thyssen and Krupp also merged, and the Swabian auto manufacturer Mercedes Benz acquired the North American manufacturer Chrysler. After a phase of euphoria it became clear that the expected synergy effect in most mergers could not be achieved to its full extent, or could only be achieved much later than expected, because the melding of differing business cultures usually led to frictions and communication difficulties. The anticipated cost reductions were also often not realized to their full extent. The new institutional economics of the 1990s tried to find the reason for this and, through empirical research into the problem of optimal business size, discovered that workers in larger firms (including subsidiary enterprises) have to spend more time to influence the top management.[9]

After the middle of the 1990s, there was a worldwide rise in the stock market. Investors most favoured the so-called 'dotcoms', companies that wanted to market business applications of the internet, and technology firms. The issuing of the so-called 'public stocks', such as the stock of Telekom AG (1996), encouraged even small investors in Germany to speculate on the market. Market fever allowed the DAX to climb to unexpected heights: in 1997 the German stock index DAX surpassed 3,000 for the first time and by March 2000 it had reached over 8,000 points. In terms of domestic politics, the electoral victory of the SPD and the Green Party in 1998 ended the 16-year 'Kohl era', during which Helmut Kohl headed a coalition of the CDU, CSU, and FDP. Gerhard Schröder became the new chancellor.

Cash-Cow: Linde in the 1980s and 1990s

The 1980s

At the end of the structural changes that began in the middle of the 1960s, Linde in 1985 as a company had four branches: Plant and Process Engineering; Gas Engineering; Refrigeration Technology; and Material Handling and Hydraulics.[10] But the relative size of the branches had shifted. Plant Engineering stayed at about the same level, but in terms of relative growth within the business, it was reduced to about 10–15 per cent of total business. Material Handling, on the other hand, rose to about 40 per cent of total business. After some strategic purchases, the Executive Board decided that they would use research to find new fields of application, in order to give the company more options in the classic Linde focus on business development.[11] Beginning in 1986, the expenditures for research and development grew at a greater rate than sales. In 1986 they reached 74.0 million DM or 12.6 per cent of sales and in 1988 they reached 109.4

Figure 6.1   Linde AG Board of Management, 1980–86. From left to right:
Gerhard Full, Dr Reinhardt Lohse, Dr Hans Meinhardt, Georg Plötz,
Joachim Müller.

million DM or 3.1 per cent.[12] The economic stagnation before the boom year of
1981 in traditional plant engineering could only be broken by new innovation.
The days of large projects with extensive detail engineering were over for the
time being. Through falling oil prices, the profitability of production became
of foremost importance. For this reason, the main task of plant engineering in
Western Europe was to find ways to bring existing plants maximal production
capacity with the lowest possible investment costs.

Market factors also changed in the area of gas engineering. The long-term
crisis in the steel industry led to a noticeable decrease in on-site business, because
major clients from the metal industry were ceasing production. New clients for
oxygen had to be found in other areas. In the area of environmental technology,
Linde was able to find new uses for oxygen. Biological process, above all in
waste water purification and drinking water preparation, and later ground
purification, were accelerated by the dosed addition of oxygen. In the automobile
industry, the use of shielded arc welding increased, causing an increase in the
sale of liquid argon, a buffer gas. Even more economically significant was the
increasing demand for highly purified gases in the semiconductor and fiber optic
industries, which were suppliers for the growing computer and communications
technology industries. In this market Linde was again able to deliver everything
from one place: modern air separation plants with computer controlled process
control systems, a specially designed distribution system, and final purifier. The
second half of the 1980s was a time of rapid growth for Linde, especially in
exports, which increased from 37 per cent in 1980 to 57 per cent in 1989.[13] The
internationalization of the Linde business was evident in the global business
yearly report, which was first published in 1989. Investors were also pleased by

the long-term positive growth: the dividends for each 50 DM stock rose from 9 to 10 DM in 1984, and from 10 to 11 DM in 1986. In 1987 discounted employee stocks were given to employees for the first time.

In 1985 the Executive Board first discussed the question of women in the company. It decided not to take any special measures to increase the percentage of women, which was 15 per cent in 1985, mostly in the lower income groups.[14] Women played no role – and still do not play a large role – in the first and second levels of management at Linde. What was the cause of this?

According to research in Gender Studies, the rise of women to leadership positions occurred more rapidly in the public sector than in private businesses, where the imperative to gender equality was less institutionalized and where quota systems could not be introduced. According to Schultz-Gambard the percentage of women in middle management across Germany in the 1980s was 4.1 per cent; the percentage of women on boards of directors was 0.3 per cent.[15] Half of the 12,000 businesses surveyed by the Institute for Job Market and Career Research (Institut für Arbeitsmarkt- und Berufsforschung) could not name a single woman in company management.[16] But the total lack of women in the first and second management tiers in the entire history of the Linde business demands a more specific explanation. Several factors came together at Linde that especially hindered the hiring and promotion of women.

During the time when the Linde family still influenced the business, there were sons and sons-in-law who wanted to be promoted. In twentieth-century family-run businesses in Germany, women usually only had an opportunity for leadership positions when there were no male alternatives, much as in a monarchy. Carl Linde's two wives played no role whatsoever in the business. This was also the case for the wives of sons, sons-in-law and grandsons. However, Richard Linde, who was a member of the second generation and had no daughters, did keep his four sisters informed of internal business dealings through family circulars. Johannes Wucherer had no sons; Hermann Linde had three daughters and a son.

The lack of women at Linde may have also been due to its technological orientation. Only a small percentage of female students studied engineering. Linde's international presence and orientation towards export may also have been a factor, because of the strong reservations in other cultures about women in leadership positions. Other factors may have included the necessary willingness to travel and the south German/Bavarian base of the business. Policies to promote women in private business were more successful in unconventional branches, branches having little to do with technology and 'Americanized' branches, where there also have been examples of affirmative action.[17]

The harvest: the 1990s

In the 1990s the positive trend continued. The business made significant profits and grew continuously. The number of employees grew from 27,676 in 1990 to 35,597 in 1999. There were some short recessionary periods, for example in 1992–93, during which company sales decreased. But the increasing focus on exports – the percentage of exports climbed to about 70 per cent at the end of the 1990s – allowed the business to weather periods of low domestic demand. It became increasingly clear that long-term profits would come mostly from the gas business in Höllriegelskreut. The company directed an increasing percentage of its capital into its gas branch. In 1996 Linde's investment resources reached an all-time high of 1,109 billion DM, more than twice as much as in 1990 (502.7 million DM). In 1998, 328 million Euros were invested in gas engineering, almost twice as much as was invested in material handling.[18] And much greater investment resources were to come.

In 1995 company sales surpassed 8 billion DM.[19] In that year the company had its largest annual surplus ever, in the amount of 358.5 million DM. In 1996 the company had sales amounting to 8.8 million DM, invested the unprecedented amount of 1.1 billion and, with a surplus of 395.9 million DM, achieved record profits.[20] In 1998 and 1999 the company grew by over 12 per cent. The stockholders also profited from this growth. The dividends grew from 16.00 DM (1996) to 17.50 DM (1997), 19.50 DM (1998) and 22.00 DM (1999). The stock prices grew even faster. With a stock value of over 10 billion DM, the annual general meeting made a decision in 1997 to end the 10 per cent maximum requirement for voting that was passed in 1973, since there no longer seemed to be a danger of a takeover.[21] There was much praise for the Executive Board during the annual general meeting: 'The balance sheet is more than a treat; it is real caviar.'[22] It again seemed possible for Linde to maintain consistent growth despite cyclical phases in the global economy.[23] With their sights set on future acquisitions, the annual general meeting decided in 1999 to enact the largest capital increase in the history of the business, in the amount of 1.84 billion DM.

The election of the CSU politician Gerold Tandler to the Executive Board in 1990 came as a surprise. He was to be responsible for refrigeration technology for ten years. He had been trained in the bank business and had been a member of the Bavarian state parliament since 1970, during which time he kept his position as branch director of the Bayerische Vereinsbank. Tandler was known as the 'foster son' of the CSU leader Franz-Josef Strauss and had held many high positions in the CSU. In the 1980s he was the general secretary and faction leader of the CSU in the Bavarian parliament, as well as domestic, economic, and finance minister. However, after the sudden death of Strauss, Tandler withdrew from politics. As early as 1989, while he was still the Bavarian finance minister, Tandler gave a speech about fiscal policy and business taxation to leading employees of Linde.[24]

At the end of the 1990s there was a debate about the financial compensation of forced labourers from World War II. These laborers had been forced to work for German companies, receiving minimal payment in return for hard labour, substandard food, and life in barracks. In 1998 the government of Helmut Kohl held the position that the moral obligations of Germany had already been met by the reparation payments to other countries. But after the change in government in that year, a serious examination of the question of individual reparations began. A foundation called 'Remembrance, Responsibility and Future' was established to compensate forced labourers. The foundation was co-financed by the government and private businesses, with each contributing 10 billion DM. At first 27 companies agreed to work with the foundation, in response to class action lawsuits of former forced labourers in the United States and the prospect of export difficulties. A broad debate about guilt and responsibility accompanied the initiative, during which many businesses disavowed their moral responsibilities. But in the end about 6,500 businesses more or less voluntarily took part in the initiative – a significant sign that was also recognized internationally.[25] Linde AG also decided in January to help fund the foundation 'to gladly support the concept of reconciliation and international understanding'. The first payments to former forced labourers began on 30 May 2001; as of the middle of 2003, more than 1.2 million victims had received modest compensation.[26]

During the annual general meeting in May 1997, one day before his 66th birthday, Hans Meinhardt switched from corporate head of the Executive Board, a position that he had held for 21 years, to the Supervisory Board of the company in Wiesbaden. With him came Josef Ackermann, CEO of the Deutsche Bank, while Hilmar Kopper and Reinhard Lohse left their positions on the Supervisory Board.[27] Gerhard Full became the new chairman of the Executive Board. Full came to the management department of Linde after receiving a degree in engineering and economics in Darmstadt. In 1969 he became the acting director of production at the Werks Kahl, owned by Matra-Werke. After a stint in the leadership of the central management department ZV in 1970–71, he left Linde for the time being. He became an instructor at the public school of engineering at Rüsselsheim in 1971 and in 1973 a professor at the University for Applied Sciences in Wiesbaden. In 1975 he returned to Linde and took charge of the technology section in the central administration office. Two years later, he entered the management board of the division for material handling and hydraulics, and in 1978–99 he was the president of the new Linde subsidiary Baker Material Handling (today known as Linde Lift Truck) in the USA. In 1978 he rose to the position of deputy member of the Executive Board of the Linde Corporation and in 1981 he became a regular member. At first Full was responsible for refrigeration technology and the hydraulics branch; in 1991 he went back to Aschaffenburg to run the division. Since 1993 Full had been in charge of the Plant Engineering Division. From 1991 to 1993 he was also director of labour relations.[28] Through these positions, Full gained years of experience in the different divisions of Linde so that he was the ideal successor to Meinhardt in 1997 as chairman of the board. He managed to combine the approaches and typical working styles of Refrigeration Technology (Sürth) Plant Engineering (Höllriegelskreuth) and Material Handling (Aschaffenburg).

The transfer of power to the new chairman of the board in 1997 went smoothly; the *Süddeutsche Zeitung* called the event 'possibly the most peaceful general meeting of the season'. Investors and investment representatives had only words of praise for the Executive Board and the Supervisory Board, because, firstly, the business development in the previous and current fiscal years left nothing to be desired and, secondly, because the meeting served to pay respect to the chairman of the board Hans Meinhardt at the end of his term'.[29] But Meinhardt continued to influence the direction of the company from his position on the Supervisory Board. He was still a regular presence in the company and kept his old office. The presence of a 'strong chair of the Supervisory Board', who was previously chairman of the board, is not unusual in Germany. This practice has advantages and disadvantages. But even by German standards, Meinhardt's position was exceptional due to his decades of personal connection to the corporation, the organization of the board, the central administration and the Supervisory Board, as well as due to his importance as a business leader above and beyond the Linde Corporation.[30] A critical article in the leading German weekly *Die Zeit* in 2003 described Meinhardt as 'a textbook case of a leader who cannot let go and continues to rule the company from the Supervisory Board', and the Linde Supervisory Board under his leadership was described as the 'most powerful senior citizens' club in Germany'.[31]

During his tenure as the chairman of the board, Meinhardt was not able to achieve his goal of developing a fifth branch for Linde that would help it weather market fluctuations. This new branch was to be in addition to the existing four branches: Gas Engineering; Plant Design and Construction; Material Handling; and Refrigeration Technology. This new branch was to be created by taking over a large foreign business, although not through a hostile takeover, since, according to Meinhardt, this was not 'the style of the company'.[32] The 'fifth branch' was mentioned as a strategic goal for many years. It was supposed to be a company and field that would allow Linde to be number one in the world, or at least 'near the top'. In a 1993 interview, Meinhardt mentioned that the company had already examined the possibility of several branches, including measurement and control engineering, certain chemical fields, as well as areas of environmental engineering.[33] Even though it was said in 1996 that a decision about the new field had already been made, the 'fifth leg' (as Meinhardt sometimes called it) never materialized.[34] The new field was researched with some intensity, but the expansion of the four branches that already existed took priority in terms of investments and corporate takeovers. In this area Meinhardt had set very high standards: by 2000 all four fields were to have a capital return of 20 per cent. In the previous century this goal was only reached in the field of Gas Engineering. Plant Design and Construction still reached 17.6 per cent and Material Handling and Refrigeration Engineering each achieved 15.5 per cent.[35]

Investments in the new states of Eastern Germany

The successful reunification of Germany gave Linde the opportunity to expand rapidly into the new states. The widespread investments that were necessary after the economic collapse of the GDR caused an economic boom in the old states. All areas of the Linde Corporation profited from this boom. The area of plant research and design, which had already undertaken many projects before reunification, saw great potential since the chemical plants of the GDR had to be 'modernized and restructured with regard to energy consumption, environmental protection, product selection and product quailty'.[36] Beginning in April 1990, Linde participated in a joint venture with the second largest engineering firm in plant design and construction from the former GDR. The firm was previously called 'Komplette Chemieanlagen Dresden', and began operating under the name 'Linde-KCA-Dresden GmbH'. Linde contributed not only capital for large investments, but also technical expertise and management know-how in the areas of cryogenics, process engineering, and environmental engineering. Linde expected 'to be better able to serve this market from Dresden than from Munich, due to the previous good relationships with the GDR and other Eastern European countries'.[37] These expectations were not met. The large former chemical combines did not have the financial means to renovate and modernize the existing plants. Many Western firms hesitated to do business with them, due to the anticipated high costs of processing existing hazardous waste. In addition, the proceedings to determine ownership often took longer than was expected in 1990. In general, the market for plant design and construction in the countries of the former Eastern block, the Soviet Union, the Ukraine and Belarus, and the Eastern European countries did not develop as favourably after the fall of state socialism as was expected. The political and economic problems in these countries did not at first offer a favourable environment for long-term investment. When it acquired KCA, Linde took over a company with 1,150 employees; within a year the number of workers was reduced to 760, through phased-in incentives. Despite this fact, the takeover of KCA by Linde was in general a successful new beginning. Without a strong Western partner, the KCA would 'not have been capable of survival',[38] in part because of the loss of clients in the chaotic Eastern European markets. Half of the management positions were filled by leaders from Munich. Positions in the fields of cyrogenics plants, processing plants, and sales were filled with personnel from Höllriegelskreuth, while positions in the areas of pharmaceutical plants, environmental technology and general engineering were filled by managers from Dresden. In order to keep the new subsidiary busy, contracts from Munich had to be transferred to Dresden. At the same time, Munich transferred many competent former employees of KCA into the Bavarian capital city. But in 1993 the situation changed. Linde-KCA-Dresden received a contract for two large polyethylene plants in the Ukraine and in Russia, as well as a contract for a pharmaceutical plant in Jena. In 1994 the environmental-technical aspects of plant design and construction were completely transferred to Linde-KCA; shortly thereafter, the construction of polyethylene plants began. In 1973 Linde erected its first polyethylene plant for Chemopetrol Litvinov using the unipole method, which was new at that time. The 'start-up' of this plant is still remembered by 'old employees'[39] and clients as the longest and most difficult start-up of a polyethylene plant.[40] In 1986 Linde made a turnkey HD-polyethylene plant using the unipole method for VEB Chemische Werke Buna in Schkopau. In November 1999 Linde KCA was able to negotiate its first independent contract for the delivery of another polyethylene plant for Chemopetrol Litvinov; 28 months later the plant was ready for occupancy. The licensor, Univation Technologies (previously Union Carbide), had already been monitoring the engineering using the virtual CAD module. Since then Linde KCA has played a leading role in the market for polyethylene plants.[41]

In the area of gas engineering, Linde was able to attain a central position in the Eastern German market through the acquisition of all parts of the Technische Gase Werken Brandenburg (290 employees) as well as the smaller gas plants Bützow[42] and Reichbach.[43] These companies were made available due to the privatization of the public combines and the breaking of the combines into smaller firms. In addition, Linde acquired an acetylene plant and a transfer plant of the energy company Schwarze Pumpe of Hoyerswerde in 1991. In 1992 Linde acquired the gas branch of the EKO-Steelworks company of Eisenhüttenstadt.

Linde's most far-reaching investment plan in the new states was the takeover of the maintenance of the chemical area Leuna in Sachsen-Anhalt. Leuna had become one of the most important chemical production centres in the GDR by the late 1950s. After the end of the GDR, unprofitable and non-typical areas were shut down. Remaining were the refinery, organic special products, and plastics. In December 1990 the newly founded Leuna Corporation and the Linde Corporation signed a contract for long-term co-operation in the area of gas engineering. Linde took control of the Leuna gas engineering branch as of 1 March 1991 and built a modern gas production centre, which would serve the Leuna plants and the east German market.[44] Linde could claim that it had built a total of 19 air separation plants in Leuna since 1916. Of six large-scale air separation plants (four produced by the Soviet Union, two by Linde) that existed in the plant area, two had to be disassembled; a large new unit was built.[45] As of 1996, Linde had invested 460 million DM in over 300,000 square metres of new constructions and renovations; by 1998 it would become 600 million.[46] Linde was one of the biggest investors in the Leuna area. The co-operation between Linde and the local authorities varied. Whereas commercial licensing procedures were 'more accommodating and faster than in the old states', going through any public procedures was 'catastrophic', because the 'Eastern authorities' tried 'not just to follow Western requirements but to surpass them'. In the area of environmental protection, eastern Germany caught up quickly with the west: 'Environmental authorities sprang up like mushrooms.'[47] Such difficulties did delay the construction and opening of gas production plants, but Linde was still able to put its plans into effect. The dedication of Linde's largest gas engineering centre in the world occurred in 1994. Pipelines from Leuna were built to supply Schkopau (the site of Bunawerk, which belonged to the olefin network of Dow Chemical) and 60-km distant Bitterfeld with hydrogen and nitrogen. Linde built dedicated hydrogen equipment for the nearby refinery of Elf Aquitaine. By the end of the decade more than half of all of Linde's pipeline transactions in Germany were directed to the gas centre Leuna, which in turn did 70 per cent of its business with on-site equipment over the pipeline net.[48] At this point, the investments were making returns.

Linde was not aggressive in investing in the new states in the area of material handling. The subsidiary STILL opened a central office in Leipzig and a sales office in Chemnitz, Dresden, Erfurt, Magdeburg, and Rostock in rental spaces. Customer service in the new states was at first carried out through contracted repair shops.[49] In 1992 Linde acquired the Hydraulik Ballenstedt firm (Ballenstedt/ Ostharz), which was once a part of the Orsta-Hydraulic combine. The company produced hydraulic valves.[50] The refrigeration technologies branch was also hesitant in the new states. Directly after reunification, the working group on refrigeration technologies and appliances opened a marketing branch and offices in Chemnitz and Leibzig, and placed importance on 'hiring workers primarily from Saxony'.[51] Directly after reunification, in the second half of 1990, the large chains of stores increased their activities in eastern Germany; because of this, the Linde plants in Mainz-Kostheim, Köln-Sürth and Bad Hersfeld were at full capacity. However, this boom resulting from reunification was only to last for a few years.

Plant design and construction: ups and downs in Höllriegelskreuth

The production planning for the location Höllriegelskreuth in the 1980s is an example for how quickly strategies can be made obsolete by real market developments. The Executive Board had already decided in the late 1970s to relocate production to Schalchen and to use the expensive location in southern Munich solely for planning, research and development. For this reason a new development centre for plant engineering was created in 1980 in Höllriegelskreuth and in 1981 the conventional equipment construction was moved to Schalchen. Only one part of serial equipment construction remained in Höllriegelskreuth, the construction of vacuum-soldered aluminum heat exchange panels. But precisely this field underwent a boom in the following years. Linde developed a production method that allowed heat exchange panels to be used at operating pressures of up to 96 bar. These heat exchange panels proved to be a complete success, and Linde became an important heat exchange supplier for other firms. Therefore, a second vacuum-soldering oven furnace was installed in Höllriegelskreuth. It was producing at capacity from day one, so a third furnace was installed there the following year.[52]

In the 1980s Linde could maintain its position in the competitive field of large plant design through its traditional strengths in the planning and construction of large plants.[53] Meinhardt used political connection for international business more than his predecessor. Franz-Josef Strauss, the governor of Bavaria, often lobbied directly for Linde in negotiations, not only for business in the east, but also for a large oxygen extraction plant in Botany Bay and a plant for creating synthetic fuels out of natural gas in Mosel Bay, both in South Africa.[54] The construction of these plants was related to attempts by South Africa, which was threatened with international isolation, to be self-sufficient. Linde's most important large project in the mid-1980s was the natural gas station for Satoil of Kaarstø, Norway, through which natural gas was to be distributed via the

Ekfisk-Emden Pipeline to Western Europe. Under the leadership of Linde project management and over 100 Linde specialists, around 1,800 construction and assembly workers were employed. The project was completed in 1985, three months ahead of schedule, and led to many other contracts, even today. Without any political cover whatsoever, Linde was able to get a turnkey-based contract for an ethylene plant for BASF in Antwerp in 1990. With a contract volume of 1.3 million DM, it was the single largest contract in the history of the business.[55] The size increase in air separation plants was taken to another level in 1997. Linde began building the four largest air separation plants in the world for the public oil company Pemex in Mexico, with a total contract volume of 150 million US dollars, to be completed in 2000. In early 1999 an air separation plant that Linde made for the largest Chinese steel company, Shanghai Baosteel Group corporation, went into operation. Because Baosteel played a leading role in that branch, Linde anticipated contracts from other Chinese steel producers.[56]

In the 1980s Linde put a great deal of effort into intensifying its business in America, which was of strategic importance. In 1983 Linde signed a contract with Georgia Pacific for the construction of the first methane plant in the US, despite a risk of up to 7.5 million DM.[57] It seemed promising to build a model methanol plant on a large technical scale. But because of the weak dollar and the tough competition – Linde still could not use its name at that time – the American market remained unprofitable in the 1980s. This was only to change in the 1990s.[58]

Linde used its expertise in plant design and construction to grow its consulting business, and created studies for industrial clients suggesting energy-saving measures and equipment modernization. As of the mid-1980s, the increased use of CAD-systems in the technical implementation of models served to improve the profitability and quality of engineering work. The systematic expansion of data processing into graphics as well as the construction of three-dimensional models of building components, modules, and complete units replaced model work, which had been standard until then. The conversion occurred gradually because the capacities of the early CAD programs were not yet able to adequately diagram complex pipe networks. For a long time, model testing remained the only way to prevent planning failures. To prevent problems on the part of the client, Linde developed a device for the dynamic simulation of plant operations and for training plant workers. This simulator could be used for experiments with better regulation strategies to train equipment users, who could be intensively trained before plant operations began and without danger of injury.

In 1993 Joachim Müller stepped down from his position as head of Plant Design and Construction for health reasons. He had succeeded Hermann Linde in this position in 1976. He wrote to his successor that the 'most pressing task' was the development of 'new techniques and products' for 'energy conservation, energy use and environmental protection'.[59] In order to highlight this broad scope of plant design, Linde followed Müller's suggestion in 1990 and renamed the working group from TVT (cyrogenic and process engineering) to VA (process and plant design). Gerhard Full unexpectedly took over the

Figure 6.2 From left to right: SASOL Manager and Linde Board of Management members Dr Hans Meinhardt, Gerhard Full, and Joachim Müller at the South African Coal, Oil, and Gas Corporation (SASOL) in Sekunda, South Africa, in 1989, standing next to a heat-exchanger.

leadership of this area. He was not a process engineer, but rather a mechanical engineer. Since 1990 Müller had been following the strategy of ensuring growth by acquiring firms, which had been a successful strategy in other working groups. Full promoted this strategy even more aggressively. In 1990 Linde subordinated the MAPAG firm, which had been independent until then, into the working group for industrial processing engineering and plant design and engineering. In 1996 a new Mapag plant was put on line in Horgau, near Aschaffenburg. In the 1990s Linde took over a series of companies in Eastern Europe. In 1992 Linde acquired the low-temperature processes division of the Gebr. Sulzer corporation in Winterthur, Switzerland and in Aldershot, England. It continued business under the names Linde Kyrotechnik AG, Pfungen, Schweiz and Linde Cryogenics Ltd., Aldershot. In 1994 Linde began a joint venture with a Chinese equipment builder in Dalian, China, with the purpose of planning and constructing air separation plants in Asia.

Business in Eastern Europe

During a board meeting in the summer of 1983, Joachim Müller, who succeeded Hermann Linde as head of Plant Design and Construction, had a negative assessment of the prospects for his branch. With the exception of Linde and Uhde, all other large German plant construction companies were running reduced shifts.[60] Müller thought that plant construction business could only be found in the Near East and Eastern Europe. For this reason, in 1993 TVT opened its own business office in Dubai (United Arab Emirates). In this difficult situation, business with socialist states, where Linde had traditionally had a good name, played an important role. The most important contract in 1982–84 was for a petrochemical plant in Kalusch, USSR. In some cases, the Soviet contractors did not tell their Western partners where the plants were actually going to be erected. Therefore, employees at Linde could only guess at their final use by examining the technical parameters. For example, through the Salzgitter-Industriebau GmbH, Linde had a contract in 1984 for a cryogenic gas purification plant for the production of highly purified carbon monoxide, which is used in the chemical industry as a basic material for syntheses. The buyer was Techmaschimport, USSR. Plant construction transactions with the GDR were usually handled through the VEB Außenhandelsbetrieb Industrieanlagen-Import. This was also the case for the construction of a turnkey HD-polyethylene plant using the unipole method in the VEB Chemische Werke Buna in Schkopau, which Linde undertook with the co-operation of its French subsidiary S.A.T. Linde S.A.R.L. The plant started operations in 1987. In co-operation with VOEST-ALPINE AG (Linz/Austria), Linde negotiated a second contract, for the renovation and modernization of an olefin plant in VEB 'Otto Grotewohl' in Böhlen. A third important contract was the steam reformer at Schwarzheide.[61] As of 1985, the RGW countries, especially the Soviet Union as an exporter of oil and gas, lost export profits through the fall of the price of oil. For this reason, they became less able to buy technology and equipment from Western countries.

In the production planning of the GDR and other Eastern Bloc states, the importance of transportation and logistics for the economy were notoriously underestimated for decades. Therefore, the GDR encountered major setbacks in these areas. The attempts to reverse this trend provided opportunities in the area of materials handling. In the late 1980s the Linde subsidiary Wagner, acting as a general contractor for the knitting plant VEB Aprotex in Limbach-Oberfrohna, delivered a complete transportation and storage system (online-controlled high-rise stackers and driverless conveyers) including the computers and software for production control, storage management, and flow of materials. But during this period, the GDR did not have the financial means for further contracts.

The social and economic situation in Eastern European countries changed with the political turn of 1989–90. This change entailed great progress in social and political freedoms for the people of the post-communist states, but actually caused a decrease in Linde's business with Eastern Europe.[62] The early euphoria caused by the partial liberalization of the planned economies did not match the reality. Breaking up the centralized economies in the former Soviet states left behind some unstable entities with less economic power. This was only to change in the mid-1990s. The EU-neighbours Poland, Hungary, and the Czech Republic were the first to experience economic growth. A few years later, most of the rest of the former Soviet states followed.

China's political system remained virtually unchanged by the dramatic political transformations in the Soviet Union and Eastern European countries. In the area of economics, the state party began a long-term strategy of modernizing the economy and introducing a limited openness to Western technology. In addition to a more rapid development of chemical production, environmental protection also played a role as of the 1980s. For example, in 1984 VR China ordered a varox and unox plant for biological waste water purification, including air separators to supply oxygen to sewage treatment plants. But the real focus was on petrochemical plants and metallurgical plants. In addition, Linde delivered equipment to China for the extraction of gaseous and liquid oxygen and nitrogen as well as all noble gases contained in air. The Plant Engineering Division was especially successful in the area of turnkey plants for the extraction and processing of the gases accompanying petroleum: ethane, propane, butane, and natural gas.

Cryogenic research pays off in the long term

Since the 1950s Linde had made major investments in the research of cryogenics, which includes the temperature ranges of liquid hydrogen (–253 °C) and liquid helium (–269 °C). But for a long time there were no fields of application. Cryogenics only experienced a significant upswing in the 1980s, through the increasing use of helium-cooled superconductors in medical technology and high-energy physics. MRI scanners used in medical diagnostics allow internal photography without radiation damage. The main component of this equipment is a superconducting magnet that requires an operating temperature of –269°C. Special helium refrigeration equipment was developed to independently provide refrigeration for MRI scanners. This helium refrigeration equipment was sold in series of about 25 per year and earned profits. DESY in Hamburg built the Hadron-Elektron-Ring-Anlage (HERA) for the purpose of pure physics research. This device allowed research into the smallest dimensions of the micro cosmos. In an underground storage system with a length of 6.2 km, particles could be accelerated through magnetic fields to speeds close to the speed of light and made to collide. Research and measurement of the collisions provide insights into the smallest structures of the construction of matter, in dimensions as small as 10^{-16} mm. Linde provided the helium purification and vacuum-isolated circular pipeline systems for helium to cool the superconducting magnets of the proton ring. To minimize insulation losses in this large project, far-reaching experiments on superinsulation in a high vacuum were conducted. Linde took part in all of the large superconducting projects, such as by providing a helium refrigeration system for a high-energy physics project at the technical university in Munich. To keep cryogenics development in pace with the increasing demands of the industry, Linde built a new cryogenics lab in the development centre in Höllriegelskreuth. Experiments could be made in this lab at temperatures as low as 1.5 K (–272.7°C).

The expertise gained through working on the HERA project could later be applied in liquid hydrogen technology. At this time, hydrogen was already one of the most important basic materials for the chemical and petrochemical industries, and Linde already had a long and successful tradition with the handling of hydrogen-containing gases. At the beginning of the twentieth century, the purification and preparation of coke oven gas for ammonia synthesis was of central importance. Later there was greater emphasis on the creation of hydrogen from the fossil hydrocarbons petroleum, natural gas, and coal. However, the resource debate resulting from the first oil price shock led to the realization that fossil fuel reserves are limited and that petroleum, natural gas, and coal would become scarcer and more expensive. In addition, in the late 1980s there was increasing criticism of the waste materials produced by the burning of fossil fuels, which had to be removed from waste gases through expensive technological methods. There was also criticism of toxins such as carbon dioxide, which could affect climate. Under these circumstances, the production of hydrogen through water electrolysis seemed to be the royal way to the future for energy technology, although questions about the economic implementation of this ambitious technical vision remain unresolved today. It is worth noting that by the middle of the 1980s Linde had already realized the future significance of hydrogen technologies for the transformation of the primary energy system, and applied itself accordingly.[63] A necessary component of a hydrogen energy system is an efficient electrolysis. Linde founded a society for hydrogen production in order to develop water electrolysis to meet the particular needs of intermediate use (for example in vehicles) with a high degree of effectiveness and produced with the most modern production methods. Its partners in this enterprise were aerospace company Messerschmitt-Bölkow-Blohm GmbH (Ottobrunn) and the power utility Hamburgische Electrizitäts-Werke.[64] Linde also formed a working group with Siemens for the purpose of developing and testing high-temperature fuel cells in which hydrogen-containing gases can be converted into electricity with a high degree of effectiveness. Linde also worked with Solar-Wasserstoff Bayern GmbH,[65] which built a testing and demonstration plant in Neuburg/ Oberpfalz near the proposed site of the reprocessing plant Wackersdorf. All the technologies and components of hydrogen electric engineering, including photovoltaic, electrolysis, hydrogen tanks, and fuel cells, could be tested in practical applications at this plant. Because these technologies were still in a conceptual phase in the 1980s, conventionally extracted hydrogen was used at first to test applications such as powering vehicles with appropriately redesigned conventional piston motors.[66]

The use of fuel cells for small power plants and street vehicles only became important in the following decades, but hydrogen had already been used for rocket propulsion for years. Linde also was active in this area in the 1980s. In 1987 Linde was contracted by the German Aerospace Research Centre (DLR) for work on the European rocket project Ariane 5. Linde planned the tank depot that provided hydrogen-fuelled rocket propulsion to the test stand. It later delivered several hydrogen tanks, each with a volume of 270,000 litres and with highly effective super insulation. The company also participated in providing cryogenic equipment for both Ariane-5 rocket test stands in Lampoldshausen and Vernon, and is still profiting from the successful introduction of Ariane as a commercial carrier rocket.

Sustained growth opportunities in environmental protection

The environmental movement of the 1980s, which was especially strong and long-lasting in Germany, was an economic blessing for Linde, since it continually brought new projects and contracts to the company. Some business areas only developed weakly, because the ecological pressure was not strong enough. An example is the unsuccessful attempt to turn purification plants into major buyers of oxygen. The oxygen content of many rivers and lakes had fallen so greatly due to eutrophication, especially during the summer months, that higher forms of life could not survive there, and the water threatened to 'turn over'. For this reason, many purification plants in Germany had to add another biological purification level, for which Linde developed many processes for adding oxygen. In 1984–85, several large German cities – Bremerhaven, Karlsruhe, Nürnberg, Munich, Peine – received 'Lindox' and 'Lindpor' equipment for biological purification with oxygen. However, modern purification plants were also able to achieve good purification results using normal air, so the plants that used expensive oxygen did not become the standard. Linde marketed these plants especially to cities with strong seasonal pressures, 'such as occur in vacation areas or during wine and sugar beet harvests'.[67] The use of oxygen for the purification of industrial waste water also had economic benefits, for example in cellulose processing. In this area, Linde produced 'Unox' plants[68] for biological waste water purification with oxygen and improved the units in the following years through new processes beyond the use of oxygen.[69]

The oxygen enrichment of polluted or manure-contaminated water also remained a small area of business. The contaminated water of the Saar rivers was enriched with oxygen until measurements of oxygen reached minimum levels. The oxygen was transported in a ship refitted for the purpose, aptly named 'Oxygenia'. The Teltow canal in the south part of west Berlin (which used to form a part of the border to the GDR) was supplied with oxygen using gassing mats (the so-called 'Solvos-B' process).[70] The growth of this business area was hindered by the high cost of liquid oxygen and the gradual improvement in the quality of water in the Federal Republic.

The purification of contaminated soil on a biological basis provided another use for oxygen in the area of environmental protection. The break-down of contaminants by special microbes is aided by oxygen. A comprehensive mapping of hazardous waste in the FRG, including the new areas after 1990, showed that the soil in many former industrial and commercial sites was contaminated. For this reason, processes for biological soil purification became increasingly important.

In the 1950s, smokestacks were still a sign of economic success. But by the 1960s, the SPD had already won elections in Nordrhein-Westfalen with the call for 'a blue sky above the Ruhr river',[71] which was, however, at first only accomplished by the construction of higher smokestacks. In the 1980s maintenance of air quality moved to the centre of public discussion. The 'dying forests' caused anxiety, and the government reacted in 1986, after several years of preparation, with stricter guidelines for the reduction of the release of gaseous and solid wastes into the atmosphere. Existing fume purification plants had to be refitted, and many new ones were necessary. Linde developed new processes for the extraction of hydrogen sulphide and sulphur dioxide from industrial gases, natural gases, methane gases, and fumes. In short: Linde played a major role in the state-mandated ecological restructuring of industry.

Beginning in the 1980s, environmental protection also played a large role in refrigeration systems. Research and development focused on the replacement of ozone-destroying coolants, the reduction of coolant amounts and the recycling of used coolants. Since then, classic coolants such as ammonia or CO_2 and new synthetic coolants have been used in the place of chlorofluorocarbons, which damage the upper levels of the atmosphere that shield UV radiation. The sale of energy-sparing equipment, such as heat pumps, was not as successful; the interest of potential clients remained low and decreased after 1985. In the early 1990s research and development in the areas of environmental protection and energy conservation remained a high priority for Linde, 'but environmental engineering did not become a major business', according to Joachim Müller, chairman of Plant Design and Construction from 1976 to 1993.[72]

Linde in the USA

In the 1990s Linde increased its activities in North America. The American market had been underrepresented in the company since 1916. In the post-war period, Linde was not able to make inroads into the US market commensurate with it significance in Europe due to the loss of the rights to its name. After reunification the post-war period was finally over, and Linde wanted to grow its business in the USA. In 1992 Linde took over T-Thermal in Conshoken, a producer of furnaces for contaminated emissions and fluids as well as specially designed burners and vaporizers for liquefied gases. One strategically important purchase was the acquisition of the Pro-Quip Corporation in Tulsa, the US market leader in ethane, propane, and elementary sulphur production plants. Pro-Quip was even the world leader in hydrogen production plants. Tulsa was also a good location. The cost of living there was only 90 per cent of the US average. In contrast, the cost of living in New York, the location of Lotepro (Low Temperature Processes, Linde's American subsidiary in plant design and construction) was 226 per cent of the US average. For this reason, Linde decided to move the New York subsidiary to Tulsa, Oklahoma. Only 30 of the 100 Lotepro employees actually moved; 70 were 'let go'.[73] Linde also expanded in the area of gas engineering: in 1996 the Linde subsidiary Holox acquired the gas company Sunox in Charlotte. Through a co-operative agreement with Millennium Petrochemicals, Linde made its entry into the American hydrogen and carbon monoxide business in 1998. 1 January 1999 is another symbolically important date. Through the purchase of the name rights from the liquidated Union Carbide, the North American affiliate of the company could finally use the name Linde, 44 years after the end of the war and 82 years after the integration of Linde Air Products into the Union Carbide Corporation.

The increasing importance of gas engineering

Since at least the end of the 1985 preliminary corporate restructuring, Gas Engineering had become increasingly central in corporate politics as one the most important branches for profits. For decades, the market percentage of gas business was more or less stunted. Linde was the market leader in only a few European countries, and this was in a business for which regional dominance was usually the deciding factor of success. Linde's importance in this branch was again increased through targeted acquisitions, first in Germany, then in Europe and overseas. In 1985 Linde took over the remaining business units of Kölner IGA-Gesellschaften (Industriegas GmbH, Industriegas GmbH & Co. KG, Azetylenfabrik Hagen GmbH, Industriegas GmbH & Co. Nord KG) and by 1986 it controlled six gas production centres, 28 gas inflation plants and 26 acetylene plants (1985: 17) in the Federal Republic of Germany and West Berlin. In the following years, several more gas production centres were put into operation.

On the European level, an order by the EG regulatory commission in Brussels in 1989 forced Linde and Messer-Griesheim to separate their joint activities in the EU in the area of gas engineering. Messer-Griesheim received all parts of L'Oxhydrique Internationale S.A. in Belgium, Fedgas (Pty.) Ltd in South Africa, as well as the Société Industrielle de l'Anhydride Carbonique S.A. (SIAC), and the Soudures Nevax S.A. in France. Linde took control of all previously jointly held parts of the Dutch gas company W.A. Hoek's Machineen Zuurstofffabriek and Linde Industriegassen B.V. In addition, the French gas market was split; Linde acquired southeast France with the takeover of an air separation plant in Salaise near Lyon. The intervention of the EU regulatory commission demonstrated a new test of power by a regulating entity; in exactly ten years Linde would have to deal with this commission again.

After reunification Linde invested heavily in Eastern Europe. To some extent these investments were necessary to make inroads into the market, as was the case with acquisitions in the Eastern European states. Another reason for the investments was the long-term expansion of on-site business; in targeted takeovers, Linde acquired the entire gas supply for several large steelworks. In general, liquid gas deliveries were to be replaced by standardized on-site plants. In 1991 Linde made an agreement with the Czech ministry of industry for the purchase of the majority of shares in the largest Czech gas company Tecnoplyn; the complete takeover occurred in 1995. In Hungary, Linde undertook a joint venture in 1992 with Répecelac, the largest carbonic acid producer in Hungary and founded Linde Repcegas AG. In 1993 Linde took over two units (Mielec and Psyczyna) of the formerly state-run gas company Polgaz and fouded – initially with state participation – the firm Linde Gaz Polska, Krakau. Linde bought the rest of the units of Linde Gas Polska in 1997. In 1999 Linde also took over the gas business of the American firm Airgas, Inc. in Poland. Through Linde's efforts, Linde Gas Polska became the largest producer and supplier of gases in Poland, with 35 per cent of the market (previously 18 per cent).[74] In Slovenia, Linde acquired a gas inflation plant in Celje in 1996. Beginning in 1990, Linde acquired Western European gas producers in Portugal and England.[75] In 1992 Linde increased its share in Hoek Loos in Ijmuiden to a majority (60 per cent). Linde also acquired the gas supply of the EKO-steelworks in Eisenhüttenstadt in 1993 as well as air separation plants (from Voest Alpine in Austria) and took over their gas supply. The expansion in the area of on-site business continued in 1994 with the acquisition of air separation plants and related plants of the Hungarian steelworks Dunaferr. In 1994 Linde acquired the Italian gas company Caracciolossigeno s.r.l. in Rome and in 1997 it acquired the hydrogen business of Air Products GmbH of Salzburg, Austria. In the 1990s there were a few sales by Linde in markets where it had a weaker position; in 1992 Linde sold its 50 per cent share in the Brazilian Aeroton Gases Industriais Ltda, as well as the entire Linde do Brasil Ltda business, to the Swedish company AGA AB, which had owned the other half since 1974. Just eight years later, the Brazilian business returned to Linde through the takeover of AGA.

Linde was also active in the area of 'lighter than air' technologies. Linde had already been using helium as a carrier gas for hot air balloons for many years, and interest in rigid airships increased at the end of the 1990s. The 'Zeppelin NT' of the Friedrichshafener Zeppelin Luftschifftechnik GmbH was first put into operation in the summer of 1997,[76] and the Linde working group on gas engineering participated as a sponsor and provided the helium, as well as the technology for gas supply and purification.[77] Linde's collaboration with the air ship manufacturer Cargolifter was unsuccessful. In 1998 Linde contributed 120,000 DM to the firm, which planned to build airships in Brand near Berlin for transporting heavy cargos. In return for this investment, Linde was promised 25 per cent of the helium delivery, which amounted to a value of 5 million DM per airship filling.[78] The only thing that was completed was a large airship hanger, and the project went into bankruptcy in 2002. But taken together, all of Linde's acquisitions paled in comparison to the most important business acquisition in the history of Linde AG, the purchase of the Swedish company AGA in 1999.

Material handling: on the way to being the leader in the global market

By 1985, the business branch Material Handling and Hydraulics had moved into third place in the global market, and first in Europe, with a market share of 27 per cent. This position was achieved through the acquisition of businesses and forklift branches. However, in the early 1980s, the worldwide market for material conveyers began to stagnate. The difficulties in this business area manifested themselves in below-capacity production and increasing concentration tendencies. In 1984 the Sparta conveyer was able to achieve growth from the previous year, but this was only achieved due to exports.[79] Unlike in plant design and construction, the USA was an important sales area, in addition to the European markets. The strongest competition in all markets came from 'the Japanese, who have flooded the markets of the Western world with cheap equipment in the last few years'.[80] Particularly frustrating for the European manufacturers was the sealing-off of the Japanese domestic market, which was more than twice the size of the German market. Up until 1990 Linde had only been able to sell one forklift to Japan, despite many years of intense efforts. The desperation about the market pressure in Japan is reflected in a statement by Hans Meinhardt in a 1990 interview: 'The Japanese want to dominate the world; nothing else is important to them. When they say competition, they mean war.'[81] But as in the car industry, the position of European manufacturers relative to Japanese manufacturers improved considerably in the 1990s. By 2000, the Europeans had made improvements in price, quality, and market share and were able to put pressure on their Japanese competitors.[82] At the end of the 1990s, Linde was also able to open up the Japanese market though co-operation with the Japanese manufacturer Komatsu.

Within the company, Linde maintained the friendly competition between the working group Güldner Aschaffenburg and STILL for strategic purposes, while at the same time creating co-operation in the areas of development and production. For example, the same computer-aided-design (CAD) systems were installed in each location.[83] Linde continued to take over competitors, such as the takeover in 1986–91 of the material handling section of Ernst Wagner KG, the leading producer of driverless transport systems. Due to the tough competition and the relatively high prices of Linde forklifts, the company became active in the area of temporary use of forklifts and in 1986 founded a 'Stapler-Leasing GmbH', along with Deutscher Bank (30 per cent share) and the IKB-Leasing GmbH (25 per cent share).

The construction of material conveyers was the most labour intensive branch at Linde. Within Germany this area employed 7,258 people in 1988; around 46 per cent of all jobs and internships were in the area of material handling. Due to acquisitions, the production was heavily networked in the European Union. Just under 20 per cent of the material conveyers sold by Linde in the Federal Republic of Germany were produced in France. Because of capacity bottlenecks, the manufacture of four-wheel stackers (1.6 and 1.75 tonne load capacity) was moved from Aschaffenburg to the newly constructed facility Châtellerault of Fenwick. At the same time, STILL began manufacturing at Asea Truck AB in Sweden; in return Asea took over the sale of the STILL series in Sweden. Another step towards internationalization of the material handling business was the purchase in February 1989 of the material handling division of the Kaye Organization, known as 'Lansing'. Lansing belonged to the largest European material handling equipment manufacturers and had a complete range of storage equipment. The acquired Lansing businesses had around 4,700 employees and achieved sales of over £200 million in the business year 1987–88.[84] After the takeover of Lansing, Linde became one of the leading producers of storage equipment in Europe.

In organizational terms, Linde divided the Material Handling Division into two groups. The Linde-WGA-Gruppe was led by the Werksgruppe Güldner Aschaffenburg (WGA) and included Fenwick-Linde (France) and Lansing-Linda

(Great Britain). The Still-Gruppe was led by Still GmbH of Hamburg and included Wagner Fördertechnik, Reutlingen and the French company Stil & Saxby. Each group had a complete range of counterweight stackers and storage equipment and competed against one another. At the same time, important components such as electrical motors, steering wheels, gears, and axles were produced in one location for the needs of all the projects.

The working group Güldner Aschaffenburg was renamed Linde-FH-Gruppe (for Flurförderzeuge Hydraulik) in 1990. Thus the name Güldner, which had a long tradition in Germany, disappeared from the German market. 1989 was a record year for this working group. The production capacity limits for conveyers and hydraulics were surpassed. The high investments in CNC workstations proved to have been worthwhile; the flexible production centres allowed an efficient transformation into a multi-shift operation. Due to the additional demand from the new German states, the sales and profits in 1990 topped those of the successful year 1989 in the FRG, the most important European market according to volume. In the new states, Linde had 'above-average participation in the market', and successfully positioned itself in sales after the Wende. In reunified Germany, the sales of material conveyers were still 40 per cent higher than in the previous year in the old states. In the area of material conveyers, FH sold 28,000 units and had sales of more than 1.6 billion DM.[85] STILL also was more successful than in the previous year, with double-digit growth rates contracts, sales, and transactions.[86] However many companies in other countries experienced losses, such as the Lansing-Linde and Baker Material Handling Corporation in South Carolina, both of which belong to the FH Group, and STILL Materials Handling in England. Baker Material Handling Corporation was able to achieve profitability in 1992 after many years of loss.

Despite the somewhat unfavourable profit situation in foreign business, Linde continued to expand. The most important purchase was in 1992, when Linde acquired a 51 per cent share in the largest Italian material conveyer producer, FIAT OM Carelli Elavatori S.p.A, Milan. At the time of the purchase, FIAT OM was the leader in the Italian market for forklifts and storage equipment, and sold around 14,000 vehicles each year in Europe, North Africa, and the Near East, under the trade name OM PIMESPO. Pimespo Carelli Elevatori S.p.A. had merged with FIAT OM the previous year. Linde increased its shares in the following years and acquired the remaining 25 per cent of shares from IVECO, Turino, at the beginning of 2003. Like Still and Aschaffenburg, FIAT OM maintained an independent status in the division. Thus, according the multi-brand strategy, there were now three groups – Linde-RH, STILL and FIAT OM – which each had a technically independent range of products.[87]

In Eastern Europe, Linde began preparations in 1992 for the founding of marketing companies in the Czech Republic and Hungary. In 1993 Linde collaborated with Jungheinrich AG to found the JULI motorworks k.s. in Brno, Czech Republic, which was to produce electric motors for material conveyers. Production began in 1994; after two years 100,000 electric motors had already been produced. With the 1997 takeover of JIPO Domoradice in Cesky Krumlov,

Czech Republic, Linde acquired a factory for the manufacture of parts for forklifts in a low-wage area.

In the 1990s Linde made large investments in an attempt to enter the growing Asian markets and founded Linde-Xiamen Staplers Ltd, along with a Chinese partner.[88] In 1996 a new production site was put into operation, at which 15,000 material conveyers could be produced per year. The co-operation experienced major difficulties in the early phases. At one point Full and Schling, the Linde directors in charge of this project, had to interfere directly in China and afterwards complained to the Executive Board about a 'leadership style strongly influenced by the state' and about the high fixed costs of production.[89] In 1999 Linde was able to take over all shares of the Linde-Xiamen Gabelstaplergesellschaft and improve efficiency. Falko Schling reported to the board that 'a relief about the new, final situation could be felt' at the Chinese plant.[90]

In 1996, Linde tried to gain access to the protected markets in Japan through the founding of a marketing company for material conveyers in Yokohama; actual access, however, only became possible through the co-operation with Komatsu in 2000. Additional marketing companies were opened in other Asian countries through Linde Material Handling Asia Pacifc Pte. Ltd in Singapore. In the following year STILL followed FH and opened its own marketing company in Singapore.

The area of material handling had achieved successful growth, but now had difficulties with the initial strategy of having each brand compete with the others in the full range of product offerings. This prevented additional gains in efficiency, and the Executive Board decided to alter the strategy in 1999. Linde conceived a different plan for the brand name FIAT OM: the group was to 'stop production of small-batch and niche products and replace them with units from the other branches of the company'. The problems that caused this change in strategy can be deduced from the claim that FIAT OM would achieve 'lower costs, stable quality and shorter delivery times'.[91] A similar assessment of the other subsidiaries in the forklift business was soon to be made.

Refrigeration: success in supermarkets

On 5 November 1981, two years after the closing of the international offices for industrial refrigeration, the Executive Board decided on further restructuring in the working group for refrigeration technology.[92] In the area of industrial refrigeration, future areas of focus were to be refrigerators for breweries, process engineering equipment, cold stores, and marine refrigeration equipment. Refrigeration Technology was to follow the lead of Plant Engineering and place more focus on engineering services. But Linde was not able to get out of the red in this area. Even reunification did not change the position of the industrial refrigeration branch. At the end of the 1990s, the size of this area shrank commensurate with its decreasing significance; 60 employees were reduced to 20.[93]

The situation was much better in the area of commercial refrigeration, which could successfully adapt to new market factors. The united European internal market of 1992 as well as the concentration tendencies of large discounters led to fewer and fewer buyers with greater and greater market power. In the field of store installations, there was a trend in two directions: to markets with expensive refrigerators for 'experience shopping' and to markets with simple, functional equipment for 'basic shopping'. A general long-term trend was a steady increase in the amount of fish and frozen foods, which required comprehensive installations of new chest and upright deep-freezers, so that customers could find a similar range of products everywhere. Linde made direct sales not only to the large grocery chains, but also to grocery producers, who in many cases delivered refrigerators along with their products and assumed responsibility for maintaining and stocking them.[94] In order to adapt to the new structural changes on the part of the buyers, Linde divided the operating company into five operating regions. Linde also had to be flexible in its product offerings because even its wide range of serially produced products often did not meet the needs of the store chains. For example, a special counter style was developed for France.

Since the mid-1990s the Schwelm location had produced refrigeration and freezer units as well as store fittings at capacity. In order to prevent capacity bottlenecks, Linde built a new factory in Bad Hersfeld in 1987 and bought back a refrigerator factory in Mainz-Kostheim from Duofrost Kuehl- und Gefriergeräte GmbH, a subsidiary of AEG. This factory had a surface area of around 95,000 m^2 and sold for around 32,5 million DM.[95] However, narrow profit margins remained a problem in the growing business. But the political change in Eastern Europe created interesting opportunities. In 1992 Linde and a partner in Beroun founded Linde Frigera for the production of ready-made refrigeration appliances.

In the early 1990s Linde used its expertise in refrigeration technology to develop refrigerators and commercial refrigeration equipment that should not cause environmental damage. In 1993 Linde installed the first HCFC-free refrigeration equipment in a German store, and ended the use of chlorofluorocarbons as a propellant for the production of polyurethane foams, using a propellant mixture that only had a small potential for damaging the ozone layer and promoting the greenhouse effect. This was done by all its competitors, but Linde claimed to have a lead after switching over to pentamethylen as a propellant in 1995.[96] This was five years before the use of ecologically harmful propellants was finally outlawed.

Linde entered the Swiss sales market in 1996 with the takeover of Frigorex AG of Luzern from Sulzer AG in Winterthur. Frigorex specialized in the area of commercial refrigeration technology. Linde also acquired the remaining shares of ZEHAG Kälte+Klima AG in Buchs, as well as EQUIPE FROID in La Conversion, where the company had owned shares since 1991. In Eastern Europe Linde was active in the Czech Republic, Hungary, and Poland. It founded the marketing company Line Polska Urzadzenia y'Systemy Chodnicze in Warsaw in 1996, because Western retail chains were investing in Poland 'at high levels'.[97] In 1997 Linde also began concentrating heavily on the Western European markets and acquired Radford Retail Systems Ltd in Great Britain. With this purchase, Linde became the market leader in commercial refrigeration in the United Kingdom. The initial plans to produce specialized equipment for the British market at that site had to be changed, probably due to financial considerations. Three years later, the factory at Chew Stoke was closed and Great Britain's supermarkets were supplied with serially produced equipment from the continent.[98] In 1997 Linde also strengthened its market position in Southeastern Europe through the acquisition of Frigel Apostolou, a Greek refrigeration technology firm. In the Western European market in 1998, Linde took over a majority of the Chief Group, one of the leading marketing companies for commercial refrigeration appliances in France, Great Britain, Belgium, and the Netherlands, and also bought the remaining 20 per cent of the well-established Italian family-owned firm Criosbanc. For these many takeovers, Linde developed its own evaluation formula for business acquisitions.[99] In the Far East, Linde began a joint venture in 1994 for the installation of refrigerators in Thailand, and in 1996 Linde founded Linde Refrigeration Philippines Inc., as a joint venture with a partner in the Philippines. In the following years many contracts were made in Thailand and the Philippines. The 1998 purchase of a 75 per cent share of Seral do Brasil, one of the leading suppliers of commercial refrigerators, store construction, and checkout systems, proved to be a strategic investment for opening up the Latin American markets because the business environment there was difficult for a while.[100] In the 1990s Linde made large investments in trying to make commercial refrigeration a profitable branch. The investments in 1999 were almost twice as high as the before-tax profits. Through its acquisition plan, Linde was able to become the European market leader, but was never able to make the branch as profitable financially as the other divisions.

Linde at the end of the 1990s

After a generally profitable decade, during which Linde was sometimes even able to achieve sales and returns records despite the economic trend, the company was very healthy and confident in 1999, even though it had not been able to profit from the stock market fever of the late 1990s. 'Mister Linde' Hans Meinhardt was on the Supervisory Board in 1997, but not yet in retirement; Gerhard Full ran the company just as Meinhardt would have done. The long sought 'fifth leg' was not found; the consolidation and profit-oriented optimization of the other four branches had priority.

Linde had already begun its transformation from an export-oriented company to a global player. The foreign percentage of sales already amounted to more than two-thirds (69.2 per cent); sale proceeds reached over 6 billion DM. Linde bought gas companies all over the world in rapid succession; in plant design and construction Linde was able to achieve a competitive basis in the USA in the long term through business acquisitions. The reclaiming of the name 'Linde' in the USA in 1999 carried significant symbolic meaning for the new strength of the company.

Business did not develop as successfully as expected in the successor countries of the Eastern Bloc, which had formerly been dominated by the Soviet Union. In particular, plant design and construction was hindered for several years, due to economic and political turmoil. In the area of gas engineering, however, Linde was able to buy companies formerly owned by the government in Eastern European states, and thus was able to achieve a favourable market position. In the new German states, Linde built up structures for all of its branches. In the area of gas engineering, the company was able to create a comprehensive supply grid, which it had not been able to do since the end of the war. Linde was one of the biggest investors in the Leuna location, which was able to supply industrial areas up to 80 km away through pipelines. In the area of material handling, Linde continued its long-standing multi-brand strategy in the 1990s, which was an 'internal competition' model. The resulting high costs and relatively low productivity in small batch production began to be a problem. For this reason, in 1999 Linde began to talk in terms of a desired 'bundling of different core areas of competence'.[101]

In the area of commercial refrigeration, Linde was the clear market leader in Europe and improved its position in Asia and Latin America, but without earning much money. In general, by 1999 Linde was following a successful business plan in the areas of gas engineering and plant design and construction, and had good prospects if large-volume investments would increase market access. In the areas of material handling and refrigeration technology, it became clear that although leading the market in Europe can be gratifying, production costs are ultimately the deciding factor.

7
Linde in the Twenty-First Century

AGA – A global Swedish group is acquired by Linde

At the turn of the millennium Linde acquired AGA, its Swedish competitor in the gas business. It was the most important acquisition in the company's history. As in its early days, the global market for industrial gases was still an oligopoly, with the seven largest producers supplying 70 per cent of the world market. As of 1998, the truly dominant companies, however, were Air Liquide (with 17 per cent market share), BOC and Praxair (each with 14 per cent), and Air Products (9 per cent). AGA (6 per cent), Linde (5 per cent), and Messer (5 per cent) were the smaller players among the big seven. In the European gas market, too, Air Liquide had the largest market share (26 per cent), followed by Linde (13 per cent) and AGA and Messer (each with 11 per cent); BOC and Air Products each had 10 per cent market share, and Praxair 5 per cent. In the North and South American markets, Praxair had the strongest presence with 31 per cent, followed by Air Products with 15 per cent, Air Liquide with 14 per cent, and BOC with 11 per cent. Here, AGA (6 per cent), Messer (3 per cent), and Linde (2 per cent) ranked among the next-smallest suppliers. These international market structures had developed gradually over many years.

In late May and early June 1999, these market structures shifted dramatically by means of mergers. Air Liquide acquired 50 per cent of BOC's shares and thus controlled 24 per cent of the global market. Air Products bought the remaining 50 per cent of BOC's shares and thus advanced to the unequivocal number two spot in the global market, followed closely by Praxair (14 per cent). These mergers significantly widened the gap between the sales volumes of the major and minor players in the global gas business. In order to compete successfully in the future, Linde now felt forced into a merger as well. For decades, profits from the thriving gas business had also financed the build-up of Linde's industrial truck business segment. Now, that had to change. 'In order to grow the gas business', Gerhard Full wrote in his review of the year 2002, 'the acquisition was vital'.[1] He was referring to the spectacular buy-out of the Swedish company Aktiebolag Gasaccumulator, or AGA for short, in December 1999.

On 16 August Linde bought AGA from Martin Ebner and two large Swedish banks and made an open offer to buy outstanding Linde shares. This surprising takeover had been preceded by months of haggling over the acquisition of Messer

Griesheim and AGA. For quite a while it looked as though Linde might even be able to acquire both companies. Linde had already announced a public takeover bid for AGA shares in August 1999. But the acquisition of Messer stalled. The shareholders of Messer Griesheim (the Messer family with 33.3 per cent, Hoechst-Aventis with 66.3 per cent) and Linde terminated the talks on 10 February, a day after the EU Commission made clear that it would approve a merger only under unattractively tough conditions (Linde would have had to give up Messer Griesheim's German activities due to its pre-eminent market position).

AGA was particularly strong in Scandinavia, the United States, and Latin America – that is, in markets where Linde was more weakly represented. A merger thus offered the opportunity to take over well-developed markets in which AGA had consistently ranked among the three largest gas suppliers. The other activities of the two companies also complemented each other, especially AGA's medical gas division and Linde's on-site business. In addition, AGA had a healthy structure following the restructuring and efficiency measures of the preceding years, so that the merger laid a good foundation for further growth. AGA was bigger and more internationally situated in the gas business, but Linde was larger overall. The strength of Linde's strategy of developing multiple business segments now became evident. They provided the mass for the takeover deal. At the end of 1999, Linde employed about 35,600 workers; at AGA it was 9,500. Together, Linde and AGA covered 11 per cent of the global market for industrial gases. With the merger, Linde would become the fourth largest supplier of industrial gases worldwide and the second largest in Europe.

The takeover at the turn of the millennium was fraught with symbolism. Linde's costly recommitment to the gas business was also a crucial decision that set the course for the company's further growth in the coming decades. The €3.4 billion acquisition of AGA represented the largest takeover in Linde's company history. Compared to other big mergers of the 1990s, the two companies blended very successfully on the whole, despite the initially sceptical reaction of the financial markets. AGA and Linde were also a good match due to the similar histories of their founding and growth. The history of AGA quickly became part of the joint corporate identity.

The company that would later become AGA was founded in 1904 as Gasaccumulator AB. Like Alfa-Laval, LM Ericsson, and the Swedish ball bearing manufacturer SKF, AGA belonged to those Swedish companies, founded circa 1900 by inventor-entrepreneurs, that were leaders in the global market even before the First World War. AGA began as a applications developer for acetylene gas, which was then relatively new. Gustaf Dalén's inventions for lighthouses launched the creative and versatile development of technologies in a variety of fields. Gustaf Dalén (1869–1937), AGA's most influential director (from 1909), was born into a family of farmers in Västergotland. He began by running his parents' farm but soon showed his dynamic disposition by adding a seed store and a dairy. In 1891 he invented a tester for milk fat, which so impressed Gustav de Laval (an important Swedish inventor of the era) that he encouraged Dalén to develop his talent and pursue a university education. Dalén took his advice, sold the farm, and matriculated at the Chalmers Technikschule in Gothenburg, where he received his diploma in 1896. This was followed by a year at the famous Swiss Federal Institute of Technology and his first job as engineer at Laval's steam turbine factory in Stockholm. He began working with applications of acetylene gas on the side and soon founded his own company, Dalén & Celsing, to manufacture acetylene lighting systems.[2] In 1901 Dalén became chief engineer at Svenska Carbid & Acetylen AB in Stockholm, which was owned by the Swedish businessman Axel Nordvall. This company was renamed AB Gasaccumulator in 1904. Its main speciality was lighting systems for coaches, railroad cars, and buildings. In the years after 1904, Dalén realized his most important inventions: in 1905 he developed a flash apparatus for lighthouses and buoys that used only a tenth of the gas of conventional ones, and that also allowed for various flash characteristics.[3] In 1906 came a compound for storing acetylene in gas cylinders that reduced the danger of explosion, and in 1907 a 'sun valve', which automatically turned the lights in beacons and buoys on and off in response to daylight. In order to generate especially bright light, acetylene must be mixed with air in a 1:10 ratio before burning. In 1909 Dalén developed a device that produced this highly explosive mixture automatically and safely, the 'Dalén mixer'. Together, these four inventions of Dalén's – the flash apparatus, the AGA compound, the sun valve, and the Dalén mixer – constituted the 'AGA system',[4] the basis of AGA's economic success.

The Nobel Prize in Physics that he received in 1912 recognized the applied significance of these inventions.[5] Reliable, economical, and automated lighthouse systems were needed in every country with a coastline. The rapid spread of the AGA system was primarily the work of Axel Nordvall, the sales director of AB Gasaccumulator, who brought back major contracts from his travels (such as the beacon system for the Panama Canal in 1911) but recognized Dalén's leadership role in the company. Dalén held the position of managing director starting in 1909.

The company's successful growth was increasingly cramped by its too-small production facilities in Stockholm, and AGA thus relocated in 1912 to the island of Lidingö, outside the city. There, Dalén injured himself so severely during an experiment to improve the AGA compound that he was left permanently blind. But with admirable strength of will, he managed to overcome this handicap sufficiently to continue leading the growing company until his death in 1937.

Gustaf Dalén was a successful entrepreneur and engineer, but he was less concerned about earning money than about establishing a company that functioned in an exemplary manner. Since 1912, when the Handelsbank Stockholm bought out the bulk of AGA's shares, the company led by the blind Nobel Prize winner had enjoyed a special status as a model of Swedish industrial culture, perhaps comparable to that of Volkswagen in Germany after the Second World War. The Lidingö site represented what Sweden's industrial future should one day look like. Admiral Arvid Lindman was appointed chairman of the Supervisory Board; he was the Swedish prime minister from 1906–11 and again from 1928–30. He kept his position on the board until his death in 1936.[6] In 1912 Nils Westberg and Rolf von Heidenstam joined the company; after Axel Nordvall's departure in 1916, the two men took over the sales of AGA products, with Westberg responsible for central and Eastern Europe and von Heidenstam for all other countries worldwide. The First World War cut AGA off from its German and North American subsidiaries which until then had supplied mirrors and lenses for beacons and searchlights. In response, AGA bought a Swedish glass manufacturer and built its own glass grinding facilities on Lidingö, the basis for its later opto-electronic developments. In 1915 there were AGA companies in seven European countries, as well as in the USA and Brazil.[7] But the war crippled international trade, and from this point forward AGA concentrated more closely on its domestic market.[8]

AGA diversified strongly during the inter-war years. In addition to beacons and buoys, it produced signal systems, welding equipment,[9] radiators, radios,[10] large film projectors, and automobiles.[11] AGA's manufacturing programme grew organically from Dalén's original inventions; the technology for beacons and buoys was developed further and used initially for rail and road traffic signals, and later for aviation signals as well.[12] Automatic welders developed in-house at AGA laid the groundwork for manufacturing welded steel radiators from 1924 onward – a line of business that remained profitable for decades. Gustaf Dalén's last invention with practical applications was the AGA cooker in 1929, an especially effective coal cooking stove that could burn 24 hours without being tended and that brought AGA into many Swedish households. It is produced to this day in England.[13] Lighthouse equipment sales dropped sharply from 1931 to 1933 due to the worldwide depression,[14] so that it became necessary to develop new markets; among these were radio buoys, short-wave transmitters, and medical devices. After Gustav Dalén's death in 1937, Rolf von Heidenstam, who had been responsible for AGA's international business since 1912, became president. The Second World War cut AGA off once again from its foreign suppliers and customers; the lion's share of the production capacity on Lidingö was used for Swedish military purposes. In 1944, only 5 per cent of AGA's products were exported.[15] From the late 1930s until the 1960s, AGA supplied instruments such as gyro compasses, artificial horizons, and bomb sights to the Swedish air force.[16] The governmental defence contracts allowed AGA to establish itself in new areas of business.[17] One more product line was added in 1947 with the battery producer Tudor, which was put up for sale due to legal conditions imposed on it as a company that had previously been under German ownership. In 1950 Gustaf Dalén's eldest son, Gunnar Dalén, became the third president of AGA and led the company until 1967. Comparable to Linde, the influence of the Dalén family in the company was not the result of its share ownership but its technological, scientific, and managerial capacities.

For AGA's many products apart from the gas business, the Scandanavian market consistently remained the most important until the 1960s, while the gas business was always highly international. For instance, AGA's electronic goods – the radio equipment and flight instruments delivered to the military, as well as its radio and television sets – were sold mainly in Sweden. This synergy was lost bit by bit during the 1960s. The gradual opening of the Swedish market made domestic production of labour-intensive electronic devices with a high vertical range of manufacture unprofitable for the relatively small Swedish sales market. Defence technology was more strongly dominated by international manufacturers, particularly North American ones, while consumer electronics were crowded out by inexpensive Japanese imports.[18] Innovative products – such as distance measurement with the geodimeter in 1953, the world's first heart-lung machine in 1954, and touchless temperature measurement with the AGA Thermovision in 1965 – guaranteed the electronics division an edge in the market again and again.[19] In 1967, AGA established AGA Innovation, a central research and development department intended to be a 'creative playground'. As an 'advanced complement' to the existing research and development departments in the respective divisions, its task was to explore projects (including quite unconventional ones) on behalf of the individual business units and to develop them to the point of marketability.[20] AGA was also engaged in environmental protection at a relatively early date.[21] But in the long run, its high-technology efforts would lead into a cul-de-sac.[22] Starting in 1970, its two smaller divisions – Electronics and Medical Technology – operated in the red, while the three large divisions – Gas and Welding Technology, Batteries, and Radiators – were (still) profitable.[23] AGA's mass production of flat radiators had become so efficient that it was able to compete very successfully in the market until the 1970s.[24] In 1983, AGA offloaded its last industrial activities. President Marcus Storch, Gunnar Dalén's successor, announced in the 1983 annual report: 'We regard AGA today as a chemicals company with gas operations as its dominant product area.'[25] At this point, gas and applications engineering constituted 75 per cent of the business volume; Frigoscandia accounted for the other 25 per cent. The acquisition of Frigoscandia in 1977 was intended to let AGA enter the market for the storage and transport of frozen foods. Following initially good business development, Frigoscandia suffered earnings setbacks with the reduction of the EU-subsidized 'butter mountain' in refrigerated warehouses. The purchase of the steel, tool, and energy group Uddeholm in the mid-1980s was justified mainly by the company's improved liquidity, for the 'treasure chest of Värmland' had almost 200 million US$ available in liquid assets.[26] But this acquisition could not stave off the long-term transformation of AGA into a pure gas company. In 1994 Frigoscandia went public on the Stockholm stock exchange; its stock was divided among AGA's shareholders. This was the last step toward AGA becoming exclusively a gas company.[27] Its diversification during the 1950s is quite comparable to Linde's development. In both companies it was the gas division that provided secure economic backing for the technological playground, though at Linde this was supplemented by its very successful plant engineering. AGA's concentration on the gas business since the 1980s allowed its returns to outstrip Linde's. But Linde AG, with its broader holdings, saw stronger sales growth and was able to play off its size in 1999.

The growth of AGA's gas business

Industrial gases belonged to AGA's manufacturing programme from its beginnings. In its first decade, welding technology was included in the production programme. In 1902 Gustaf Dalén demonstrated gas welding for the first time in Sweden at the Finnboda shipyard in Stockholm. Starting in 1906, AGA used an acetylene welding process to produce buoys. After 1910, welding and cutting with acetylene and oxygen steadily grew more important for industrial processes; the demand for these two gases rose accordingly. In 1914 AGA in Lidingö began producing oxygen for use in welding.[28] With the First World War, petroleum supply became problematic in Sweden, which turned out to be beneficial to the carbide and acetylene business, since petroleum lamps were often replaced by carbide lamps.

Beginning in the 1930s, oxygen was used increasingly in hospitals, generally as a mixed gas together with laughing gas and 'Carbogen' (oxygen with about 5 per cent carbon dioxide), to treat respiratory disease, to anaesthetize, and to alleviate pain. AGA also developed the medical devices that used these gases: respirators, incubators, anaesthesia equipment, oxygen therapy equipment, and heart-lung machines.[29] The Second World War brought growth in acetylene purchases and the development of new application areas for the gas. The defence industry became a quantity buyer of argon for welding. From the 1940s onward, a new market emerged with the use of oxygen in steel plants. In order to supply large customers via tank trucks, in 1951 AGA built in Lidingö the first facility in Sweden for producing liquid oxygen; it was modelled on a German plant that made fuel for the V2 rocket.[30] Despite these initial hesitant steps, AGA remained an 'acetylene company' until the early 1970s.[31] An alliance with L'Air Liquide begun in 1971 mainly affected AGA's subsidiaries in Germany and the Benelux. This partnership led to an expansion of the German AGA's product range. In addition to acetylene, it produced and sold liquid atmospheric gases, inert gases for welding, and speciality gases. Gas production at AGA increased sixfold between 1970 and 1980; from that year on, the company produced its entire supply of commercial gases in-house and no longer had to rely on external vendors.[32] In 1986 AGA acquired the Rommenhöller group, which specialized in carbon dioxide gas. In 1999 AGA was a highly specialized gas company that produced and supplied industrial gases for the food processing industry, medicine, welding, environmental protection, metallurgical processes, process chemistry, and the production of electronic components.

Corporate strategies

Like Linde, AGA had its eye on the international market from the start. Lighthouses and buoys were sold in Europe and Latin America. In order to ensure a stable supply of acetylene for these devices, AGA set up acetylene gas factories in many countries, which later also supplied other customers with dissolved acetylene.[33] This established the company's essential structure in the gas area: relatively autonomous subsidiaries in Europe and Latin America produced acetylene gas in cylinders and sold it regionally. This decentralized organization[34] existed until 1996.[35]

Figure 7.1 International business: AGA as the world's leading maker of beacon buoys. Rio de Janeiro 1915

AGA's broadly diversified production programme with a large vertical range of manufacture similarly grew out of the company's history.[36] For a long time, the breadth of its production spectrum and its international orientation proved advantageous in smoothing out the fluctuations of its various markets.[37] But starting in the late 1960s, when AGA's industrial product lines achieved unsatisfactory results in the relatively small Swedish market and dynamic developments in the industrial gas market became apparent, a change of course became inevitable. The group policy formulated in 1968 was meant to ensure that all AGA divisions would operate profitably in the future. Particularly in the gas area, AGA was to become an aggressive growth company in the 1970s, with a small number of larger plants.[38]

AGA was traditionally set up to produce bottled gases for small and medium-sized customers.[39] In the long run, this led to a poor strategic decision. Starting in the 1940s and especially in the 1950s, there was rapid growth in the so-called tonnage market, the production of large quantities of atmospheric gases delivered by pipeline to industrial customers. AGA overslept this important trend and later had trouble catching up.[40] Starting in 1970, managing director Sven Ågrup initiated a restructuring of the group to orient it more toward the gas business. In 1971, with the goal of expanding quickly in the European market, AGA entered an alliance in the industrial gas area with L'Air Liquide S.A. in the Benelux countries and Germany; 50 per cent of the German AGA-Gas GmbH shares were handed over to Air Liquide, partly in exchange for cash, and partly in an exchange of shares.[41] The 1970s saw an accelerated expansion of AGA's gas division's production capacity and a strengthening of its market position. The era of heterogeneous products coexisting peacefully within the corporate group was finally about to end. In the middle of the decade, long-term strategic decisions were made that essentially remained in effect until the merger with Linde in 1999. In the future, the AGA group's activities were to concentrate on business segments in which AGA had sufficient competence to enable expansion in international markets.[42] In the future, return on total capital was to be consistent at 15 per cent, and two-thirds of the earnings per share would be retained within the company to build up capital reserves for continuous expansion. At that point the gas business accounted for less than 50 per cent of total sales. It was correctly understood at AGA that the international gas market had the greatest long-term potential for profitable growth,[43] and the decision was thus made to concentrate the company's strength here. This brought the market situation into the spotlight: starting in 1977, each of AGA's company reports contained a description of its competitors. In that same year, AGA bought Burdox, the seventh-largest North American gas producer, in order to enter the world's largest gas market.[44]

The gas market has certain peculiarities that distinguish it from industrial markets. The high capital requirements for building new air separation plants act as a barrier to the market entry of new competitors. For decades, a relatively small number of companies has dominated the global gas market; the eight largest ones account for more than 75 per cent of the total sales volume.[45] Production methods and the products offered vary only slightly. Despite their international orientation, gas companies must operate mostly on the local level.[46] All large gas companies can rely on a strong domestic market. The relatively small Swedish market in which AGA operated constituted something of an exception; its Latin American markets compensated for this.[47] The production of acetylene from calcium carbide was technically simple, so it required only low investments. The high costs in this process were due primarily to the raw materials. In contrast to this, production of atmospheric gases entails no outlay for raw material, but the capital costs for building such plants are immense.[48] Air separation plants are thus particularly subject to economies of scale, because their high capital costs force their operators to achieve optimal plant utilization, and thus to tap additional clienteles beyond large industrial customers. Another adverse factor is that the composition of the atmosphere by per cent does not match production requirements. The demand is primarily for oxygen and, to a lesser extent, argon; but the atmosphere comprises 80 per cent nitrogen, only 20 per cent oxygen, and less than 1 per cent argon. The large quantities of nitrogen make it necessary to develop applications that will reduce production costs.[49]

AGA's first air separation plant went online only in 1980. Two more plants were built in the following year in co-operation with the French L'Air Liquide group. Despite fierce competition, AGA rapidly became a contender in the air separation market.[50] The increasing importance of the gas business was also evident in the appointment of the head of the gas division, Marcus Storch, as AGA's new president in 1980.[51] In 1981, AGA became the fifth-largest gas producer worldwide and held this position until its merger with Linde. The political changes in Eastern Europe after 1989 allowed AGA to return to markets where it had been present until 1945: Hungary, East Germany, Estonia, Latvia, Lithuania, the Czech Republic, Slovakia, Poland, Russia, and Rumania. Starting in 1997, Lennart Selander led the gas group as president and CEO. AGA's further business development after 1997 was less favourable.[52] The 1998 results were unsatisfactory; the lion's share of extensive investments in Russia and the Ukraine had to be written off.[53] AGA gave up its involvement in industrial gases in Great Britain in 1999, because it had been unable to build up a strong position in the gas business there in ten years. Comprehensive efficiency measures improved the cost structure at AGA.[54] After a 15-year phase of 'aggressive' investments,[55] restructuring, and rationalization, AGA was actually now looking forward to an uneventful epoch: 'We are now entering a period of comparatively low investment requirements and with a strong positive cash flow.'[56] But things were to turn out differently.

Co-operation with Linde before the takeover

Linde's early history resembles that of AGA: on the basis of his own inventions, an extraordinarily creative engineer and scientist established a globally active industrial corporation circa 1900. In Linde's case, the basic inventions remained dominant in the later course of the company's history. At AGA, the gas business was initially established to supply the lighthouses it built throughout the world; only the displacement of acetylene light by electric lighthouse systems made it necessary to develop new markets. Linde developed and produced air separation plants for industrial gases starting in 1913; plant engineering dominated, while the gas business played a subordinate role for a long time.

Both companies expanded their business segments into fields of mechanical engineering. At AGA, vehicle manufacturing remained only a short episode during the 1920s, while the tractor and diesel motor segment lasted from 1938 to 1969 at Linde. AGA, especially its subsidiary AGA Baltic, developed new business segments from the 1920s into the early 1960s with opto-electronic equipment (film projectors, aviation instruments) as well as with home electronics (radio and television sets) and medical equipment. Its main sales area for these products was always the relatively small Swedish market. Only its heating division's products (flat radiators, valves) – and of course its industrial gases – were also successful abroad. By contrast, apart from its Güldner tractors, Linde did not attempt to enter mass production of consumer goods until relatively late. Small compressors came out of commercial refrigeration, and commercial refrigerator construction served as the basis for producing household refrigerators and freezers. But Linde was involved in the manufacture of 'white goods' for a mere two decades.

AGA's and Linde's company histories also show parallels in their continuity. Over nearly a century, the destiny of both companies was determined by only a relatively small number of long-term leaders. In both cases, company policy can be described as tending toward conservatism, whereby AGA after 1970 devoted itself exclusively to the gas business from 1983 onward. In contrast, Linde, in addition to selling industrial gases, was able to expand its traditionally strong market position in refrigeration technology, plant engineering, and material handling over the previous decades and finally turned out to be the bigger and financially more powerful company.

Until a few years ago, Linde's gas business focused mainly on Germany and its neighbouring, German-speaking countries. Here, one of Linde's competitors was AGA's German subsidiary, AGA Gas GmbH, which for many years was involved primarily with the production and sale of acetylene.[57] The German subsidiary AGA Gas resulted from the merger of two companies in which AGA had been the majority shareholder for several decades. The Hanseatische Acetylen-Gas-Industrie-Gesellschaft m.b.H. and the Autogen-Werke in Berlin (the latter founded in 1906) were renamed Autogen-Gasaccumulator AG (AGA) after the takeover by the Swedish parent company. 'While the Hanseaten devote[d] themselves exclusively to acetylene and put a new plant on stream in Wilhelmsburg in 1916, the Berliners [were] involved mainly with the production of welding equipment'[58] but also produced a small automobile after the First World War in the Lichtenberg district of Berlin. The acetylene business at Berlin's AGA grew rapidly; the first plant was commissioned in 1916 near Breslau. Additional production facilities were built near Chemnitz and Königsberg. In 1926, an acetylene plant in Gelsenkirchen began production. In 1928, one plant was added in Dortmund and another in the Adlershof district of Berlin. In 1929 came a plant on the Saar River and one in Duisburg; in 1930, a plant was established near Leipzig. Additional plants were leased near Zittau and in the Vogtland in 1936. By 1929, the Hanseatische Acetylen-Gasindustrie AG had built plants in Hanover, Cassel, Rostock, Neu-Isenburg (near Frankfurt am Main), Ludwigshafen, and Kiel. During the Second World War, the German AGA subsidiary built plants in Bromberg (Silesia) and Bützow (Mecklenburg). Since foreign exchange laws in National Socialist Germany now prohibited the transfer of profits abroad, in 1939 the German AGA bought a site for the construction of a new administration building in the Zehlendorf district of Berlin. At the end of the war, the plants in the areas governed by Poland and the Soviet Union were lost, while the factories in the Soviet occupied zone of Germany were first sequestrated and only later expropriated. In 1953, the German AGA was transformed into a GmbH (a limited liability company). Both companies continued to produce acetylene gas exclusively, which was for the most part delivered in returnable cylinders. Only at its shipyard locations did the Hanseatic AGA also maintain facilities to supply large customers by pipeline.

Prior to AGA's entry into air separation technology, there was no overlap of business segments between AGA and Linde where plant construction was concerned. Although AGA did design and build its own plants, it bought the equipment from other manufacturers. Linde and AGA had enjoyed cordial business relationships since even before the First World War, and AGA gave Linde preference in awarding subcontracts; however, there were no exclusive agreements.[59] On the other hand, Linde utilized the AGA compound developed by Gustav Dalén to fill explosive acetylene cylinders.

Even the Second World War, during which the other European countries experienced Germany mainly as an aggressive military power, did not harm the good relationship between AGA and Linde. As early as October 1947, AGA director Gunnar Dalén asked Friedrich Linde whether Linde would be able to deliver oxygen installations.[60] Since AGA did not build plants itself and thus did not come into question as a competitor, AGA engineers were allowed to intensively inspect Linde's plant facilities and operating gas production stations.[61] Between 1949 and 1999, Linde delivered a total of 45 air separation plants to AGA. In addition, the two companies informed one another when potential competitors planned to build oxygen installations.[62] Yet, throughout this the business integrity of the two companies was constantly upheld.[63] Contacts existed on a personal level between high-ranking personnel in both companies.[64] Upon AGA's 50-year jubilee, the plant engineering group in Höllriegelskreuth sent a vase to Lidingö in the name of the Executive Board. It was displayed there in the conference room and, in Gunnar Dalén's words, it was supposed to 'constantly remind' the AGA management 'of the many years of good collaboration with your esteemed house, a collaboration that we very greatly appreciate'.[65]

A joint venture of AGA and Linde was the founding of the TEGA subsidiary in Austria in 1958 to produce and sell industrial gases. In Steyr, the centre of the Austrian steel industry, Linde built a large air separation plant in 1974, in which AGA and Linde each had a 50 per cent stake.[66] In 1974, AGA and Linde launched a joint venture to build and operate an air separation plant for the COSIGUA steelworks in Santa Cruz near Rio de Janeiro. With a capacity of 650 tons of oxygen and nitrogen per day, the plant went online in 1976 and represented the largest plant operated by a gas company on the South American continent at that time.[67] To design, build, and operate this plant, Linde founded a separate company, Aeroton Gases Industriais Ltda, in which AGA had a 50 per cent stake. In addition, a joint sales company was founded, into which AGA S.A. integrated its business activities in the Rio de Janeiro area. Here, too, each partner corporation held half of the shares; the joint sales company offered AGA's product range.

Takeover

Under pressure from the break-up of BOC and its subsequent acquisition by Air Liquide and Praxair, Linde was now prepared to enter into negotiations with the Swiss financier Martin Ebner, who had bought about 40 per cent of AGA's shares just a few months earlier.[68] In doing this, he made use of the BZ Bank of Zurich (later Wilen), which he had founded in 1985 and which grew into one of the world's largest investment banks after the mid-1990s, as well as the investment funds in the BZ Group, including Gas Vision AG.[69] Especially in the late 1990s, Ebner had very good Swedish contacts.[70] Ebner sold his stake in AGA to Linde on 16 August 1999. The sale was only completed after the final permission of the European Commission in February 2000. At the time of the purchase, the market valued Linde at 8.8 billion Swiss francs, and AGA at 5.6 billion Swiss francs. The Deutsche Bank, which also served as a consultant on this merger, provided the €3.5 billion needed as a bridging loan. The successful takeover of AGA by Linde at that time surprised the employees of both companies as well as the market. After all, Linde had been expected instead to take over its German competitor, Messer-Griesheim.[71] On 16 August 1999, the minority shareholders were tendered a buy-out offer of 141 Skr per share. On 8 September 1999, the AGA Supervisory Board issued a statement on Linde's public bid describing the merger as an 'industrially sound solution'.[72]

For AGA, the merger meant the end of autonomy after nearly 100 years. A week after Linde and AGA's joint press conference, AGA's fifth president, Lennart Selander, contacted AGA's employees and expressed understanding for the sorrow many of them felt at AGA losing its independence after almost a century. But he emphasized that the merger of the two companies in the gas business with its difficult conditions was beneficial to both, and that it guaranteed continuity for the employees. Moreover, the corporate cultures of Linde and AGA would harmonize especially well; in this respect, he was glad that Linde was the purchaser, Selander said.[73] At Linde's express wish, Selander joined the board of management of the newly formed gas group at Linde, which consisted of Gunnar Eggendorfer, Rainer Goedl, and Volker Metzler, along with CEO Gerhard Full. However, some AGA employees – especially in middle management – saw no future for themselves at Linde and resigned.[74] In addition, cost-cutting was, of course, one reason for the merger. A total of 1,500 employees became redundant through the Linde–AGA merger. Among them were 400 of the 700 positions at AGA in Germany, where Linde already had a well-developed infrastructure for producing and distributing industrial gases.[75]

In light of a certain accumulation of takeovers of old-line Swedish companies by foreign firms, the Swedish public reacted with dismay. Many AGA employees were also rankled by the fact that AGA had made great efforts in the preceding decades to modernize and rationalize its structures. Until the 1960s AGA had been a heterogeneous conglomerate in which its separate divisions grew (mostly independent of one another) on the basis of their own traditions and underwent only incremental change. The eventual decision to limit itself to industrial gases (and their medical applications), accompanied by modernization and efficiency measures, had helped AGA become the fifth-largest gas supplier in the world, with a decentralized and effective international structure. But AGA and Linde's industrial gases segment were each too small on their own to compete in the future with the large suppliers. In the tight oligopoly system, growth seemed possible in 1999 only through mergers.

The European Commission, which must approve mergers of this magnitude, analysed the probable impact of the merger of Linde and AGA on the EU gas markets, and on 9 February 2000 it authorized the acquisition of AGA under two conditions: first, Linde and AGA had to withdraw from selling bottled gases in the Netherlands and bottled and liquefied gases in Austria. But the European Commission also made clear that it took a generally dim view of mergers of this magnitude – both the acquisition of AGA and the takeover of BOC by Air Liquide/Air Products. Therefore, the takeover of Messer Griesheim by Linde, planned for the same time as the AGA acquisition, fell through.[76] The Commission was unlikely to approve it, or if so, only with unacceptable conditions.[77] On 5 May 2000, trading of AGA stock was officially stopped; from that point forward, AGA was no longer a public company.[78] The acquisition of AGA occurred at a very high market price. The international stock markets reached their all-time high in March 2000. Since then, as of 2003, stock prices have fallen to about a third of this level. In light of the competition's acquisition activities, there was not a large window of opportunity for purchasing AGA. The acquisition was a lucky success for Linde, but its high capital investments will burden the company for some years to come.

AGA and Linde since the takeover

After the EU competition department gave the go-ahead for the AGA acquisition in February 2000, Linde optimized purchasing, production, distribution, sales, and administration. A total of €100 million were to be saved annually by 2002 within the industrial gas segment through the integration of AGA.[79] These cost savings were to come partly from eliminating 1,500 jobs, and partly from the now denser customer network that allowed more efficient use of transport resources (capacity utilization of tank trucks; simpler routing and accordingly fewer empty runs).

In many European countries, Linde was able to become the market leader as a result of acquiring AGA; its gas business in the North and South American markets was significantly expanded. Linde experienced change in the wake of the merger with the separation of its gas business into industrial and speciality gases. In 2000, the industrial gas area – including the cylinder, tank, and on-site business – comprised about 88 per cent of gas sales. The strongest sales growth in the industrial gas area was achieved in the on-site business, that is, supplying customers by pipeline from plants located directly on the customer's property.

With the takeover of AGA's medical division, Linde has ranked among the leading suppliers within the gas industry of medical gases worldwide since 2000. Linde established a self-contained business area in 2000, AGA Linde Healthcare, which is mainly involved in supplies for patients with respiratory diseases. Linde provides gases, therapeutic equipment, and services for patient care in hospitals and in private homes. In 2000, AGA Linde Healthcare achieved sales of €450 million, a 25 per cent increase over the previous year.[80] This growth can be attributed in part to the successful introduction of the Inhaled Nitric Oxide (INO) process. INOmax is a gas mixture that contains nitric oxide (NO) and dilates pulmonary blood vessels; it is used as a pharmaceutical to treat newborns with respiratory failure.[81]

The integration of the Swedish gas company AGA AB into Linde's Gas and Engineering segment was led by Gerhard Full. In a joint interview in 2002 with his successor as CEO, Wolfgang Reitzle, Full summarized: 'The AGA integration has been very successful and is largely completed.' Reitzle, too, pronounced the buy-out an important step for the company: 'The acquisition of AGA was an absolute stroke of luck and strategically key for Linde.'[82]

However, by acquiring AGA at the height of the stock market boom, Linde incurred exceptionally high interest charges, which analysts viewed very critically. In 2000, nonoperating charges grew by €119 million to €172 million; the main causes of this were interest on the AGA financing, provisions for restructuring measures, and goodwill amortization. The assumption of AGA's loans increased Linde's debt by €1.44 billion to €4.18 billion in 2000.[83] Following the AGA acquisition, Linde has thus 'taken pains to rapidly reduce our financial liabilities as a proportion of total capital and to decrease financial expenditures'.[84] Since then, cost-cutting has been ubiquitous within the company.

The AGA takeover has also given further impetus to the international integration of the Linde group as a whole. As of 2002, the five members of the board of management for Linde's gas segment came from four different companies; their common commercial language is English. This international orientation is expected increasingly to spread to the other areas of the company as well.[85] Since the AGA takeover, the majority of Linde employees work outside Germany; the ratio of employees working abroad to those in Germany is about 60:40 today. This proportion is even more striking when it comes to sales: in 2001, 75 per cent of sales revenue was earned abroad, and only 25 per cent in Germany. Today, Linde is represented in 50 countries around the globe and supplies 1.5 million customers.[86]

Since 2002, AGA has been completely integrated into Linde Gas AG. Through the merger, sales for the industrial gases division rose by 140 per cent to €894 million in the first quarter of 2000; without AGA, the division would have achieved a gain of a scant 18 per cent. The increase in sales was even more apparent in America: from €643 million in 1999 to over €1.747 billion.[87] Total corporate sales rose by 36.4 per cent to €8.45 billion, and pre-tax profits rose from €417 million to €481 million, a gain of 15 per cent. The foreign portion of total group sales increased to 76 per cent, compared to 69 per cent in the previous year.

In addition to expanding Linde's international business, the acquisition of AGA also promoted further internationalization within Linde AG, which – despite its strong international relationships dating back to Carl von Linde's era – had long remained a 'German' company internally. Group Headquarters was located in Germany, and upper management was staffed almost exclusively by Germans. Its production sites were located in Germany, and about a half of sales were generated domestically. Even the company newsletter existed only in German until 2002. But today this is no longer sufficient to remain internationally competitive.[88] Thus, corporate strategy aims not only to strengthen significantly Linde's international business, but also to achieve more English communication, exchange of personnel, and meetings.

Linde in the twenty-first century

In 2000, the world economy set another record with 4.5 per cent growth – only to register one year later the most sluggish growth since 1991 with a mere 1.7 per cent. Even before the 9/11 terror attacks in the United States, a worldwide economic slowdown had begun, which was only exacerbated by the dramatic events in New York.[89] The German stock index, the DAX, plunged from its all-time peak of 8,136 points on 7 March 2000 to 2,769 points in March 2003, though it has rallied somewhat since then. Linde stock followed this trend, too; during 2001, it fell to a yearly low of €36.04, or 37 per cent below its high for that year of €56.90 in February. Linde stock performed significantly worse than the German sector index of publicly traded machine manufacturing companies (CDAX-Machinery), which slipped only 7.4 per cent. However, it did better than the CDAX-Chemical index, which tumbled 21.7 per cent. This was presumably due in part to Linde being viewed increasingly as a chemical company by the most influential stock analysts, since it now earned more than two-thirds of its operating profits in the Linde Gas division.[90]

Analysts' attitudes and stock market movements have become increasingly important in recent years for the strategy of publicly traded companies. For Linde this was even more true after 2000. For one thing, the company had to carry a significant debt burden for the first time due to the AGA acquisition.[91] For another, the three reliable institutional investors recruited by Meinhardt gave notice that they would reduce their industrial holdings in the medium term.[92] Linde has consistently had to accept a 'conglomerate discount' in its market value – that is, its industrial trucks and refrigeration technology reduce the attractiveness of Linde stock in the view of buyers, and its shares thus trade at a lower price than the stock of its international competitors in the industrial gas area. Linde cannot completely escape this pressure from the stock market. To be sure, discussion of the primacy of shareholder value – the single-minded orientation of corporate strategy toward stock market prices and yields – has somewhat abated since narrowly return-oriented strategies often brought only a short-term advantage: while they put a lot of money in investors' pockets, in the long run they ruined the substance of the company concerned, and obstructed strategic investment and acquisition decisions that made sense with a longer time horizon.

In the future, Linde is likely to come under even more pressure to conform to international practices – especially North American ones – and, when making business decisions, to consider mainly how they will affect stock prices and returns. The focus will thus be on 'a market return that is satisfactory in the long term' – and that is indeed a novelty in the company's 125-year history. The yardstick for this is return on capital employed (ROCE).

The crisis of the so-called New Economy after 2000 indirectly gave Linde a boost, for after years of neglect the classic industrial values were once again considered solid investments with relatively good returns. But the international economy, stock market movements, and business decisions are not the only factors that influence the development of Linde AG: for example, the change-over of certain rating criteria to Anglo-Saxon standards had serious consequences for Linde's stock price. In mid-February 2003, the rating agency Standard and Poor's said the Linde's credit rating could fall. As a reason the rating agency claimed that the future pension costs and other debt were too high. In doing so, the rating agency simply equated pension reserves with other debt, which is not usual in Germany, where pension reserves are not comparable to other financial obligations due to their long maturity times and their non-cancellable nature. In the agency's view, the Linde group's debt rose by a billion Euros at a single stroke, from €3.3 billion gross to €4.3 billion.[93] This led not only to a downgrading of the company's rating for long-term credit from A– to BBB+, with corresponding consequences for the amount of interest it must pay, but also led to a drop in the share price. In the meantime, the strong position of the rating agencies, whose ratings influence the stock prices and refinancing options of publicly traded companies, has come under heavy criticism, in part because there is as yet no effective 'control over the controllers', as became clear with the collapse of the 'new market' of the (often insubstantial) dot-coms.

In recent years, Linde has begun to strengthen its international position more through strategic alliances and not just through acquisitions – a corporate policy that has been popular since the 1990s in a variety of sectors. Especially in the sales area, strategic partnerships are interesting for Linde in overcoming market entry barriers more easily and meeting market-specific needs more quickly.[94] In this regard, the alliance begun in 2002 with Komatsu Forklift Ltd in Japan was especially successful in gradually obtaining access to the Japanese market for industrial trucks, which had been tightly closed against imports up to that point.

The change of the millennium brought important further changes to top management. In January/February 2000, three new members joined the Executive Board: Aldo Belloni, Hubertus Krossa, and Hans-Peter Schmohl. Falko Schling had been brought onto the board a year earlier to replace Rudolf Mundkowski, who had suffered a stroke; Gerhard Full, Gerold Tandler, and

Hero Brahms continued to serve on the board. Then, in September 2000, Dr-Ing. Peter Grafoner (who had just stepped down as CEO at Mannesmann VDO AG) joined the board as vice-chairman and designated successor of Gerhard Full as CEO.[95] Grafoner was expected gradually to take over the duties of the CEO. But the change of top leadership did not proceed as smoothly as planned and Grafoner stepped down from the board only eight months later.[96] Schling, too, left the board after a short time, Schmohl followed at the end of 2003 and Brahms in 2004. In comparison with earlier times, the current changes to the board were much greater.

Figure 7.2 The Linde AG Board of Management, 2003. From left to right: Hubertus Krossa, Dr Hans-Peter Schmohl, Dr Wolfgang Reitzle, Dr Aldo Belloni, Hero Brahms

On 10 May 2002 the Supervisory Board appointed Wolfgang Reitzle as a regular member of the Executive Board and named him CEO effective as of 1 January 2003.[97] With Wolfgang Reitzle, an engineer with a doctoral degree again took the helm of the company.

After completing his doctorate in manufacturing engineering at the Technical University of Munich in 1976, Dr-Ing. Wolfgang Reitzle began his career at BMW and advanced to its Executive Board within 11 years. In 1999 Reitzle moved to Ford, where he headed up the newly created Premier Automotive Group, a new division that brought together the previously autonomous luxury models of the Ford Motor Company.[98]

At his first balance-sheet press conference as CEO, Wolfgang Reitzle presented his corporate strategies in the tradition of Carl von Linde. The 'Lead-Ing.' model he conceptualized tries to make explicit the company's technological orientation and the German engineering tradition (echoing 'Dr-Ing.', or Doctor of Engineering) and to link these values with the will to market leadership. He thus emphasized that Linde – despite its increasingly international orientation – continues to be committed to Germany as its base of operations. The Supervisory Board also underwent a similarly sweeping generational change to the Executive Board. Its long-standing chairman, Wolfgang Schieren, died in March 1996 while in office. His successor was Hermann J. Strenger from 1996 to 1997; Strenger left the Supervisory Board in May 2001. Hans Meinhardt followed him as chairman in 1997. Following the successful changeover from Full to Reitzle as CEO, the 72-year-old Meinhardt left the Supervisory Board of Linde AG by the end of May 2003 at Linde's annual general meeting.

Meinhardt served Linde for a total of 48 years. He is thus in the familial tradition of the leadership team stemming from the Linde family, which attained similarly long tenures in office. In retrospect, the Meinhardt era from 1970(76) to 2003 was a very successful time for the company, which during these years increased its sales 25-fold in nominal terms. After-tax profits increased even more dramatically, and the number of employees grew from 13,500 in 1970 to more than 46,500.[99]

Plant engineering and industrial gases

As has so often been the case in Linde's history, plant engineering was able to withstand the economic recession that began in 2000. While many plant engineering companies had to accept economic losses, Linde's contracts and sales did not significantly decline. The division continues to enjoy international confidence and has managed to hold a technological lead for decades. Its sustained success surely has something to do with the identity of this business segment and its roots in the tradition of the development-oriented engineering firm. This has been coupled with a successful strategy of concentrating on a few closely interlinked core fields in petrochemical, air separation, hydrogen, and pharmaceutical plants. Linde anticipates good future prospects in the growing hydrogen market, because for decades the company has been involved in all the essential research and application projects in the hydrogen field.

Linde continues to distinguish itself in hydrogen technology; as a developer and builder of plants for producing and distributing hydrogen, and of vehicle equipment as well, the company is able to offer complete hydrogen application systems. In the area of alternative-energy drive systems, Linde is involved in projects with nearly all the leading vehicle makers. Together with MAN, Siemens, and Ludwig-Bölkow-Systemtechnik, Linde developed a fuel-cell bus in 2000 that made its debut in Erlangen in parallel to conventional diesel-driven buses.[100] In 2000, the company filed multiple patent applications for vehicles with hydrogen fuel-cell drives. The future demand for hydrogen will increase dramatically, not only due to its use in vehicle power trains but also for new applications in (for example) the chemical and electronics industries, food processing, and refineries.[101] In the coming years, Linde expects the hydrogen market to grow by more than 8 per cent annually.[102]

Figure 7.3 Entering the hydrogen era: Mercedes-Benz Necar 4 with fuel cell and Linde hydrogen tank, 1999

Natural gas plays an important role in the global energy supply: as an environmentally-friendly energy source for power generation, as a low-emission vehicle fuel, and in the production of cleaner diesel fuels. According to experts, total demand for liquefied natural gas will rise by 60 per cent by the year 2010. Linde has at its command all the technologies for treating and separating natural gas, and is also involved in development projects to make natural gas usable as a clean fuel. Together with Erdgas Südbayern GmbH, Linde built a combined natural gas service station for ARAL in Munich, at which customers can fill their vehicle's tank with either liquefied natural gas (LNG) or compressed natural gas (CNG).[103] Linde has also retrofitted one of the switching locomotives of the Deutsche Bahn AG to run on natural gas and built the necessary filling station.[104] Particularly in public transit, natural gas has a bright future as a fuel, and this environmentally friendly energy source will also find broader use as a fuel for large corporations' vehicle fleets – especially since the infrastructure for natural gas already exists today. Liquefied natural gas is a promising energy source for the future due to its high energy density, constant heating power, and high purity. The demand for liquefied natural gas is expected to grow by about 60 per cent by the year 2010.[105]

Natural gas from previously uneconomical sources that are inaccessible to pipelines can be made usable by converting natural gas into sulphur-free diesel fuel ('gas to liquid'). Since this process requires 'enormous quantities of oxygen', Linde can expect good business here, too.[106] The breakthrough in this area is represented by Linde's contract with the Norwegian company STATOIL to build a LNG plant in Hammerfest. The geographical location of this natural gas field presented special challenges; transport of the natural gas out of this remote development area via a pipeline on the ocean floor would not have been economical; only liquefaction makes it possible to deliver the gas profitably with tankers to its destination.[107] Besides natural gas liquefaction, natural gas separation is also creating new fields of work; the size of these plants continues to increase steadily. Between 2000 and 2003, Linde built the largest natural gas separation plant of its kind in the Persian Gulf, which extracts ethane from the South Pars natural gas field as raw material for two petrochemical complexes. The Iranian National Petroleum Company has commissioned Linde to build another plant of this type in Bandar Assaluyeh to produce pipeline-quality natural gas as well as ethane and heavy hydrocarbons, which will be processed in an olefin plant of Linde's that is also currently the world's largest.[108]

Linde's plant engineering segment is also highly active in the petrochemical area. In 2000, the Mexican oil company Petroleos Mexicanos Corportivo (Pemex) in Cantarell, Mexico, put on stream the world's four largest air separators, built by Linde. These plants deliver nitrogen, which is placed under pressure in the producing formation in order to increase the recovery of crude oil while simultaneously reducing the danger of explosion. More contracts can be expected here in the future, as more and more petroleum formations are tapped that were previously not considered exploitable because the pressure was insufficient for extraction.

BOC and Linde began negotiations in 1998 on a partnership in plant engineering; BOC subsequently turned its plant construction activities over to Linde altogether.[109] In the USA, too, Linde Process Plants (LPP) in Tulsa, Oklahoma – founded in late 2001 through the merger of the former subsidiaries Lotepro and Pro-Quip – took over BOC's plant-building activities in the USA in March 2002. In return, BOC received a 30 per cent stake in the company, which now operated as Linde BOC Process Plants (LLC).

A much more far-reaching plan, the founding of a 'Deutsche Anlagenbau' (a 'German plant construction consortium'), in which Linde would have participated along with other major German manufacturers, was rejected by the Executive Board under CEO Gerhard Full.[110] Full took the position that the loss of autonomy for Linde's plant engineering would not be offset by corresponding advantages, because in his view the Deutsche Anlagenbau 'offered no convincing concept'. Full also emphasized that Linde's plant construction business had never been in the red during its 100-year history: 'Where else can you still find a plant builder that has operated so economically?' In addition, he said that plant construction plays an important role for the Linde Gas division.[111] Linde Gas depends on plant engineering for expanding both the on-site business and the hydrogen business. In this context, Full also referred to the supportive function of Linde's plant designers' expertise for the gas business. Wolfgang Reitzle agreed with this view at Linde's 2003 balance-sheet press conference: 'My predecessors Mr. Full and Mr. Meinhardt decided quite far-sightedly a year and a half ago to combine Gas and Engineering into a single business segment.' This has increasingly proven to be an excellent starting position, he said; the two areas complement one another perfectly and are increasingly interlocked. At least in this segment, there will be no change of strategy in the foreseeable future.

Industrial gases

A successful marriage of plant engineering with the industrial gas business is represented by the €90 million helium liquefaction plant that Linde will utilize in Algeria as a joint venture with the state-owned Algerian company Sonatrach, one of the world's leading suppliers of natural gas and oil.[112] The bidding process for developing the helium source on the North African coast began in 1998, and up to the end all the major international competitors vied for access to it. In January 2003, an initial contract was concluded for the construction and operation of the helium liquefaction plant; a second joint venture regulates the marketing and sales of the helium. Linde will command 50 per cent of the quantity produced, and Sonatrach will market the other half worldwide. The amount reserved for Linde, with a transaction value of $50 million (in US dollars), will meet about two-thirds of its sales requirements from 2005 onward. Linde will thus rise to the rank of fourth-largest helium supplier worldwide, after having previously purchased helium mainly from American gas companies. For a 20-year period, Linde will command about 10 per cent of global helium production.

Gas and Engineering now play the most important role in the group – due, above all, to the AGA acquisition. This segment brought in 56 per cent of group sales and a full 80 per cent of operating profits in 2002.

Material handling

This segment achieved yet another new profit and earnings record in 2000 and 2001. Sales rose by 14 per cent to 105,000 units.[113] In line with the strong economy, Linde brought on stream a plant in Merthyr Tydfil, Wales, which it had acquired a year earlier, and which produces heavy-duty trucks and container handlers. At the start of 2001 the segment could still assume that its strategy of expanding its market position with a wide and ever more varied product range would continue to succeed in the future. But the worldwide downturn starting in the second half of 2001, especially for counterweight trucks – in the USA demand collapsed, in Asia it diminished slightly, while it declined continuously in Europe from the middle of the year onward – called for a realignment.[114] The 'consistent technical product differentiation'[115] of the three brands Linde owns – a strategy with which Linde hoped to stand out from its competitors – proved problematic with the fall-off in the economy and in sales since 2001. In light of dwindling margins and keener competition, it is no longer possible to sustain an independent, complete programme of warehousing equipment and counterweight trucks for each of the three brands, along with 'diligence and care' even for products 'which, though they serve only small market segments, are important to customers – for example, tow tractors and special-purpose vehicles'. In a 2002 conversation, Gerhard Full saw 'the most difficult market situation since 1993' for the material handling segment. In the same interview, the incoming CEO Reitzle skated around its failings very diplomatically:

> As long as the market is growing, potential improvements don't come so much to the fore. They are also often not pursued with the necessary vigour. It's a little like a sea – at high tide all the shoals are invisible, but as soon as the water level drops, the rocks appear that one must sail around.[116]

Despite the slump in sales and profits, Linde continues to be the most profitable forklift manufacturer in the world and has a market share of a good 37 per cent in Europe.[117]

Since a real economic recovery that would boost industrial truck sales is not to be expected at present or in the foreseeable future, Linde's material handling segment is prepared to achieve better results again through reorganization and comprehensive cost-cutting. Part of this effort is the so-called TRIM.100 project, which should save about €100 million in costs in its first phase through to the end of 2004.[118] It has met or exceeded this goal until now.

The material handling segment is also changing its structures for the development of new material handling and hydraulic products. In the future, products will be developed at the locations where they will later be produced. Warehousing equipment will be concentrated in the Châtellerault plant in France, electric four-wheel and reach trucks in the Basingstoke plant in England, and heavy-duty trucks and container handlers in the new Merthyr Tydfil plant in Wales. The parent plant in Aschaffenburg forms the centre for developing and mass-producing electric and combustion motor forklifts as well as hydraulic drive systems.

Linde sees growth opportunities primarily in North and South America and in East Asia.[119] In order to compete the previously hermetically sealed Japanese market, which is still hard for European vendors to enter, Linde has fostered increasingly close relationships with Komatsu Ltd, one of the world's largest construction machine manufacturers. First, agreements were reached on selling Linde's industrial trucks through the Komatsu organization. In 2000, Linde founded a joint venture with Komatsu in Tokyo, the Komatsu Forklift Co. (KFL), in which Linde originally had a 10 per cent share and then, starting in April 2003, a 35 per cent share (with the option to gradually continue increasing its stake). At the beginning KFL had a 10 per cent share in the Linde group company FIAT OM Carrelli Elevatori S.p.A., which handles the import and sale of Komatsu products in Europe. In addition, FIAT OM produces forklifts for KFL. On the other hand, KFL sells Linde equipment in Japan through its own sales organization; Linde's Xiamen plant in China produces Komatsu forklifts and components for KFL. In North America, the Linde and KFL product ranges complement one another. Komatsu Forklift is the second-largest industrial truck supplier in Japan and number seven in the world. In Southeast Asia, Linde has consolidated the sales organizations of KFL and STILL in order to now offer all warehousing equipment and counterweight trucks from a single source.

The hitherto sportsmanlike competition between the five forklift brands in the Linde group has been a marketing success, but in the future it will lead to cost savings through closer co-operation, especially in manufacturing.[120]

Refrigeration technology

While Gas and Engineering has been successful despite the poor economy and Material Handling continues to lead the global market after some very good years with a series of new unit records, Linde Refrigeration has not achieved satisfactory performance despite significant investments and acquisitions. Although Linde, with its various brands, continues to lead the European market for commercial refrigeration and is number two in the global market, Refrigeration is generating losses or at best slight profits, which probably cannot be significantly increased in the foreseeable future. This is due in part to the international competitive situation, the market power of the relatively small number of major retail chains which are increasingly globalized, the shift of the market toward Eastern Europe, Latin America, and Southeast Asia, and the low margins in retailing.[121] Up to now, Linde could command higher prices in Germany and Europe, thanks to its long-standing existing good business relationships with large customers and its good reputation, than is likely to be possible in these new markets. In addition, bid invitations are increasingly for equipping entire supermarkets, including the uncooled displays and shelves; Linde has naturally reacted to this.

Linde is also trying to become more competitive in this area through technical innovation and new design solutions. But in today's business of commercial refrigerators and refrigeration systems for supermarkets, the money is less in innovative new technologies than in complete set-ups that are as inexpensive as possible and quickly deliverable. Linde's typical strategy of trying to satisfy every customer desire – even those requiring elaborate custom production – can no longer be sustained in the present economy. Linde was already forced at the end of 2000 to close its Chew Stoke plant in Great Britain, which specialized in expensive custom refrigerators, because cheaper standardized refrigerators are increasingly in demand in England, too. Moreover, this will permit the other Linde plants to run closer to capacity.[122]

In the long run, a comprehensive restructuring programme should lead to a concentration on a few large production sites, which chiefly means the gradual relocation of production volume to the Czech Republic. The €50 million cost savings achieved in 2002 under the current restructuring programme 'were absorbed by the market' – that is, prices fell faster than costs. Although Linde Refrigeration is making great efforts to reduce costs, it was currently unclear when the segment would consistently contribute appreciable profits to the group's results. Therefore until the end of the year 2003, all three options still were possible: to continue the refrigeration business, to merge it with a competitor or to sell it.[123] However, on March 15, 2004, Linde sold it to an American company, the Carrier Corporation, a subsidiary of United Technologies Corporation in Hartford/Connecticut. The sale ended an era in Linde's history. At the same time, the sale guarantees and is a sign of continuity. Carrier, like Linde and AGA, is an old technology company, founded by a research oriented engineer. The refrigeration business remains part of a company with a strong emphasis on technological leadership and high quality production.

In light of the ongoing globalization of the markets and worldwide competition, no one is able to make a responsible prediction today as to where this corporation will be in 10 or 20 years, as Reitzle observed to its leadership: 'Is Linde in the year 2020 a pure gas company? Or should Linde position itself as a technology company? And what argues against retaining some sort of hybrid structure?' This question of strategically weighting and mixing the disparate business segments has run through the company's history since the 1920s. Only a review of the conglomerate's successes and failures in its impressive history reveals the long-term effects of strategic decisions. The blend of business segments, which can only be understood historically, has often made life difficult for the Executive Board in the short term – there was always a crisis somewhere – but in the long run it has expanded their options and scope of action.

Resumé: looking back to the future

History, as the Swedish historian of technology Svante Lindquist has written, is the discussion of the present with the past about the future. This is also true for an analytical retrospective of the 125-year history of the international, Germany-based technology company Linde, which is influenced by questions about the company's future and the perspectives of the economic style it represents. What constitutes the identity of Linde AG? Which factors have influenced and moulded its unquestionable economic success? Are there trends and lessons that the company and its employees can use as a compass?

1. Technology and research orientation

In the refrigeration industry that arose after 1870, Linde's Ice Machines was the most successful developer and system vendor in the European market. It is traditional in the history of technology to ascribe this success to Linde's close ties to academic research. Yet this assessment is not nuanced enough. Linde and his company fought hard to achieve market leadership in fierce competition with older and sometimes more successful refrigerator producers. Linde's scientific reputation and proximity to research were an important factor and argument in this struggle. Just as decisive were Linde's contacts with several leading machine factories, especially the Maschinenfabrik Augsburg, the Gebrüder Sulzer in Switzerland, Carel Frères in Belgium, CAIL in France, and many other licensees. Perhaps more important yet was the Linde Company's organization as an engineering firm with licensees, which was flexible and capable of growth. It offered the machine factories opportunities for participation and profit-sharing, and made them partners of the company. To be sure, in the medium term there was a potential for conflict over customer business. But the initial choice to dispense with production facilities of its own, together with high-powered partners, led to a competitive advantage for Linde, which was able to develop and for a long time sustain the unbureaucratic and development-oriented style of an international engineering firm.

Linde's fundamental decision in favour of the compression refrigeration machine was motivated by thermodynamic considerations, but it was based on premises that thermodynamics later repudiated. For several decades, the configuration Linde chose – a horizontal, double-action ammonia compressor – was the most widely used and frequently copied refrigerator design in Europe. Only the Americans initially favoured single-action, vertical compressors; they were cheaper to produce and easier to service. But Linde failed in his attempt (in the 'system dispute') to elevate free competition among refrigerators onto a scientific basis and to replace it by neutral testing in an experimental station for refrigerators that he founded – a precursor to the Stiftung Warentest (the independent consumer watchdog in Germany) – but that was financed by one of his competitors. For, starting in the 1890s, the variety of designs was again growing, first with regard to coolants, then in a renaissance of absorption technology, and finally with a radical change in compression refrigerators: from low-speed motors similar to steam engines, to high-speed single-acting motors resembling Otto engines.

Since its founding, the technologically leading Linde company employed primarily university-trained engineers, first in Wiesbaden and later also in the gas liquefaction and separation division in Höllriegelskreuth. The German refrigeration industry modelled itself on the Linde Company and in international comparison had an unusually large number of engineers on board. In this way, the entire industry was launched on a technological path and soon took the lead even vis-à-vis the much larger US refrigeration market. This changed only in the 1920s, when American conglomerates entered the refrigeration sector and took the lead (also in technological terms), at least in small refrigerators, while in Germany small and mid-sized businesses moved up in the refrigeration industry.

But history has also shown that there is no clear correlation between technological success and the developer's theoretical background. Many engineers got by on intuition, even when it came to the highly complex phenomenon of heat transport. There were only a handful of technical problems in development that were fundamentally inaccessible to practically trained engineers without thermodynamic expertise. Inventors that exemplified this intuitive, practical, ingenious approach in gas liquefaction and separation at Linde include Mathias Fränkl and Paulus Heylandt; at the competition, the highly innovative Abraham Freundlich in Düsseldorf (an autodidact) and also Adolf Messer in gas separation.

Nevertheless, close ties to science at Linde were also important for its economic success. These connections shaped not only individual developments but also the overall spirit and strategy of the company – the constant search for innovations, the desire to compete on the basis of technological performance rather than price, and thus the quest for fields of technology in which this was possible. Carl Linde and his family went into low-temperature processes for precisely this reason. They sought a technologically challenging area in which thermodynamic process engineering developments and inventions would count, and that is where they shifted the company's economic emphasis. Production of refrigerators, which offered challenges only in manufacturing engineering, took a backseat after the First World War at the latest. A series of other leading refrigerator makers met a similar fate though to a lesser degree, such as MAN, Humboldt, Borsig, and Sulzer, who either reduced or halted refrigerator production, or drifted into challenging niche areas.

Since at least the 1960s, the development-oriented model has been not just an asset for the growing Linde group, but a problem as well. It continued to be very suitable for motivating and focusing the Höllriegelskreuth factory groups, but it was less appropriate for the production-oriented areas, refrigeration technology and materials, which suffered from a lack of perspectives and strategies for their business segments. The change in strategy pushed through by the economist Hans Meinhardt in the 1960s and 1970s, which aimed at economic market leadership through market power, acquisition of the competition, and efficient processes on all levels, offered a new, attractive model for employees in these areas. At the same time, the engineering division in Munich managed to preserve its tradition and its development-oriented corporate style and growth, and the company leadership in Wiesbaden respected Munich's special status. With the founding of the Linde Academy at the Technical University of Munich in 2004, the company reinforced this element of its corporate tradition.

2. Co-operation with the competition

Besides its scientific and technological leadership, Linde followed a second path over many decades to mitigate the economic struggle for survival: securing market share by forming cartels. From the turn of the century until the end of the 1930s, Linde used marketing agreements to try to retain its position as the clear market leader. In those years – unlike today – syndicates, cartels, and marketing agreements were still considered reasonable and even necessary instruments for structuring the economy. Particularly in Germany, they sometimes even had the character of state-sanctioned coercion. In refrigeration technology, however, attempts to form cartels met with little success over the long run, because the number of new entrants was too large. In industrial gases, by contrast, agreements were successfully reached from the beginning, not least because there are only a few companies in this market to this day. The initial investments in infrastructure – gas plants, mains, bottle parks – were so great for each region as to deter new entrants. Only in the special situation of the two world wars could new companies grow up and join the ranks of industry leaders. Otherwise, on an international level we are still dealing predominantly with the 'prime movers' from the early twentieth century. For example, the most important European competitors Linde and Air Liquide divided the world market between themselves, preferring to pay royalties to each other rather than accept competition in 'their' regional markets. This system continued through both world wars, because both sides knew that they would be dependent on a good relationship after the wars. However, after the Second World War, Linde was almost completely stripped of its international properties and assets. Re-establishing its network of international subsidiaries took 20 years. On the other hand, compared to markets with many players, competition within oligopolies is fundamentally harsher and more threatening to survival. Thus, the takeover of British Oxygen Company by Air Liquide and Air Products in 1999 virtually forced Linde to acquire a large partner as quickly as possible, too, even if the turn-of-the-millennium stock boom might have instead compelled Linde to wait and see. But Linde had to take action and acquire Messer Griesheim or AGA. Only one purchase was accepted by the German Anti-trust Office and Linde opted for AGA.

3. Family and identity

About 100 years, or four-fifths of the history of Linde AG so far, have been closely linked to the history of the Linde family. In his analysis of family capitalism, Hervé Joly distinguishes between five forms of familiy influence in big business from an absolute to a more symbolic influence and refers to share ownership.[124] This was different at Linde. The family members have made up the majority of the decisive Boards of Management and Supervisory Boards, and the second tier of management has also been drawn in part from the family or was related to them without much stock ownership. Nearly all of the Lindes came out of the research and development-oriented university milieu and thus reinforced the company's engineering culture and orientation. Many spectacular technological breakthroughs were achieved by family members. It is impressive to see how the family members put company interests above their personal ones in central decisions, agreeing to – for example – capital widening in the early phase (which diluted their importance in the shareholders' meeting); the abolition of registered shares with 20-fold voting rights in the early 1960s, which had been almost exclusively in the hands of the family; and also the elimination of the position of the chairman of the Executive Board in 1972. Hermann Linde's departure as spokesman of the Executive Board and Johannes Wucherer's resignation from the Supervisory Board in 1980 marked the end of this familial tradition, which however had been dwindling since the 1950s and was increasingly confined to the Plant Engineering Division in Munich. An indicator for this generational separation process was the lack of targeted grooming of the fourth generation for company leadership in the 1960s. But the displacement of the family in upper management had less to do with a 'Buddenbrooks syndrome' (weak performance of the third generation) than with a lack of consideration of management strategy issues. To this day, the Linde family continues to have friendly relationships with the company and is present in third-tier management. What the future will bring here remains open, for the family's influence was from the very beginning grounded not in shareholding but in networking, personal authority, and technical performance. In the Plant Engineering Division, there was also a series of other families who were especially close to the company for many decades. Additionally, there was the 'familial' closeness to the Technical University in Munich and an impressive number of 25-year and even 40-year service anniversaries in the company, the feeling of belonging, and the investment in minds. This family tradition lives on.

4. Debates over the portfolio

Since the 1920s, Linde's top management has discussed the strategic sale and later also the acquisition of factory groups. At first, only Refrigeration Technology was in question. Should it be hived off to permit full concentration on low-temperature processes? The decision fell in favour of expanding Refrigeration Technology and the company's manufacturing competence. Until the 1960s, the argument for this expansion was – besides reverence for the founder – the (partly trumped-up) technological overlap with the gas area. In contrast, the purely economic rationale for very different business segments belonging together was expressed for the first time by Hans Meinhardt, with different investment cycles and complementarities not only in technical but also in financial terms. To be sure, this argument of economic balance between the business segments functioned only imperfectly in practice due to the varying size and profitability of the factory groups, but as a mobile model it has had a sustained impact on acquisitions. In practically every market, even where Linde was not the market leader, it has led to Linde outstripping its competitors in sheer economic size and thus facilitated the acquisition of competitors. For 20 years, from 1976 to 1996, the guiding light of the balance model, Hans Meinhardt, considered buying an entirely new additional factory group, but he finally refrained from this in favour of strengthening the existing business segments through targeted acquisitions.

Wolfgang Reitzle, however, sold the refrigeration business in March 2004 and therewith shifted the symbol of the hanging mobile for the connectness of the branches towards the model of a walking beam between gas and engineering and material handling.

5. Internationalism and German style

From its beginnings, Linde has been an internationally operating corporation, yet at the same time it has remained a typical German engineering firm in many respects. One factor in this was that engineers – who moreover came overwhelmingly from just a few technical universities, particularly the Technical University Munich – set the tone with some interruptions. Another factor was the family tradition in top management, the manners, the academic titles that remain very important at Linde to this day, and the predominance of the German language at all management levels despite an international orientation. Linde has profited greatly from this German engineering culture in the past, and it is an open question as to which elements of this culture the globally active corporation should preserve and which it should leave behind. With the 'Lead-Ing.' model, Linde's new CEO is trying to develop further a style that is global yet rooted in German engineering tradition, knowing that a globally active corporation has a particularly acute need for a unifying identity and unifying guiding models.

A technological orientation and close ties to research, co-operation with the competition, family tradition and homeland, diversification and/or focusing of the business segments, internationalism and recognizably independent tradition – these are strategic questions and decisions not only for Linde AG but also for technology companies around the world. Thus, looking back at the history of Linde AG lends inspiration not only for the future of Linde but also for the future of international technology groups worldwide.

Notes

Preface

1. Pierenkemper, *Unternehmensgeschichte*; Pierenkemper, *Was kann*, ZfU 44 (1999), 15–31; Pohl, *Zwischen Weihrauch* ZfU 44 (1999), 150–63; Pierenkemper, *Sechs Thesen*, ZfU 45 (2000), 158–66.
2. Berghoff: *Zwischen Kleinstadt und Weltmarkt*.
3. Examples: Feldman, *Die Allianz*; Lorentz, *Industrieelite*; James, *Deutsche Bank*.
4. Example: Gall, *Deutsche Bank*.
5. During the last few years, company histories have belonged to some of the most important historical publications from Germany's best-known historians. See the history on von Krupp written by Gall, Tenfelde, Abelshauser and Pierenkämper: Gall, *Krupp*.
6. Borscheid, *Der ökonomische Kern*, ZfU 46 (2001), 5–11.
7. North, *Institutions*; Williamson, *Die ökonomischen Institutionen*; Richter/Furubotn, *Neue Institutionenökonomik*.
8. Wilhelm von Siemens, Allgemeines undated (probably 1900–10). Siemensarchiv Munich, SAA, LD 144,2. See Dienel, *Herrschaft*, p. 110.
9. Feldman, *Hugo Stinnes*, p. IX.
10. Important examples: Heuss, *Robert Bosch*; Manchester, *Krupp*.
11. One example is the history of the Quandt family, a major BMW shareholder: Jungbluth, *Die Quandts*.
12. Heller's study was certainly contracted by the company in order to contrast the portrayal given by the 'Independent Expert Commission of Switzerland in the Second World War'. Heller, *Zwischen Unternehmertum*.
13. Seidl's study neglected the internal conflicts that affected the networks involved with industrial reconstruction. Seidl, *Die Bayerischen Motorenwerke*.
14. Mommsen/Grieger, *Das Volkswagenwerk*.
15. Tilly, Großunternehmen, *GuG* 19 (1993), 531.
16. Linde was ennobled in 1897 for his efforts. His name changed to Carl von Linde. In the following, I will use the name 'Carl von Linde', although Linde himself rarely made use of the title (cf. Linde: *Aus meinem Leben*, S. II (Notes to new edition 1998)).
17. Linde, *Aus meinem Leben* (1979).
18. Linde, *Eine Münchener Familie*.
19. The American economic historian Alfred Chandler sees stronger family connections in German companies compared to US companies, and, at the same time, more readiness to relinquish selfish family interests compared to England. Chandler/ Takashi, *Scale and Scope*, pp. 451, 463 and 500.

Chapter 1

1. See Linde, *Aus meinem Leben*, p. 48.
2. According to Rosenberg, *Große Depression*.

3. Stürmer, *Das ruhelose Reich*, p. 83; Erker, *Dampflok*, p. 82.
4. Landes, *Der entfesselte Prometheus*, p. 186.
5. Kiesewetter, *Industrielle Revolution*, pp. 75–90; Hentschel, *Wirtschaft und Wirtschaftspolitik*, pp. 205–10.
6. Landes, *Der entfesselte Prometheus*, pp. 231–2; Fischer, *Deutschland 1850–1914*, p. 433.
7. Wehler, *Deutsche Gesellschaftsgeschichte*, Bd. 3, p. 581 f. This estimate assesses the share of the exchange of the net domestic product in the secondary sector. If the third sector was calculated into the net domestic product this takeover would be moved up to the beginning of the 1890s; compare Fischer, *Deutschland 1850–1914*, p. 393.
8. Wehler, *Deutsche Gesellschaftsgeschichte* Bd. 3, p. 610.
9. Born, *Wirtschafts- und Sozialgeschichte*, pp. 46–7.
10. Ott and Schäfer (eds.), *Wirtschafts-Ploetz*, p. 146 and p. 263.
11. Radkau, *Technik in Deutschland*, pp.120, 125 and 149.
12. Hoffmann, *Das Wachstum*, p. 522.
13. Jaeger, *Geschichte der Wirtschaftsordnung*, pp. 107–18. Kaelble, *Industrielle Interessenpolitik*, p. 102.
14. In 1908 there were 5,100 refrigeration machines exported from Germany, at 1914 there were 7,945 machines exported valued at approximately *ca*. 300 million Marks. Stetefeld: *Die deutsche Kälteindustrie im Auslande*, Z-VDI 63 (1919), 932. Außerdem zum Export: *Die Kälteindustrie* 11 (1914), 5–7; EKI 21 (1919), 123–5. Wegener, *Kälte-Industrie*, p. 19.
15. In 1901 the customs tarifs were raised for US refrigeration machines from 10 Marks per 100 kg to 30 Marks per 100 kg. BayHSTA, MH, 10489.
16. Catrina, *BBC*.
17. Wengenroth, *Unternehmensstrategien*, p. 296.
18. Radkau, *Technik in Deutschland*, p. 122.
19. Weckerle, *Kältemaschinen-Industrie*, p. 26.
20. Pfeiffer/Niebergall, *50 Jahre Borsig Kältemaschinen*, p. 76.
21. DMMS, APV 102.
22. *ZgesKI* 11 (1904), 12 and 34.
23. Weckerle, *Kältemaschinen-Industrie*, p. 24 f. Out of the 55 factories only 47 were completely built, 9 only ordered parts.
24. Only 2 of 52 manufacturers covered all three areas in 1935; 8 manufacturers concentrated solely on large cold stores, 20 on small cold stores and 5 on refrigerators, 7 on refrigerators and small cold stores, 10 on large and small cold stores. Weckerle, *Kältemaschinen-Industrie*, p. 61.
25. GFLE, *50 Jahre Kältetechnik*, p. 183.
26. Weckerle, *Kältemaschinen-Industrie*, p. 22.
27. Chandler, *Scale and Scope*, p. 8.
28. Chandler, *Scale and Scope*, pp. 395 and 497. See also: Siegrist, *Deutsche Großunternehmen*; Jaeger, *Geschichte der Wirtschaftsordnung*, p. 119. Summary of the new literature: Schmitz, *Review Article*, 91–103. Schmitz criticizes Chandler's ideal development of 'modern co-operation' as too linear, and for negating alternative successful paths of German business development (p. 102).
29. Chandler sees that Germany as well as the United States had 'close ties with universities', however: 'Germany's organizational capabilities can hardly be attributed entirely to the symbiosis that developed in Germany [...] between the country's industrial enterprises and its financial and educational institutions.' Chandler, *Scale and Scope*, p. 499 f.
30. Chandler argues that the diversification in the machine plant was probably due to risk management. Ibid., p. 459.
31. Ibid., pp. 451, 463 and 500 f.

32. On the American natural ice industry, see Anderson, *Refrigeration*, pp. 14–71; Everson, *Tidewater Ice*; Cummings, *American Ice Harvests*; Jones, *America's Icemen*. The best contemporary source is Hall, *Ice-industry*. On the history of marketing natural ice for private customers, see Jones, *American Iceboxes*. For breweries see Appel, *Artificial Refrigeration*. For the meat industry see Walsh, *The Rise*; Wilde, *Industrialisation*. For citrus fruits see Hopkins, *50 Years*.

33. Rohbeck, Die Bedeutung, *KA* (1928), 79–81.

34. Meidinger, Die Aufbewahrung, *BG* 4 (1870/71), 51.

35. Regulations for building ice basements have existed in Germany since the medieval period. The first books, however, were published in Germany starting in the 1860s. These include, Oncken (1864), Menzel (1867), Swoboda (1868), Meidinger (1883), Schatteburg (1893), Nöthling (1896). A huge theoretical advancement can be found in these handbooks. See the reviews from Swoboda, Anlegung and Benutzung *DBZ* 9 (1875), 180; Hans Lorenz' review of Menzel's Bau der Eiskeller, *ZgesKI* 1 (1894) 180; and Hans Lorenz' review of Schatteburg's Eiskeller, *ZgesKI* 1 (1894), 21, and *ZgesKI* 8 (1901), 75 (2nd Edition).

36. Schwarz, Eisproduktion, *EKI* 3(1901/02), 129–32.

37. Artmann, *Lehre von den Nahrungsmitteln*, p. 177.

38. Heinrich Meidinger purposefully aimed his article at America. Compare Meidinger, Die Aufbewahrung des Eises, *BG* 4 (1870/71), 49–64, 57.

39. For the history of refrigeration technology in German breweries see: Hård, *Frozen Spirit*, pp. 193–209. Heckhorn/Wiebe, *München und sein Bier*; Behringer, *Löwenbräu*.

40. Ganzenmüller, Über den gegenwärtigen Stand, *ZgesKI* 16 (1909), 190; Heinel, *Bau und Betrieb*, p. 23; Wegener, *Kälte-Industrie*, p. 31.

41. Meidinger in *DPJ* 217 (1875), 471.

42. Hoffmann, *Das Wachstum*, pp. 652–3; Struve, *Entwicklung des bayerischen Brauwesens*; Tillmann, *Einfluß der Kältetechnik*, pp. 88 f. In Bavaria, however, beer use rose – although from a higher base amount – more slowly, from 12.9 million hectolitres to 17.9 in 1900.

43. In Germany there were approximately 11,000 breweries, of which 40 per cent were in Bavaria: 4,996 breweries in Bavaria (1874), 5,471 (1880), 5,186 (1890), 4,063 (1900), 3,796 (1910).

44. Holzner, Mitteilungen über den verstorbenen Gabriel Sedlmayr, in *ZfdgB* 15 (1892) 5.

45. Bavarian beer needed a few decades to catch up with Bohemia's popularity.

46. Lintner, *Lehrbuch der Bierbrauerei*, pp. 259 and 265.

47. Meidinger in *DPJ* 228 (1878), 240.

48. Philippe, *Bierbrauen einst und jetzt*, pp. 8–17.

49. Linde, *Aus meinem Leben*, p. 41.

50. Linde, Wirtschaftliche Wirkungen der Kältetechnik, *VDI-Z* 50 (1906), 1036. Heinel calculated even greater savings, from 7.04 Marks to 2.8 Marks per Hectolitre of beer. Heinel, *Bau und Betrieb von Kältemaschinenanlagen*, p. 23.

51. Philippe, *Bierbrauen einst und jetzt*, pp. 8–17.

52. Letter from R. Banfield to R. Diesel, 7 December 1880. MAN, Diesel Estate, Correspondence Diesel/Gesellschaft für Lindes Eismaschinen 1880–94.

53. Hård, *Frozen Spirit*, p. 225.

54. Jacobsen, Über die Fortschritte, *ZfdgB* (1884), p. 49. Schwarz, Zymotechnische Reiseskizzen, *ABuHz* 25 (1885), 1596.

55. Prussian law on the exclusive use of public slaughterhouses from 18 March 1868. GStAPKM, 89H, XIII, 11. In 1880 the law was somewhat relaxed. GStAPKM, Zivilkabinett 2.2 1 , Nr. 28016. Linde, Schlachthofgesetzgebung, *IIR* Vol. 2 (1908), 845–7. In addition, see Häutle, *Schlacht- und Viehhof*.

56. Schwarz, Die Bedeutung der Kühlhäuser, *ZgesKI* 8 (1901), 170–2.

57. Göttsche, *Kühlmaschinen*, p. 112.
58. Krabbe, *Deutsche Stadt*, pp. 116 and 121 ff.
59. Linde, Schlachthofgesetzgebung, *IIR* Vol. 2 (1908), 845–7.
60. Nimax, Über Kühlanlagen, *DBZ* 26 (1892), 602–4, 614–16, 628–30.
61. Schwarz, Die Bedeutung der Kühlhäuser, *ZgesKI* 8 (1901), 170–2; Heiss, Schlachthofkühlanlagen, *IIR* Vol. 2 (1908), 924–30.
62. Kögler, Fleischkühlhäuser, *IIR* Vol. 2 (1908), 935–50.
63. Linde, Das Berliner Werk, *ZgesKI* (1903), 105–9.
64. Nipperdey, *Gesellschaft*, p. 321; Wegener, *Kälte-Industrie*, pp. 115–17.
65. Schwarz, Die Fleischversorgung Deutschlands, *ZEKI* 8 (1916), 102–9,105.
66. Bützler, Die Notwendigkeit der Gefrierhäuser, *ZEKI* 10 (1917), 1 ff.
67. Plank/Kallert, Behandlung und Verarbeitung. Plank/Ehrenbaum, Konservierung von Fischen. Martin Krause writes in his review: 'The achievements of these authors have only been possible because of the war. During peace, they would not have been possible.' *ZgesKI* (1916), 64.
68. Bützler, Die Notwendigkeit der Gefrierhäuser, *ZEKI* 10 (1917), 1 f.
69. Verband deutscher Kühlhäuser, *ZgesKI* (1917), 24.
70. During the DKV's general meeting in 1916, Rudolf Plank gave the plenary speech on 'Wartime Duties of the Cold Storages' (*ZgesKI* 23 (1916) 57–64). Cattaneo, Die Bedeutung der Kältetechnik, *ZgesKI* (1915), 42.
71. *VDI-Z* 63 (1919), 370.
72. Wegener, *Kälte-Industrie*, p. 64.
73. The port cold store in Cologne (1921), the herring cold store in Hamburg (1929), and the egg cold in Berlin (1929). Schipper, Bedeutung der Kältemaschinen, *KA* (1929), 65. Moderne Deutsche Kühlhäuser, *KA* (1934), 58.
74. Schipper, Bedeutung der Kältemaschinen, *KA* (1929), p. 71.
75. Hellmann, *Künstliche Kälte*, pp. 109–19.
76. Heiss, Einrichtungen und Aufgaben des Reichsinstituts, Plank, *Festschrift*, 55–63.
77. The institute was directed by the committed National Socialist Heiss and had 25 employees in 1940. In 1944 it moved under the name Research Centre of the Army (Forschungsstelle der Wehrmacht) to Munich. BayHSta, MK 40108.
78. Mosolff, *Der Aufbau*, pp. 11–16.
79. Heiss, *Aufgaben der Kältetechnik*, pp. 87–91.
80. Kasdorf, *Eis und Kälte*; Schweizer Milchwirtschaftlicher Verein, *50 Jahre*, pp. 119–22.
81. Thévenot, *A History*, p. 102.
82. Carles, Le Froid en oenologie, *IIR* Vol. 3 (1908), 3, 29–49.
83. After the turn of the century, industrial American bakeries used refrigeration to slow dough fermentation to avoid a night shift. Thévenot, *A History*, p. 104.
84. Göttsche, *Kühlmaschinen*, p. 249.
85. Porges, Über die Verwendung der künstlichen Kälte, *IIR* Vol. 3 (1908), 3, 263–71.
86. Hoffmann, *Acht Jahrzehnte Gefrierverfahren*, pp. 513–36.
87. Large refrigeration machines and air dryers were used in American steelworks in 1904, and their implementation was also discussed in Germany. Linde, Trocknung des Hochofenwindes, *S&E* 25 (1905) 3–13.
88. Doederlein, Künstliche Eislaufbahnen, *ZgesKI* (1898), 77–8.
89. Chandler, *Scale and Scope*, pp. 430–3.
90. The following information quoted from Linde's autobiography, Linde, *Aus meinem Leben*, pp. 3–35 and 145–8.
91. According to Kocka, the percentage of entrepreneurs with degrees in higher education lay below 10–15 per cent until 1870. It was 'für den Erfolg als Fabrikant offenbar wichtiger, in einem der wenigen großen und hochqualifizierten Unternehmen empirische Erfahrungen gesammelt als eine technische Fachschule absolviert zu haben' (Kocka, *Unternehmer*, pp. 51, 64). Kocka's comparison of academic and industrial experience does not apply in particular cases, as Linde's biography

illustrates. For more on the educational background of employee-entrepreneurs: Kocka, *Die Angestellten*, pp. 110–14; Schumann, *Bayerns Unternehmer.*

92. Linde, *Aus meinem Leben*, p. 73.
93. Threat to resign to Buz from 5 March 1888. PAHL, Cb. 4, p. 229.
94. Linde, *Aus meinem Leben*, p. 73.
95. Supervisory Board meeting on 11 February 1893: Linde proposes again to establish a branch in Munich, but is rejected. Linde 1, IV (Supervisory Board meeting minutes, 1879–1916).
96. The following data quoted from: Linde, *Aus meinem Leben*, pp. 130–45; ATUM, PAA, Personalakt Carl v. Linde.
97. GFLE, *50 Jahre Kältetechnik*, p. 13.
98. Linde's influence was, in the beginning, significant. Klein and Böttinger were friends of his. The first three professors at the Göttingen Institute for Technical Physics (Mollier, Meyer, Lorenz) were all at least 50 per cent refrigeration specialists. Linde, *Aus meinem Leben*, pp. 135–7.
99. He spoke resolutely against building a zeppelin (letters from 14 February 1897, 21 March 1898, p. 29 December 1901 and 13 January 1902. PAHL, Cb. 10, p. 22, 149, 233 and 245), but he was outvoted. His support of dynamic flight and early helicopter projects was just as resolute as his rejection of the zeppelin. On 21 November 1904 he applied for 40,000 Marks from the Jubilee Foundation to support this research and later in 1912 he applied for another 15,000 Marks. PAHL, Cb. 10, p. 403 and Cb. 11, p. 421.
100. This was followed by membership in the Academy of Sciences in Vienna, Berlin, and Göttingen.
101. In 1892, Linde founded in Berchtesgaden the Lutheran Kirchbauverein, in Munich he became the chairman of the Mission Association and was a board member of the Munich Adult Education Association. Linde, *Aus meinem Leben*, p. 145.
102. Like the German Chemical Society (Deutsche Chemische Gesellschaft), the Prussian Commercial Association (Preussischer Gewerbefleißverein, the Technical Associations of Frankfurt and Augsburg, the Polytechnic Association, the Institute for Brewery Experimentation and Education in Berlin (Versuchs- und Lehranstalt für Brauerei in Berlin), the German Refrigeration Association (Deutsche Kälteverein), the Lower Austria Commerce Association in Vienna (Niederösterreichischer Gewerbeverein in Wien), the Association of German Engineers (Verein deutscher Ingenieure).
103. On 28 August 1889 Linde rejected Silbiger's proposal. PAHL, Cb. 4, p. 465. However, after resigning from his position as chairman, he agreed to Silbinger's proposal and corresponded more intensively with him. PAHL, Cb. 6, pp. 1, 21, 32, 70, 79, 108, 133, 143, 151, 152 and 187.
104. Letter from Linde to Sulzer-Steiner from 22 October 1890. PAHL, Cb. 6, p. 158.
105. Letter to 'Most esteemed colleagues' from 28 November 1893. Further correspondence on 12 December 1893, 27 December 1893 and 18 January 1894. PAHL, Cb. 8, pp. 262, 271, 289 and 335. Linde wanted to contribute 20,000 Marks five times. Given this history, it is astonishing that Linde did not provide any support of Diesel's motor project in 1893.
106. PAHL, Cb. 8, 459. PAHL, Cb. 9, p. 109. The project failed, Knoch fled in October owing Linde 40,000 Marks and later was jailed for a few years.
107. Linde oversaw particular major or innovative projects abroad, such as a cooling system for a hospital in India, meat cold-storage plants in Australia and the refrigeration of fermenting cellars in the world's largest brewery: Guinness in Dublin.
108. Linde wanted to prove that carbon dioxide machines could not achieve the same efficiency as ammonia machines with these experiments.
109. The letter exchange spanned 53 years, from 1876–1929, although the regularity thereof is inconsistent.

110. As of 1895 he had a telephone connection in Munich.
111. The correspondence came predominantly from him. Linde's business correspondence took up much of his private time, as well as that of his employees and business associates. Over 400 of the 3,000 received letters discuss primarily personal issues; in 150 cases association business is discussed. In his letters, Linde writes of his business style and about his university work (in the laboratories particularly) about a dozen times.
112. Guestbook in PAHL. A number of letters mention the summer house providing this function.
113. In his autobiography, Linde characterizes this passivity as a sin of failure. Linde, *Aus meinem Leben*, p. 146.
114. Subscription to Cologne's daily newspaper. Letter from the Linde Company to Linde in Berchtesgaden, 10 September 1886. PAHL, Cb. 2, p. 331.
115. Linde asked Moritz Schröter in 1895 to present the air liquefaction installations at the VDI shareholders meeting and to introduce him at the start of the presentation. Linde admitted 'how embarrassing it would be for me to hear words of praise while sitting at the Board's table'. PAHL, Cb. 9, 196. Linde turned down a request to make a statement about the German Emperor for a publication 'because I do not consider him to be one of Germany's best men'. Letter to publisher Rhode, 18 February 1897. PAHL, Cb. 10, p. 32. He rejected numerous requests for biographies. In response to a request for a lecture on air liquefaction in 1900 at the German Brewer's meeting, he stated: 'One would gain the impression that I had the need to open my tired jaws – a thought most uncomfortable.' Letter from Linde to Henrich, 5 February 1900. PAHL, Cb. 10, p. 181. On 15 December 1902 Linde requested his Berlin director Krüger to advertise liquid air to the Emperor. 'I've suffered a loss in those circles – a loss that pains me and that I've not earned.' PAHL, Cb. 10, p. 269.
116. Letter to Heike Kamerlingh Onnes, who later held a Noble Prize in Physics, 6 March 1907. PAHL, Cb. 11, p. 233.
117. Letter from Linde to staff auditor Zenk in Würzburg, 14 October 1890. Zenk had suggested dubious business to Linde. PAHL, Cb. 6, p. 145.
118. Linde attributed his hesitation in many letters to illness. He appears to have suffered often with migraines for weeks on end, making work nearly impossible. He tried to avoid extended train rides in the summer.
119. Linde, *Aus meinem Leben*, p. 5.
120. Ibid, p. 4.
121. Linde commented on dogmatic, intolerant discussions held by pietistic theologians with 'nescimus'. Information provided by Professor Hermann Linde.
122. Linde to his brother Siegmund, 12 June 1893: 'You know that according to the mores in which we were raised consider divorce to be unjust and that your siblings will likely do so as well. I at least, do not see it this way.' PAHL, Cb. 8, p. 152.
123. Linde founded and led the Protestant Kirchenbauverein in Berchtesgaden; 200 letters to farmers in the Berchtesgaden region in Linde's business correspondence indicate that there was a financial benefactor with a strong moral purpose in the area from 1885 to 1905.
124. The 19-year-old Linde justified his choice of career as an engineer on 19 January 1861 as a calling to serve the general public. He hoped 'that I will find in this vocation the opportunity to be a benefactor more so than in any other career and to achieve something of higher value, whether it be by spreading Christianity in China or elsewhere, by providing relief to the poor slaves of America by working for their freedom, by bringing science to Africa as a member of an expedition, or as a mechanic or chemist to find a useful invention.' PAHL, Schattenmann, Integrität der Seele.
125. As the son of a minister, Linde was well-acquainted with biblical images and terms that he employed often with secular meaning in his letters. Writing to the

Supervisory Board chairman Lang-Puchhof in 1893, he refers to having 'survived the seven lean years in England'. PAHL, Cb. 8, p. 208.

126. Chandler sees this attitude as typical of German businesses in mechanical engineering after 1870. Chandler, *Scale and Scope*, p. 395 and 497. Similarly Jaeger, *Geschichte der Wirtschaftsordnung*, p. 119; Siegrist, *Familienbetrieb*, p. 1981.

127. This was made clear by the various increases to capital encouraged by Linde, which effectively shrank his relative share of the AG to an insignificant amount. On 28 September 1889 Linde tried to convince his largest shareholder, Moritz von Hirsch, to increase capital and buy the Rudloff Grübs Company, writing: 'Nobody, least of all myself, is comfortable with such an increase.' Letter from Linde to von Hirsch. PAHL, Cb. 4, p. 458.

128. Linde considered personal toughness as a character weakness. Writing to his brother in 1893: 'I believe it is worth noting that we Lindes on the one hand tend to be inconsiderate in expressing our judgment of others, however, on the other hand, we cannot stand hearing others' judgment of us.' PAHL. Cb. 8, p. 206.

129. Richard Tilly emphasizes German businessmen's honesty as a factor in their economic success in the nineteenth century as compared to England. Tilly, *Unternehmermoral*.

130. Examples: Linde went in 1877 to Paris, Trieste, Vienna, Winterthur, and Düsseldorf (Cb. 1, pp. 196 and 250), 1878 to Trieste, Düsseldorf, Mainz, Elberfeld, Dublin (Cb. 1, pp. 386 and 475), 1880 to Prague, Vienna, Pest, Florence, Amsterdam, and Odessa. Letter from Lindes to Buz, 12 May 1880. PAHL, Cb. 1, p. 623.

131. For more on this and a general overview of cooling processes, see: Dienel, *Ingenieure zwischen Hochschule und Industrie*, p. 34 ff.; Thévenot, A History of Refrigeration, p. 36 ff.; Lütgert, *Eiskeller, Eiswerke und Kühlhäuser*, p. 93 f.

132. Also: Kirk, the Scottish William Rankine, one of the founders of modern thermodynamics or the Director of the Karlsruher Gewerbehalle Heinrich Meidinger, who also had constructed a refrigeration unit had done this; cf. Dienel, *Ingenieure zwischen Hochschule und Industrie*, p. 94.

133. Cf. Linde, C., *Über die Wärmeentziehung bei niedrigen Temperaturen*; idem, *Verbesserte Eis- und Kühlmaschine*; Dienel, *Ingenieure zwischen Hochschule und Industrie*, p. 91 ff.

134. Linde saw himself here in the tradition of his Zurich professor Gustav Zeuner, who also over-simplified cyclical processes which ignored losses; cf. Dienel, *Ingenieure zwischen Hochschule und Industrie*, p. 92.

135. Linde, C., *Über die Wärmeentziehung*, p. 365.

136. Idem, pp. 323 and 325; idem, *Verbesserte Eis- und Kühlmaschine*, p. 272.

137. Hård, *Machines are Frozen Spirit*, p. 92; Dienel, *Ingenieure zwischen Hochschule und Industrie*, p. 93.

138. Letter from Lindes to Klein, 7 July 1895. PAHL, Cb. 9, p. 205.

139. Hughes, *Erfindung Amerikas*, pp. 30–3 and 62–4.

140. Hughes is mistaken here.

141. As Elmar Sperry did with Prof. William Anthony (Cornell), Rudolf Diesel with Carl Linde, Charles Holl with Prof. Frank F. Jewett (Oberlin College), Nicolas Tesla with a professor at the TH Graz. Hughes, *Erfindung Amerikas*, p. 73. Edison employed 'a number of chemists who had received their doctorate in Germany'. Ibid., p. 42.

142. 'No other nation had illustrated such inventive power and produced such brilliant inventors as the United States of America for 50 years after 1870.' Ibid., p. 23.

143. Ibid., p. 93.

144. Die Entwicklung der modernen Kompressionskaltdampfmaschine, *KA* (1929), June, 50–2.

145. 'In 1873 he could set up the first proper functioning refrigeration system at Gabriel Seldmayrs' brewery [...] perfected versions thereof were to follow soon

thereafter.' In: Kurzer Lebenslauf Carl von Lindes (Linde Press Release). Linde-Archiv Höllriegelskreuth.

146. For example, Linde's patent application at HStA, correspondence and bills for Maschinenfabrik Augsburg (MAN-Archive). The history of the Spaten brewery from Fritz Sedlmayr is based on Sedlmayr's diaries and the Spaten brewery's books, which were partially destroyed by fire during the Second World War. The remaining material is to be found in the Munich Municipal Archive.

147. HStA, Min.d. Inn. 38240.

148. There are two designs of this patented machine at the Maschinenfabrik Augsburg Archive. Sedlmayr wrote his comments and signature on one of the documents. MAN Historical Archive, Zeichnungen Eismaschinen, Fach 15.

149. On 18 January and 4 February 1873 the machine works confirmed the contract, sending Linde a detailed drawing on 4 March noting that new dimensions were 'smaller than the first machine'. Another 11 letters followed. MAN Historisches Archiv, Allgemeine Geschäftskorrespondenz.

150. Sedlmayr: Die Geschichte der Spatenbrauerei, p. 369 and HStA, Min.d. Inn. 38240. Parts of machines were calculated until the beginning of December 1873; in January 1874 a transmission followed, presumably a new construction for the engine. In: MAN Historisches Archiv, Hauptbuch 1872–74, p. 311.

151. This and the following information was gathered primarily by Fritz Sedlmayr (Sedlmayr, *Die Geschichte der Spatenbrauerei*, p. 369 f). Sedlmayr based his information on the diaries of his grandfather Gabriel Sedlmayr, discussions with his father and the report of an intern, Christian Weymar, who worked in the brewery in 1873.

152. Oil and ammonia consumption was an important operational cost factor.

153. Loss due to leakage increased waiting time. Pressure loss due to refrigerant exhaust raised only (in the best case) refrigerating temperatures, in the worst case it led to a halt in operations.

154. The commonly used refrigerants were either highly poisonous or explosive.

155. Linde also overestimated the importance of his stuffing boxes. In his autobiography he places the stuffing box drawings as number one in his index. At the end of the nineteenth century in a letter exchange with August Osenbrück, he insists on the priority of his invention. In: Lorenz: Die erste Benutzung der Laternenstopfbüchse, ZgesKI 4 (1897), 69–71. Lorenz points to descriptions of stuffing boxes previous to Linde, namely those from Karl Swoboda. Swoboda, *Die Anlegung und Benutzung*, p. 13.

156. Description of ice and refrigerating units from C. Linde, 17 January 1873. MAN Historisches Archiv, 3.37.295.

157. The machine built in 1873 for Sedlmayr was different in design from the patented machine, but the sealing principle remained the same. Without practical experiments, there would have been no clues as to how to improve the design.

158. MAN Historisches Archiv, Hauptbuch 1872–74, pp. 300–2.

159. In his autobiography, Linde claims that the theoretical predictions for rotation speed were not achieved, making the pump impractical. Linde, *Aus meinem Leben*, p. 38.

160. That is what Linde wrote in 1876 on his supplementary form for the Bavarian patent in 1873. HStA, Min. d. Inn. 38241.

161. Sedlmayr, *Die Geschichte der Spatenbrauerei*, p. 369 f.

162. Linde, *Aus meinem Leben*, p. 39.

163. Glycerine had two positive qualities. It was a lubricant and could be combined with ammonia.

164. Erecting old machines and refilling their mercury/water was, as seen in the old drawings, a difficult problem. The new design had its own expandable filling cylinder for water. See drawings in: Linde, *Aus meinem Leben*, Appendix, p. 1.

165. Maschinenfabrik Augsburg's prices were calculated exclusively according to weight at that time. According to the records, the machines weighed 4550 lbs and cost 1933 florins.
166. HStA. Min. d. Inn. 38241.
167. MAN, Historisches Archiv, Vertragsbuch Nr. 5 (1875–80). The consortium was established somewhat earlier, before May 1875, according to Linde. The first written reference is in the Supervisory Board protocol for Maschinenfabrik Augsburg, 28 February 1876 (MAN Historisches Archiv, Box 3–37, Akt 1).
168. 2 November 1877. In: MAN, Historisches Archiv, Vertragsbuch Nr. 5 (1875–80), p. 417.
169. DRP 1250, 9 September 1877.
170. Contract between Anton Dreher in Klein Schwechat and Maschinenfabrik Augsburg, 1 September 1876. In: MAN Historisches Archiv, Vertragsbuch Nr. 5 (1875–80); Linde, Aus meinem Leben, p. 39.
171. Linde wrote on the double-acting horizontal compressors, they were, according to him a 'close proximation of the usual layout of gas pumps'. Linde, Aus meinem Leben, p. 38. See also: Rheinmetall-Borsig, Deutscher Maschinenbau 1837–1937, p. 309: Linde's design 'was an addition to the common design for horizontal double-acting gas pumps'.
172. In the 1880s, Linde, Osenbrück and Pictét argued over priority and patent rights for this invention. Lorenz: Die erste Benutzung der Laternenstopfbüchse, ZgesKI (1897), 69–71.
173. Brief Lindes, 6 December 1877 to Wilhelm Schipper in Triest. PAHL, Cb. 1, p. 237.
174. Linde's notice to the Physikalisch-Technische Reichsanstalt, 29 July 1888. PAHL. Cb. 4, pp. 276–81.
175. Maschinenfabrik Augsburg's products were sold simply by weight in the 1870s.
176. In a letter to his client Gustav Jung, Linde pointed out the price difference for effectively the same machine; his machines cost 36,000 florins, compared to 37,200 (Vaas & Littmann), 50,000 (Mignon & Rouart) and 84,000 (Pictét). Letter from Linde to Otto Jung (Mainzer Aktienbrauerei), 29 June 1878. PAHL, Cb. 1, p. 483.
177. Low-running horizontal double-acting ammonia compressors were made in the German Democratic Republic (GDR) until 1963 at the Maschinenfabrik Halle. Jungnickel, Carl von Linde, LuK 28 (1992), 57.
178. This has been overlooked by many. Classifying the refrigerating unit as a 'thermodynamic invention' has led to developments with the machine being celebrated as a triumph of heat theory. For example, Mikael Hård writes of the development of a 'thermodynamically sound compressor'. Hård, In the Icy Waters, p. 198.
179. Simple vertical compressor (1871/73), interconnected three-piston system (1873), interconnected single-acting double compressor (1876) and the horizontal double-acting compressor (1877).
180. This was in and of itself not a disadvantage. Clients wanted ice that lasted. Cloudy ice had a specific weight of 0.85 kg/dm^3, crystal ice of 0,9. Göttsche, Die Kühlmaschinen, p. 582.
181. Letter from Linde to his director in Hamburg Flint, 4 June 1893. PAHL, Cb. 8, p. 145.
182. Göttsche, Die Kühlmaschinen, p. 581.
183. On 22 August 80 in a letter to Feltmann in Rotterdam, Linde views clear ice production possible by doubling generator size for the same production unit. PAHL, Cb. 5, p. 51.
184. Spring of 1880, Schipper puts an end to work on drum to extract air from water via vacuum. However, Linde advises against this in a letter, 2 March 1880: PAHL, Cb. 3, p. 435.

185. Linde negotiated with the British Bannister (Price Co.) on 6 January 1893 over a patent for finned apparatus for clear ice production. PAHL, Cb. 7, pp. 407 and 451.

186. Copper or sheet iron cans were hung in rows in a brine bath and were lifted and rotated by crane after freezing, allowing a block of ice to fall out.

187. Ice cans are mounted upon a rotary drum that is cooled from the inside. The rotary drum rotates in water, allowing for ice to form in the cans. Water overflow flushes air bubbles away. The device produced clear ice, however, use and construction are complicated.

188. Letters from Linde to Maschinenfabrik Augsburg, 12 December 1878 and 14 December 1878. PAHL, Cb. 1, pp. 248 and 253.

189. Letter from Linde to Sulzer Brothers, Winterthur, 12 December 1878. PAHL, Cb. 1, p. 245.

190. Linde thus asked the Sulzer Brothers on 14 January 1878 for 'entirely new arrangements for the Paris Exposition'. PAHL, Cb. 1, p. 278.

191. PAHL. Cb. 1, 300). Sulzer presented the drum at the Paris World Exposition in June (Linde to Heinrich Buz, 10 May 1878), PAHL, Cb. 1, p. 403.

192. Letter from Linde to Sulzer, 2 September 1878 informing of Sedlmayr's agreement to allow tests. PAHL, Cb. 1, p. 498.

193. Letter from Linde to Fa. Carels Frères, 8 February 1879: 'Ce generateur rotatif a fonctionné parfaitement', PAHL, Cb. 3, p. 52. This was presumably purposeful optimism, as Linde had complained to Sulzer on 11 February 1879 about the Carel Frères' drum being slow. PAHL, Cb. 3, p. 59.

194. Letter from Linde to Schipper, 1 August 1879. PAHL, Cb. 3, p. 257.

195. Linde considered whether 'doubling the generator dimensions would produce the desired clarity'. Letter from Linde to Schipper, 9 December 1879. PAHL, Cb. 3, p. 321.

196. Linde to Schipper, 18 December 1879. PAHL, Cb. 3, p. 331.

197. Carl Linde: calculation for an ice factory with daily production of 24 tons of crystal ice. PAHL, Cb. 3, pp. 178–81.

198. Letter from Carl Linde to Sigmund Linde, 22 October 1881. PAHL, Cb. 1, p. 662.

199. Letter from Lindes to Negele, 4 September 1890. PAHL, Cb. 6, p. 103.

200. PAHL, Cb. 5, p. 17.

201. Letter from Linde to Samuel Geoghegan (Guinness), 16 January 1893. PAHL, Cb. 7, p. 425. Also: DRP 60368 der Gesellschaft Linde aus dem Jahr 1891 für die Kondenswassergewinnung und Reinigung.

202. Linde could speak German with a number of foreign brewers, such as Geoghegan in Dublin or Moritz in Barcelona. PAHL, Cb. 7, pp. 1 and 334.

203. Letter from Linde to Robert Banfield on 2 June 1879. PAHL, Cb. 3, p. 194.

204. Four Reaumur are given in the contract. MAN Historisches Archiv, Contract record book Nr. 5 (1875–80). The machine cost 12,000 Austrian Florins and weighed 12,000 kg. It was used successfully in operations for over 30 years at the Dreher Brewery and was later put on display at the Vienna Technical Museum.

205. Linde, *Aus meinem Leben*, p. 42.

206. PAHL, Linde-Cb. 3, p. 128. After having conducted comparisons, the Sulzer brothers decided upon radial ventilators to be built into the fermenting cellar at the Hürlimann Brewery. On 14 May 1879, Linde wrote to Sulzer stating that he would continue with axial ventilators for his four projects (PAHL, Linde-Cb. 3, p. 170 f), and then ordered on 30 May screw ventilators from Dresden. PAHL, Linde-Cb. 3, p. 170 f.

207. Linde attempted to encourage and cheer Friedrich Schipper up in a long, sympathetic letter written in November 1877. PAHL, Linde-Cb. 1, p. 189.

208. Letter from Linde to Sulzer, December 1877. The machine failed due to frequent interruptions during daily operational use. PAHL, Cb. 1, p. 220.

209. Letter from Linde to Heimpel, 30 April 1879. PAHL, Cb. 3, p. 125.
210. Letter from Linde to Heimpel, 30 April 1878: 'I am most concerned about how the Sedlmayrs view the situation and what sort of assessment may result. A damaging impression will be left.' PAHL, Cb. 3, p. 125.
211. Linde accused Pfitzner of an arbitrary act for Spaten's air cooling 'which is the least efficient of all our installations'. PAHL, Cb. 3, p. 236.
212. Letter to Pfitzner, 27 March 1880. PAHL, Cb. 3, p. 457.
213. Letter from Linde to Carels Frères, 16 November 1877. PAHL, Cb. 1, p. 171.
214. Letter from Linde to Feltmann, November 1877. PAHL, Cb. 1, p. 182.
215. Letter from Linde to Heimpel in Barmen, 3 March 1880. PAHL, Cb. 3, p. 441.
216. Letter from Linde to Feltmann in Rotterdam, 25 March 1880. PAHL, Cb. 3, p. 454.
217. Linde suggested pipe sleeve joints with right-left threads to Feltmann, 20 April 1880. PAHL, Cb. 3, p. 485.
218. On 7 March 78 Linde thanked Feltmann for the order of a large installation from Jacobsen. PAHL, Cb. 1, p. 511.
219. Letter from Linde to J.C. Jacobsen, 4 July , 7 July and 13 July 1880. PAHL, Cb. 5, pp. 26 and 31.
220. Letter from Linde to the brewer J.A. Dietz, 22 August 1880. PAHL, Cb. 5, p. 47.
221. Gesellschaft für Geschichte und Bibliographie des Brauwesens (ed.): Carl von Lindes Kältemaschine, p. 41.
222. Linde clearly emphasized to Zeuner, that he 'came to our current directions through extensive and multifaceted experience'. PAHL, Cb. 5, p. 245; Göttsche, Die Kühlmaschinen, p. 323.
223. The given values for fermenting and storage cellars (2.25 m piping pro m² floor space) are twice as high, as suggested in later handbooks. PAHL : Cb. 7, p. 11.
224. Letters from Linde to Lightfoot, 7 April and 29 December 1892 with sketches for air ducts. PAHL, Cb. 7, pp. 100 and 387.
225. In Löwenbrauerei. Gesellschaft für Geschichte und Bibliographie des Brauwesens (ed.): Carl von Lindes Kältemaschine, p. 49.
226. Linde offered Theissen to patent the idea, before he was shown the drawings. 'because you are a direct competitor'. Letters from Linde to Theissen 28 November and 8 December 1890. PAHL, Cb. 6, pp. 196 and 209.
227. Letter from Linde to Theissen, 2 January 93. PAHL, Cb. 9, p. 390. Theissen offered his research services frequently to Linde, but was greeted either with refusal or disinterest. PAHL, Cb. 9, pp. 39 and 386.
228. Letters from Linde to T.B. Lightfoot, 2 December 1893 and 3 January 1894. PAHL, Cb. 7, pp. 266 and 294.
229. Letter from Linde to his Berlin branch director Sorge, 10 July 1893. PAHL, Cb. 8, p. 167.
230. Letter from Linde to Friedrich Schipper, 19 March 1892. PAHL, Cb. 7, p. 38.
231. These 220 letters are for the most part written to clients so as to defame competing refrigeration processes and machines, not to introduce solutions to technical problems.
232. Up until 1878, there were licensing agreements made with MA for Germany, with Sulzer for Switzerland, Italy and Spain, with Sartre & Averly for France, with Carels Frères for Belgium and Holland, and with Morton for Great Britain.
233. Linde, Aus meinem Leben, p. 44.
234. PAHL, Cb. 1, Brief 57 and 78; cf. also Dienel, Ingenieure zwischen Hochschule und Industrie, p. 150.
235. Moritz von Hirsch, born* 9 December 1831, died† 20 April 1896.
236. PAHL, Cb. 1, Brief 395.
237. Dienel, Ingenieure zwischen Hochschule und Industrie, p. 151; PAHL, Cb. 1, Brief 483.

238. PAHL, Cb. 1, Brief 530.
239. Ibid., Brief 524; Linde, *Aus meinem Leben und von meiner Arbeit*, p. 46.
240. PAHL, Cb. 3, Brief 57.
241. PAHL, Cb. 3, Brief 161.
242. Cf. Moll, *Aktiengesellschaften*, pp. 141–60, here p. 155.
243. Linde, *Aus meinem Leben*, p. 48.
244. Ibid.; money was the trade-off for Linde's patent rights that he transferred to the company. The interest made from this was to balance losses due to pension rights as well as to insure the family in the event of an accident.
245. PAHL, Cb. 1, Brief 804.
246. PAHL, Cb. 1, Brief 808.
247. PAHL, Cb. 4, Brief 458 and 464, Cb. 6, Brief 2 and 6.
248. About Moritz von Hirsch, cf. *Deutsche Biographische Enzyklopädie* Bd. 5, p. 63.
249. Dienel, *Ingenieure zwischen Hochschule und Industrie*, p. 152 f., Anm. 57.
250. PAHL, Cb. 1, Brief 653.
251. Ibid., Brief 742.
252. Linde, *Aus meinem Leben*, p. 48.
253. Ibid.
254. Ibid., p. 73.
255. Friedrich Linde, * 1870, † 1965.
256. Richard Linde, * 1876, † 1961.
257. Rudolf Wucherer, * 1875, † 1966.
258. Since 1898 with Maschinenbau-Actien-Gesellschaft Nürnberg, fused into Vereinigte Maschinenfabrik Augsburg and Maschinenbaugesellschaft Nürnberg A.-G., since 1908 Maschinenfabrik Augsburg-Nürnberg Aktiengesellschaft, acronym: MAN.
259. Wilhelm von Oechelhäuser, * 4 January 1850, † 31 May 1923. Oechelhäuser was general director of the German Continental Gas Company in Dessau since 1890. In 1898 he designed the first German industrial strength gas machine using blast furnace gas. He was chair of the Association of German Engineers, amongst other duties; cf. *Deutsche Biographische Enzyklopädie* Bd. 7, p. 462.
260. Johannes Hess, * 8 November 1877, † 3 February 1951. Hess was technical director of business for Wacker Chemie, which he turned into one of the top German chemical companies; cf. *Deutsche Biographische Enzyklopädie* Bd. 4, p. 671.
261. Cf. Dienel, *Ingenieure zwischen Hochschule und Industrie*, p. 154 ff.; Marsch, *Zwischen Wissenschaft und Wirtschaft*, pp. 68 and 396.
262. Cf. Dienel, *Ingenieure zwischen Hochschule und Industrie*, p. 155 ff.
263. This department was given the letter B while the engineering department, which stayed in Wiesbaden and continued with refrigeration engineering, was given the letter A for a title.
264. Linde AG, *1879–1979*, p. 85.
265. Linde, *Aus meinem Leben*, p. 54.
266. Protokolle der Generalversammlung. Linde-Archiv Wiesbaden, IV.
267. For more on the economic development of Gesellschaft für Linde's Eismaschinen see appendix.
268. On Diesel: Sass, *Geschichte*; Diesel, *Diesel*; Thomas, *Diesel*. Recently: Knie, *Diesel*. None of these texts satisfactorily explore Diesel's 13 years of experience with refrigeration in his early years.
269. The Technical University Munich rarely awarded graduates diplomas with distinction after 1868. ATUM, RA, Diplomverzeichnis 1868–1900.
270. The employment contract with Linde is dated 25 January 1883. On 26 February 1893 Diesel quit and left the Linde company on 15 April 1893. MAN Historisches Archiv, NRD, XI3b.
271. On 31 December 1884 he founded together with Louis Philippe Cohen a company Cohen & Cie. In 1886 he signed service and consultation contracts with the Société

Anonyme des Glacières, Paris and the company Grillon & Cie. In Berlin, Diesel became the representative of Sandt's malt machine for Belgium and France on 23 September 1891 (contract with J. Sandt). MAN Historisches Archiv, NRD, XI3b.

272. MAN Historisches Archiv, NRD, XI3b, K4, Folders Notes on crystallization and concentration 1883, extraction of lactose and glycerine via cooling.

273. MAN Historisches Archiv, NRD, XI3b, K4, Folder Description of ammonia engine. 1880–87; Folder Ammonia engine (there: Théorie du moteur d'ammoniaque (160 pages), 1888).

274. MAN, NRD, XI3b, K4. Folder Various ideas from Rudolf Diesel.

275. A list of the necessary requirements for a refrigeration test laboratory is at the archive. MAN Historisches Archiv, NRD, XI3b, Folder Notes on crystallization.

276. He corresponded with Moritz Schröter on steam pressure and sent him a manuscript of 'Saturated Vapours of Carbon Dioxide' (20 pages). Diesel had asked for vapour tables of sulphuric acid, ammonia and carbon dioxide from Moritz Schröter. MAN Historisches Archiv, NRD, XI3b, K4, Folder Zur mechanischen Wärmetheorie (Gesättigte Dämpfe der Kohlensäure, 1889). Also there: correspondence with Zeuner, Schröter.

277. Notes, often titled 'Essais à faire' are in abundance in the following folders: MAN Historisches Archiv, NRD, XI3b, K4, Folder Notizen zur Kristallisation.

278. MAN, NRD, XI3b, K4. Folder Extraction of lactose through cooling (1881–83).

279. His handicap kept him from working with refrigeration machines, as he wrote to his parents and wife on 25 December 1887: 'I sense that my relationship with Linde is definitely coming to an end, although I do not know what is to come.' Quoted in Thomas, *Diesel*, p. 14 f.

280. Rudolf Diesels notes, April 1893. MAN Historisches Archiv, NRD, XI3b, Folder Schriftwechsel Rudolf Diesel/Linde.

281. Linde's criticism was rather constructive and carefully encouraging. He tried to express to Diesel 'that in the best case scenario, only one-third of his calculations for theoretical efficiency can be expected to be so'. Such success is possible 'if you could work with an efficient machine factory or other person ... as much as I'd like to be of help'. But the Linde company was not in a position to do so and therefore Diesel would have to leave if he were to decide 'to truly pursue things'. Letter from Linde to Diesel, 20 March 1892. On 7 February 1893 Linde added that there was 'something uncomfortable about the thought of a possible collision with the interests of the Linde Company' with regards to the engine issue. PAHL, Cb. 7, pp. 41 and 485.

282. Linde's business correspondence shows letters regarding diesel engines were sent only to Rudolf Diesel himself. On 30 June 1893 he wrote to Diesel, attempts to establish a support network for the engine were of no success. PAHL, Cb. 7, p. 187.

283. Letter from Linde to Diesel, 28 February 1893. PAHL, Cb. 8, p. 8.

284. Letter from Linde to Lang (Supervisory Board), 14 June 1882. PAHL, Cb. 1, pp. 680–7, 793; Cb. 2, p. 190.

285. Assistants in the department of mechanics-technology at the Technical University Munich earned between 2,000 and 2,200 Marks in 1881, in 1892 between 1,720 and 2,538 Marks. ATUM, RA, IV (1881, 1892). Mollier's assistants at the TU Dresden earned between 1,200 and 2,200 Marks from 1900 to 1914. StAD, Min Vobi, 15706 (Institut für theoretische Maschinenlehre).

286. Machinists' income for ice works in Milan was at 1920 Marks p.a. PAHL, Cb. 6, p. 228.

287. The annual income of employees in the industry increased in Germany from 617 Marks in 1870 to 1190 Marks in 1911. An average worker's wage was of course much lower than this. The top 1 per cent wage earners comprised nearly 20 per cent of total income in the empire. In Prussia, less than 3 per cent of the population

earned over 3,000 Marks a year in 1896 and in 1912 less than 5 per cent. Only a solid 1 per cent of all income-earners received over 6,000 Marks p.a. in Prussia in 1896, in 1912 it was just 1.5 per cent. Nipperdey, *Deutsche Geschichte*, p. 289.

288. Jäckel established that engineering civil servants usually earned more than their colleagues in industry, whose wages in some cases fell below that of qualified workers; 18.7 per cent of those civil servants analysed earned more than 3,600 Marks p.a. Jäckel, *Statistik*, p. 51; Kaelble, *Industrielle Interessenpolitik*, p. 74 f. See Pierenkemper for another view: Pierenkemper, *Einkommensentwicklung*.

289. Supervisory Board Resolution, 4 December 1907 on hiring five engineers for 3,600 Marks each p.a. and on 3 November 1908 on hiring six engineers for 3,300 Marks to 5,400 p.a. KHDA, XII, 5.

290. Letter of hire Reichenwallner, 6 November 1882. PAHL, Cb. 4, p. 10.

291. Employment contract, Linde Company with August Krebs, 20 May 1882. PAHL, Cb. 1, 672. Krebs was previously an assistant to Grove, the best-payed professor in the mechanics-technology department at the Technical University Munich (8,000 Marks p.a.), receiving an annual pay of 2,646 Marks from the university. ATUM, RA, IV,

292. PAHL, Cb. 4, p. 32. A number of engineers and businessmen without shares in profits received in 1885 a bonus of 1,000 to 15,000 Marks. The company distributed dividends of 104 per cent in 1884.

293. Supervisory Board protocol, 25 February 1888 (Power of procuration to Reichenwallner and Scharnberger, 1 per cent of profits) and 2 October 1890 (Banfield returns to Wiesbaden from England with 10,000 p.a.) Linde-Archiv Wiesbaden, IV, Protocol Supervisory Board meetings 1879–1916.

294. Linde counted on a net profit of 500,000 p.a., that is 50,000 Marks Tantieme. In actuality, net profit was higher, but Michaelis was nonetheless let go. Letter from Linde to Michaelis, 25 March 1906. PAHL, Cb. 11, p. 11.

295. Suicides were also reported..

296. PAHL, Cb. 8, Brief 123, Letter from Linde to Robert Banfield, 10 May 1893.

297. Jaeger, *Aus der Geschichte der Werksgruppe TVT*, p. 189.

298. He later had himself be addressed as 'Highly esteemed Mr Privy Councillor'. Cf. Dienel, *Ingenieure zwischen Hochschule und Industrie*, p. 164 f.

299. Ibid., p. 173.

300. Carl letter from Linde to Friedrich Schipper, 27 July 1893. PAHL, Cb. 8, Brief 191.

301. Private Archive Hermann Linde, Cb. 6, Brief 111, letter from Linde to Banfield, 5 September 1890; Dienel, *Ingenieure zwischen Hochschule und Industrie*, p. 181.

302. Letter from Linde to Maschinenfabrik Sulzer in Winterthur, 24 April 1882. PAHL, Cb. 5, Brief 128.

303. Letter from Linde to Fenzl, 23 August 1883. PAHL, Cb. 5, Brief 204.

304. Richard Linde, Meine Tätigkeit bei der Abteilung B der Gesellschaft Linde. Ms. from 1954/1955, p. 6. PAHL.

305. Along with the Spaten breweries, Linde also had the Aktien brewery in Mainz and Rotterdam's Heineken brewery for risky projects. Linde 4, Cb. 3, p. 242 (Jung/Mainz), p. 437 (Sedlmayr/Spaten), p. 454 (Feltmann/Heineken).

306. The atmospheric condensers were tested and developed at cold stores in Hamburg during the early 1890s.

307. Letter from Carl Linde to Fritz Linde, 7 February 1892. PAHL, Cb. 7, p. 488.

308. Supervisory Board meeting, 11 February 1893: founding of a Munich subsidiary is suggested again by Linde but denied. Linde-Archiv Wiesbaden, IV.

309. Linde, Carl: Versuchsstation der Gesellschaft für Lindes Eismaschinen AG in Höllriegelskreuth bei München, in: Z-VDI 48(1903), p. 1362 f. The station had air liquefaction and separation devices (rectifiers, engines, compressors, countercurrenters and air dryers) that were tested and improved.

310. Wiesbaden headquarters had run a small lab in Wiesbaden since the 1930s that was integrated with Sürth. Tettinger, *Aus der Arbeit*; Schnabel, *Aufbau des Prüfwesens*.
311. Linde, *Aus meinem Leben*, pp. 62–8; Hesselmann, *Wirtschaftsbürgertum*, pp. 312–14.
312. Maschinenfabrik Augsburg, the most powerful licensee, tried in vain to acquire right of sale as well. Linde threatened with direct competition for this 'alarming and despicable situation' and appealed to Buz with a 'certain bitterness'. MA responded by withdrawing their request. Letter from Linde to Buz, 24 August 1885. Linde-Archiv Wiesbaden, Cb. 4, p. 53.
313. Regarding the company Mignon & Rouart. PAHL, Cb. 1, p. 42. Regarding his friend Feltmann in 1882, he 'did not value patents in terms of security ... very highly'. Value lay more in 'the experiences and contracts with Sulzer and MA'. 'There are things that one cannot have patented, which not everyone understands.' PAHL, Cb. 1, pp. 708 and 712.
314. Lossen Brothers (Darmstadt) refused 11 March 1879. PAHL, Cb. 3, p. 78. Dr Proell (Dresden) refused on 19 February 1879. Cb. 3, 60. Schambach & Craemer (Koblenz) refused on 13 April 1879. Cb. 3, 107. Manlove, Alliot & Co (Nottingham) refused on 17 December 1878. PAHL, Cb. 1, p. 578.
315. Sum of patent fees. PAHL, Cb. 3, p. 195.
316. Frequently to Linde British Refrigeration Co.
317. Frequently to Sulzer.
318. PAHL, Cb. 1, p. 42.
319. Letter from Linde to Armengaud & Fils, 21 November 1877. PAHL, Cb. 1, p. 202.
320. PAHL, Cb. 1, pp. 167, 540 and 582.
321. Letter from Linde to Diesel, 28 July 1879 PAHL, Cb. 3, p. 195. Linde congratulated him on the success of this founding, 30 December 1879. PAHL, Cb. 3, p. 340.
322. Letters from Linde to Sauvan, 19 October, 24 October, 27 October, 5 November, 7 November, 10 November 1879 with technical details and so on. PAHL, Cb. 3, pp. 291–301, 307 and 311.
323. Letter to Ducretet & Legenne in Paris. PAHL, Cb. 9, p. 386.
324. PAHL, Cb. 1, pp. 172, 239, 266, 410 and 413. On 25 May 78, letter from Linde to his employee Banfield on Morton: 'Can't we break the contract with him?' PAHL, Cb. 1, p. 443. On 24 June 1878 Linde ended his contract and insisted that the drawings be burned with a witness watching. PAHL, Cb. 1, p. 448.
325. PAHL, Cb. 1, pp. 506 and 591; Cb. 3, pp. 53 and 106.
326. Delivery bill, 7 January 1881. PAHL, Cb. 5, p. 78.
327. Letter to Le Vino, 10 December 1881. PAHL, Cb. 5, p. 117. Letter to the Austro-Bavarian Lagerbeer Brewery, London, 4 June 1883. PAHL, Cb. 5, p. 150.
328. Linde's suggestion to Feltmann, to buy patent rights from the Austro-Bavarian Lagerbeerbrewery and to start a business. Foundation plan, 23 October 1882–22 March 1883. PAHL, Cb. 1, pp. 704, 708, 712, 722 and 735. The foundation was planned for 7 March 1883, but vetoed by the Linde Executive Board. Ibid. Cb. 1, pp. 739 and 742.
329. Letter from Linde to Thomas Pigott, 26 August 1883. PAHL, Cb. 5, p. 207. The licence contract was sent on 17 February 1884. Cb. 5, p. 265. Further negotiations, followed which eventually led to the founding of the Linde Britisch Refr. Co. Cb. 5, p. 268.
330. Carl von Lang-Puchhof invested £20,000, his brother £20,000, Linde £10,000, Feltmann £10,000, v. Hirsch £10,000 and a group of British investors with £20,000. PAHL, Cb. 4, pp. 61 and 65.
331. In 1892 the company made its first big profit. Lightfoot was an inventor businessman. His air refrigeration machines sold very well in England in the 1880s, however they were phasing out. On 4 August 1890 he wrote to Feltmann, the new business

director in London Lightfoot had 'proven himself invaluable'. PAHL, Cb. 4, p. 409 and Cb. 6, p. 81.

332. With respect to increases, Linde himself did without the benefit of Lightfoot, granting him a higher percentage of profits for his services as a 'consulting engineer'. In 1898 Lightfoot was already receiving a set annual pay of 40,000 Marks plus 20 per cent share of the net profit. Letter from Linde to Lightfoot, 14 January 1898. PAHL, Cb. 10, p. 133.

333. The first partner is unknown; he was however, so bad that a number of clients were dependent upon Sulzer for delivery orders. PAHL, Cb. 7, p. 237. In August 1894 Lightfoot signed a licence contract with Davey Paxmann. PAHL, Cb. 8, p. 493.

334. Although engineers with a few years' experience in Wiesbaden under their belt such as Statham Jr and Lightfoot Jr ran the business, new co-operation was never resumed, perhaps precisely because of the aforementioned experience. Attempts towards new co-operation failed, according to Schipper, due to 'Mr. Lightfoot and Mr. Statham's arrogance'. Letter from Schipper to F. Linde, 26 May 1929. Linde-Archiv Höllriegelskreuth, Korrespondenz Schipper.

335. Letter from Linde to Carels Frères, 16 November 1877. PAHL, Cb. 1, p. 171. In January Carels could announce the first machine to the patent office. Linde congratulates, 28 January 1878. PAHL, Cb. 1, p. 293.

336. Linde, *Aus meinem Leben*, p. 66.

337. Linde referred to the licensed manufacturer in a letter in November 1876 to the English client Tarnet. PAHL, Cb. 1, p. 47. Noback & Fritze often sold refrigeration machines for the Linde Company in the years following, but did not, presumably, produce them themselves. In 1878 Noback was at the Paris Exposition as a member of the testing commission and evaluated Sulzer's Linde machines. Letter from Linde to Noback, 23 June 1878. PAHL, Cb. 1, p. 464.

338. Schröter worked as an engineer for Sigl as of 1877.

339. Letter from Linde to Sigl. PAHL, Cb. 1, p. 405.

340. Linde wrote to Schipper on 8 May 1892: 'After three negotiation attempts which failed ten times, the contract with Skoda and Ringhofer was signed.' PAHL, Cb. 7, p. 146. Ringhoffer had by 1889 already ordered their first refrigeration unit. Linde responded on 2 July 1889. PAHL, Cb. 4, p. 456. Contract draft 11 May 1892. PAHL, Cb. 7, p. 150, 2nd draft with price lists (ibid., p. 163). Ringhoffer was allowed to supply only Bohemia. Approval of a special delivery to lower Austria on 1 December 1901. PAHL, Cb. 10, p. 231. F. Ringhoffer (ed): Firmenkatalog. NMAH-Archives, EC, Refrigeration Collection, Trade- Catalogues.

341. Maschinenfabrik AG (ed): *Eis- und Kühlmaschinen System Linde*. NMAH-Archives, RC.

342. On 1 March 1913, MAN, Skoda and Königsfeld-Thanner-Letsch signed the Austrian Refrigerating Machine Treaty, which was expanded in 1914, 1916, and 1921, falling apart in 1922. MAN originaly received a share of 19.5 per cent of contracts in Austria and 14.3 per cent in Hungary. MAN, 3/37 (Folder Verträge Österreich-Ungarn).

343. Letter from Linde to Lang, 6 January 1879. PAHL, Cb. 3, pp. 6, 148 and 183.

344. On the history of the Maschinenfabrik Augsburg in the 1870s see: Büchner, *100 Jahre*. MAN Historisches Archiv, SF, Goebel, Das Werk Augsburg; MAN Historisches Archiv, MAN FS, Bitterauf, Maschinenfabrik Augsburg; Feldman, *Weltkrieg*, pp. 161–82.

345. To avoid direct competition, the two companies divided the world market amongst themselves: MA supplied in Germany, Sulzer the rest of the world, so long as no other licence agreements existed. He openly admitted the positive effects of 'the exploitation of the competition between Sulzer and MA' to his friend Feltmann. Letter to Feltmann, 17 November 1882. PAHL, Cb. 1, p. 708. Negative aspects of the competition: MA refused to give the Sulzer Brothers, 'their toughest competitor'

a payment for the Paris Exposition. Letter from Linde to MA, 14 November 1877. PAHL, Cb. 1, p. 162.

346. 'The Linde machine owes much of its great success to excellence in mechanical detail.' Statement by J.A. Ewings, quoted in GFLE, 75 Jahre Kältetechnik, p. 15. After MAN relinquished the manufacture of refrigeration machines to Maschinenfabrik Esslingen in 1928, complaints on the quality of the products increased and requests to have MAN supply them once again are on record at the MAN archives. MAN Historisches Archiv, 3–37, Korrespondenz (1928–42).

347. Linde could speak openly of the problems with Krauss in his regular correspondence with Buz. During Linde's tough battle with the Supervisory Board members Frommel and v. Lang-Puchhof, who were pushing to phase out the refrigeration machine department for the sake of gas liquefaction, Buz, in response to Linde's plea, ensured that the Supervisory Board expressed their trust in Linde. MAN Historisches Archiv, 3–37, Kontroverse (1909).

348. The choice of Krumper is possibly due to Linde. Krumper worked in 1867 as Linde's assistant at the locomotive engineering office. Linde, *Aus meinem Leben*, p. 24.

349. Lucian Vogel studied mechanics at the Munich polytechnical from 1872 to 1876, including under Linde. He worked from 1876 to 1897 at Maschinenfabrik Augsburg, and had been Chief engineer for the department of ice since 1888. MAN Historisches Archiv, ZA, Personalakt Lucian Vogel.

350. MAN Historisches Archiv, NRB, HB.

351. The Linde Company drew a fixed premium from sales revenue even with original costs and cut-rate prices. This meant that Linde made earnings even if MA experienced losses with machine sales. Setting a minimum price for refrigeration units was therefore a constant topic of correspondence between the two companies.

352. MA maintained the right to sell machines directly to their 'old customers'. MAN Historisches Archiv, 3–37, Verträge mit GFLE; Verträge zwischen MAN und Linde, 1910 and 1916.

353. Linde agreed to supply 88 per cent of the compressors, 66 per cent of the apparatuses and 66 per cent of the drives. MAN paid 7.5 per cent premium. MAN, 3–37, Verträge MAN/Riedinger/Linde, 1920–24.

354. On Paul Reusch's influence on business policy at MAN: Feldman, *Weltkrieg*, p. 178.

355. Contract between MAN-Riedinger and the 'State Moscow Machine Manufacturing Trust' . MAN-Riedinger supplied design drafts and potentially engineering help for the manufacture of refrigeration machinees at the Moscow machine manufacturing trust in exchange for 5 per cent commission ('normal German prices destroyed'), which was not to be under 20,000 dollars per year. The contract was terminated in 1932 by the Soviet Union. A. Borsig conducted similar major 'Russian business'. MAN Historisches Archiv, 3–37, Staatlicher Moskauer Maschinenbautrust.

356. On11 May 1925 Schipper supported the acquisition of Güldner Motorenwerke in Aschaffenburg in a letter to Fritz Linde, which 'definitely would secure the future of Department A'. 'The current status of MAN is without doubt directed at destroying us and this is our only vital defense against this activity.' Linde-Archiv Höllriegelskreuth, Korrespondenz Schipper.

357. MAN Historisches Archiv, 3–37, Überleitung an Maschinenfabrik Esslingen, Vertrag, 5 October 1927. MAN Historisches Archiv, NRB 253, 1–8, Kältetechnik Linde-MAN-LAR.

358. At the MAN historical archive, there are a total of 36 names given for erecting engineers for the years 1886–95. They were the most well-paid employees at MAN.

359. The erecting hall had two overhead cranes for 15 and 20 tons, 1 metal circular saw and 22 machine tools. MAN Historisches Archiv, ZA, Schriften, Goebel, Das Werk

Augsburg, p. 36. (In contrast to this, parts manufacturing had over 250 machine tools in 1900.)

360. MAN Historisches Archiv, ZA, Schriften, Möller, Berufserinnerungen, p. 46. The department of ice had Hall A3 on the plant premises and a test room in Building C7.

361. MAN Historisches Archiv, 3–37, Bericht der Abteilung Kältemaschinen, 11 March 1921.

362. MAN Historisches Archiv, Möller, Berufserinnerungen, p. 46.

363. Schipper worked in the machine department at Maschinenfabrik Augsburg from 1 January 1878 to 15 June 1880. A separate ice department was founded first in 1886. PAHL, Cb. 1, pp. 250 and 626.

364. Otto Hippenmeyer worked as Negel's assistant near Carl von Linde from 1904 to 1913. In 1926, after refrigeration manufacturing had been put to a stop at MAN and Riedinger, Hippenmeyer returned to GFLE, entering the Executive Board in 1929. GFLE, *50 Jahre Kältetechnik*, p. 19f.

365. Letter from Linde to Heimpel, 6 May 1885. PAHL, Cb. 1, p. 793.

366. Confidential letters from A. Scholler to F. Linde, 29 April 1917, 17 May 1917 and 30 September 1918 and 2 October 1918. Includes: information regarding offers from competing companies (Borsig, Schwartzkopff). Linde-Archiv Höllriegelskreuth, Akten Friedrich Linde.

367. Salomon Sulzer founded the company in 1784 as Messinggießerei; in 1834 the company moved to a new area with an iron foundry. The first steam boiler was built in 1841, the first steam engine in 1851. From 1851 to 1871 Charles Brown was director of steam engine construction. The construction of pumps was taken into production in 1869, in 1877 refrigeration machines. In 1881 the Sulzer Brothers opened a branch in Ludwigshafen. Attempts to go into large turbine construction failed after the turn of the century, instead, as with MAN, it was the manufacture of diesel engines which became the primary economic pillar in the company. Refrigeration machine manufacturing continued, and transferred eventually to the Escher-Wyss, then the subsidiary Sulzer-Escher-Wyss, being today at Axima. For more on the company's history: Matschoß, *Geschichte*; Sulzer Brothers, *Historischer Rückblick*; Sulzer Brothers, *100 Jahre Sulzer Brothers*, pp. 88–91; Sulzer Brothers, *125 Jahre Sulzer Brothers*; Oederlin, *Lebensbilder*; Oederlin, *Geschichte der Sulzer Brothers*.

368. Sulzer Brothers, *100 Jahre Sulzer Brothers*, p. 90.

369. Carel Frères and Satre & Averly built the Sulzer ventilated steam engines as of 1873. Matschoß, *Geschichte*, p. 92.

370. 'According to a 1929 summary, 49 per cent of total production for all non-European Sulzer refrigeration machines went to Argentinia and Uruguay.' This 'provides the company with one of its largest markets, as the plants will need not only compresors and so forth, but also boilers, steam engines, pumps and later diesel engines'. Sulzer Brothers, *100 Jahre Sulzer Brothers*, p. 89.

371. Matschoß, *Geschichte*, p. 69.

372. PAHL, Cb. 8, p. 250.

373. Contracts between Sulzer Brothers and Gesellschaft Linde, 10 October 1883, 15 November 1890, 16 May 1896, 24 February 1904, 12 July 1906, 27 June 1912 und 30 June 1917 (dismissal). ASAG, 92, Verträge.

374. Hirzel-Gysi studied at TU Karlsruhe. In 1897 he left Sulzer and started his own business.

375. At the end of the 1870s, Sulzer consulted the inventor Vincent, who had suggested the refrigerant Methylflorur for compression machines. Linde rejected Sulzer's suggestion to use it. Letter from Linde to Sulzer 5 September 1878. PAHL, Cb. 1, p. 503.

376. Linde's first large project proposal for the Argentinian slaughterhouse Bertuch & Co., 29 September 1884 and 18 February 1885. PAHL, Cb. 5, pp. 252 and 259.
377. Letter from Linde to Sulzer, 5 October 1886. PAHL, Cb. 1, 919 and of 22 September 1887. PAHL, Cb. 4, p. 214.
378. The first major air conditioning project was at a hotel in Cairo in 1914. Sulzer Brothers, 100 Jahre Sulzer Brothers, p. 89.
379. Linde's first writings on liquefying air were sent to the Sulzer Brothers, 30 November 1894. PAHL, Cb. 9, p. 65.
380. The engineer E. Schechtlin became the department head, his superior was Emil Huber. Huber left the company on 17 April 1917, and Junod became branch head of the ice machine department.
381. ASAG, Box 92 c. Report from Junod and Hans Sulzer on 'L'activité de notre département frigorifique'.
382. Report from Rudolfo Landoldts to Dr Hans Sulzer, 8 October 1929. ASAG, Box 92a.
383. Landoldt's observation was similar to Möller's in Augsburg.
384. In 1946 Department 6, as it was then called, had 33 employees in the acquisitions and engineering office. In 1950 Departments 6 (refrigeration installations) and Department 2 (piston compressors) were combined; in 1956 apparatus engineering was added. Research in the successful department gravitated towards plant engineering.
385. Barmen: PAHL, Cb. 1, p. 430; Cb. 3, pp. 185 and 246. Stuttgart: PAHL, Cb. 4, pp. 155 and 391; Cb. 5, pp. 69 and 154. In 1890 he sold the ice works to Ed. Kober, who went bankrupt in 1910. Handbuch der süddeutschen Aktiengesellschaften, 1912, p. 579 f. GFLE, 50 Jahre Kältetechnik, p. 45. Schiltigheim: PAHL, Cb. 5, p. 38. On 30 March 1885, Linde offered the cold store to Schmederer for the brewery owner for 450,000 Marks. PAHL, Cb. 4, p. 40.
386. Linde, F., Die wirtschaftliche Entwicklung, p. 175. His father had written in his autobiography that the ice works were built only as 'model installations for further research'. Linde, Aus meinem Leben, p. 49.
387. Letter from Linde to Siegmund Linde, 1 November 1882 (Toulouse and Barcelona) and 1 September 1889 (Eiswerkprojekt in Lissabon). PAHL, Cb. 1, p. 707 and Cb. 4, p. 476.
388. Letter from Linde to Gottfried Linde, 26 July 1890. PAHL, Cb. 6, p. 58.
389. Both the Straßburg and Stuttgart ice works went bankrupt in 1900 and 1910 respectively. Linde's ice works sent to Paris (1880), Bombay (1880) and Milan (1887) also failed to turn profits.
390. GFLE, 50 Jahre Kältetechnik, p. 46 f.
391. A Linde relative, the manufacturer Ingenohl, had invested in founding the Société Anonyme des Frigorifères d'Anvers. GFLE, 50 Jahre Kältetechnik, p. 43; PAHL, Cb. 1, p. 938; Cb. 4, pp. 99 and 186 (Founding and Construction), pp. 272 and 420 (Expansion); Cb. 7, p. 199 (complete takeover). Shortly before the war's end in 1918, the Linde Company sold the business, fearing dispossession.
392. Cutler's ice works, built in 1879 with Linde machines, went bankrupt shortly thereafter. Letter to Cutler, 9 February 1880. PAHL, Cb. 3, pp. 393. Linde sought investment from Moritz von Hirsch (Cb. 3, pp. 365 and 409) and the Sulzer Brothers (Cb. 3, pp. 387 and 425). The ice works were later taken over by Linde British Refrigeration Co.
393. Letter from Carl Linde, 20 July 1884. PAHL, Cb. 5, p. 240.
394. Letter to Sauvan in Paris, 20 February 1880. PAHL, Cb. 3, p. 415.
395. Letter to Sulzer Brothers, 4 July 1880. PAHL, Cb. 5, p. 17.
396. Linde agreed to a 50,000 Franc investment for the ice works to be headed by Carlo Vogel. Letter from Linde to Carol Vogel, 15 July and 18 August 1888. PAHL, Cb. 4, pp. 269 and 441.

397. Letter from Linde to Matheson & Grant in London, 18 July 1888. PAHL, Cb. 4, p. 287.
398. Società Anonima per la Fabbricazione del Ghiaccio Artificiale. PAHL, Cb. 4, p. 86. The shares were sold again in 1897. GFLE, *50 Jahre Kältetechnik*.
399. Linde invested in the ice works and corresponded extensively with the founder Heinrich von Ritter-Zahony, 26 March 1893, 3 September 1894, 12, 27, 29 September, 3 October, 7 December 1894, 8 February 1896 and 18 August 1896. PAHL, Cb. 8, p. 400 and Cb. 9, pp. 1, 6, 7, 8, 21, 57, 297, 366 and 470.
400. GFLE, *50 Jahre Kältetechnik*, p. 43.
401. The Linde British Refrigeration Co. put major installations into operation at their own cost, 1885 in Shadwell (PAHL, Cb. 5, p. 268), 1894 in Birmingham (Cb. 8, p. 442) and 1896 in Grimsby (Cb. 10, p. 43).
402. In 1916 the Linde Company and Carl Linde sold Linde British Refrigeration Co. shares at 43 per cent to a Zurich bank. Letter, 11 April 1916. PAHL, Cb. 11, p. 514.
403. *Handbuch der süddeutschen Aktiengesellschaften*, p. 579 f. GFLE, *50 Jahre Kältetechnik*, p. 45.
404. Max Schwarz (1848–1917) had invested in the Augsburg Private bank P.C. Bonnet (Linde's banker) and was on the Supervisory Board at Maschinenfabrik Augsburg. Hesselmann, *Wirtschaftsbürgertum*, p. 104.
405. In 1893 Linde sent the new Hamburg cold store director Klimt on a 'spying trip' through the English cold stores, prodding him to discuss American cold stores with the Darmstadt professor Gutermuth. PAHL, Cb. 8, pp. 9 and 268.
406. Letter to Klimt, 8 January 1903. PAHL, Cb. 10, p. 275.
407. Linde, *Aus meinem Leben*, p. 107.
408. Until 1909 Sürth Maschinenfabrik, previously Hammerschmidt.
409. Sürth bought a Linde machine in 1892, presumably to study it. Bill to MF Sürth, 30 June 1892. Linde-Archiv Wiesbaden, Cb. 7, p. 183.
410. In 1909 the company had only 36 employees, yet the number grew rapidly. GFLE, *50 Jahre Kältetechnik*, p. 168.
411. Proposal to the Supervisory Board, 1920. Linde-Archiv Wiesbaden, XII, Finanzen, Aufkauf der DOAG.
412. Confidential questionnaire to Director E. Volland of MF Sürth. Linde-Archiv Wiesbaden, XII, Finance, Purchase of DOAG.
413. GFLE, *50 Jahre Kältetechnik*, pp. 163–74.
414. GFLE, *50 Jahre Kältetechnik*, p. 32.
415. Letter from Schipper to F. Linde, 9 January 1926. Linde-Archiv Höllriegelskreuth, Akten Schipper.
416. Walb, *Fünfzig Jahre Walb*, p. 20. GFLE, *75 Jahre Linde*, p. 86 f.
417. Resume Friedrich Walb. Linde-Archiv Wiesbaden, II, Geschichte der Firma Walb.
418. They sent Volland to Linde. Walb, *Fünfzig Jahre Walb*, p. 20.
419. From Petroll & Co. refrigeration manufacturing in Nordhausen, from machine manufacturer Siller in Rodenkirchen and Lanz in Mannheim and the machine factory in Karlsruhe. Walb, *Fünfzig Jahre Walb*, pp. 17–19.
420. In May 1925 Schipper supported the purchase of Güldner so that Linde could be armed for negotiations with Riedinger-MAN ziehen würde. Linde-Archiv Höllriegelskreuth, Akten Friedrich Schipper.
421. Aschaffenburg's success cast doubt on Linde's identity as a refrigeration engineering company and project oriented engineering firm.

Chapter 2

1. Carl von Linde suggested on 29 May 1897 to the Supervisory Board of GFLE the founding of a Department B with an independent budget.

2. Linde, F., *Messung der Dielektrizitätskonstanten*. Linde had measured the dielectricity constants of carbon dioxide and dinitrogen monoxide. The measurements for ammonia and sulphuric acid failed because of their high conductivity.
3. Linde's carbon dioxide refrigeration machine was tested in the experiment laboratory by the refrigeration technique commission of the Polytechnischer Verein in September 1894. Letter Linde to Schipper. PAHL, Kb 8, p. 485.
4. Linde, C., *Über die Verflüssigung der Gase, BIGB* 26 (1893). Linde asked his Paris reperesentative Desvignes to send him a brochure of a French company, which brought 'compressed oxygen in steel bombs' onto the market. He wanted to have the brochure for his lecture. PAHL, Cb. 7, p. 476.
5. The compressor codensed 440 cbm from 22 to 65 bar. GFLE, *50 Jahre Kältetechnik*, p. 80.
6. Linde described how his ideas progressed. Linde, *Aus meinem Leben*, pp. 83–9. Hausen, Gedanken und Erkenntnisse Carl von Lindes, *ZgesKI* 42 (1935), pp. 220–5.
7. Linde, *Aus meinem Leben*, p. 84. The regenerative method had a long theoretical tradition in referigeration technology. It was not applied. Regenerativkaltluftmac hine, *VDI-Z* 2 (1858), 287.
8. This idea would have faced major factory material problems, since lubrication at 190 degrees in a cylinder is very difficult. The French pioneer of gas liquefaction Georges Claude ultimately pursued this idea. In the 1930s Linde built the first gas liquefier with external pressure release in the turbines. Mendelsohn, *Nullpunkt*, p. 21.
9. Linde to Friedrich Schipper, 25 December 1894. PAHL, Cb. 9, p. 103.
10. 'From Sulzer I obtained the design drawings for a work cylinder for the "oxygen machine", as we perhaps want to call it among ourselves, but I don't agree waith that.' Linde's 25 December 1895 to Friedrich Schipper. PAHL, Cb. 9, p. 103.
11. Letter, Linde to Sulzer. 27 December 1894. PAHL, Cb. 9, p. 106.
12. Clausius, *Abhandlung*, p. 236. Kirchhoff, G.: *Mathematische Wärmetheorie*. Lectures at the University of Berlin SS 1884 (ms. by Karl Schmidt), p. 84. On Joule and Thomson's experiments with pressure release through a boxwood valve: 'The cooling of a gas is inversly proportional to the sqaure of the absolute temperature.' I thank Prof. Wladimir Kirsanov, Moscow, for allowing me to look at the manuscript.
13. William Siemens (patent No. 2064) in 1857 and Solvay had already taken the patents for this method. Only Solvay had tried to also practically liquefy the air and failed. Linde, *Aus meinem Leben*, p. 84. In December Linde thanked Friedrich Schipper for the forwarding of the Solvay's patent document, 'through which the question of priority is indeed so fundamentally solved, that I no longer think of a patent for myself'. PAHL, Cb. 9, p. 78.
14. Pursell, *Science and Industry*, p. 234.
15. DMMS, Lorenz, *Praxis, Lehre und Forschung*, p. 47.
16. Linde did not immediately recognize problems posed by the factory material itself. Steel has very low viscosity at low temperatures and therefore breaks easily. In 1896 there was a large explosion in the experiment laboratory, which shut down experiments for three months. Copper bars were used thenceforth.
17. During this period numerous improvements on the air liquefaction units were made, such as the replacement of iron pipe heat exchangers with copper bars, the exchange of the carbon dioxide condenser with a four-step high pressure condenser (200 atmospheres) and the interposition of a refrigeration machine to pre-cool suctioned air. GFLE, *50 Jahre Kältetechnik*, pp. 81–5.
18. Ibid., pp. 88–92.
19. Linde-Archiv Höllriegelskreuth, Akten Dr Friedrich Linde, Akte Oxyliquit.
20. The rectification method was different from distillation in that a certain part of the distillant flows back through the vaporizing liquidity. In a special rectification

column different equilibria between the two involved elements are acheived on individual levels. Hausbrand, *Rektifizier- und Destillierapparate*, pp. 5–14.

21. Meidinger, *Fortschritte in der künstlichen Erzeugung*, p. 141; *DPJ* 195 (1870), p. 40.

22. In retrospect the Linde Company itself wondered about this long search phase, as 'similar tasks faced the chemical industry already'. GFLE, *50 Jahre Kältetechnik*, p. 93.

23. The 'separation attempts of 30/31 August 1900, report' still worked with factioned vaporization. Linde-Archiv Hoellriegelskreuth, Akte Forschung und Entwicklung.

24. On 14 August Friedrich Linde reported to his father in Berchtesgaden of 'favourable results' and suggested 'sieved cinder, permeable clay, thin bars of wood and cork' for the filling of the rectification columns. Linde-Archiv Höllriegelskreuth, Akten Dr Friedrich Linde, correspondances with Carl Linde, 14 August 1902 and 15 August 1902.

25. The partners were the oxygen works, formerly Elkan, and the carbon dioxide works of C.G. Rommenhoeller (Linde, *Aus meinem Leben*, p. 105). In 1905 Elkan and on 15 May 1906 Rommenhöller withdrew again from the partnership. Linde to Michaelis. 15 May 1906. PAHL, Cb. 11, p. 29.

26. Stinnes responded to the director of the factory who attempted to withhold experiment results from the Linde Company by attesting to the Lindes' professionalism: 'One can rely on the loyalty of Linde father and son.' Letter, Linde to Carl Linde. 18 September 1907. Linde-Archiv Wiesbaden, IX, Correspondances F. Linde/C. Linde.

27. Letters, Linde to Stinnes, 7 June, 25 October and 11 November 1907; letter, Carl Linde to Beninghaus. 23 November 1909. PAHL, Cb. 11, pp. 167, 198, 216 and 335. The new company had 9,000,000 Marks in share capital. The shareholders were: RWE, Thyssen, Hugo Stinnes.

28. Verkaufsvertrag zwischen der Gesellschaft für Lindes Eismaschinen und der Sauerstoff-Industrie-AG vom 25 September 1913. Linde-Archiv Höllriegelskreuth, Akten Friedrich Linde. The sale price noted in the Linde Company's files was a nominal value of 500,000 Marks or 700,000 Marks in cash. Auflösung der Sauerstoff-Industrie-AG, in *ZgesKI* 20 (1913), 187.

29. Letter, Linde to Meier. 13 December 1908. PAHL, Cb. 11, p. 401. Costs at 150 m³/h were suddenly only 1/5 of the costs in 1908.

30. Schenck, Die Verwendung von Sauerstoff, *S&E* 48 (1924), 521–6.

31. DMMS, Hauttmann, *LD-Verfahren*, Bd. 12. Steel factories like the Ilseder Hütte put oxygen installations for steel production into operation for the first time in the 1950s. Treue, *Geschichte der Ilseder Hütte*, p. 676.

32. Linde-Archiv Höllriegelskreuth, Akte Versuche zur Gewinnung reinen Wasserstoffs aus Leuchtgas.

33. Water gas consists of approximately equal amounts of carbon oxide and hydrogen, and is produced as luminary gas, as steam is led over the incandescent carbon.

34. Tungsram, *57 Jahre*.

35. Chloric gas liquefaction for the chemistry factory Rhenania. PAHL, Cb. 10, p. 31; Separation of Naphthalin from city gas. Heinzerling, Austreibung von Naphtalin aus dem Stadtgas, in *ZgesKI* 51 (1898), 157; Extraction of sulphuric acid from rust gases. Linde-Archiv Höllriegelskreuth, Korrespondenz mit Etablissement Kuhlmann, 1932/33.

36. Examples: Until 1908 the extraction of oxygen through gas separation competed with electrolytic oxygen production. (Petz, Elektrolytische Sauerstoffgewinnung, *DPJ* 89 (1908), 414–16.) For more on chemical methods for producing nitrogen, oxygen, argon, crypton and neon, see: Scurlock, *A Matter of Degrees*, p. 491.

37. The arc method was developed by BASF together with Norwegian engineers and was very energy consuming. No installations were put into operation in Germany. The

physicist Walther Zenneck left BASF in 1911, as the Haber-Bosch method prevailed
over the arc method. Plumpe, *I.G. Farbenindustrie*, p. 206.

38. The nitrogen obtained from gas separation is combined with calcium carbide,
which is produced from coal (or charcoal) and lime in an electric oven, into
calciumcyanamide. Kürten, Die Calziumcarbidindustrie, *Acetylen in Wissenschaft
und Technik* 28 (1925), 17–47. 'The calcium carbide industry grew in the 1890s. In
1895 Frank and Caro found a method for manufacturing calcium cyanamide. Since
1908 nitrogen was mainly provided through gas separation. One can manufacture
ammonia with calcium cyanimide through steam under pressure. Ammonia thus
produced provides in turn an exit material for various explosives.

39. The calcium-nitrogen method required oxygen and nitrogen. In contrast the
synthesis of ammonia according to the Haber-Bosch method required hydrogen
and nitrogen. Later the BASF developed a chemical method for the production of
hydrogen, which was dependent on oxygen. Plumpe, *I.G Farbenindustrie*, p. 208.
GFLE, *50 Jahre Kältetechnik*, p. 112.

40. Linde-Archiv Höllriegelskreuth, Akten Friedrich Linde, Korrespondenz F. Linde/
Caro on 16 November 1907. Caro warns against the leaking of information to
the outside, 2 January 1910, 12 January 1911 – meeting with the director of the
Berlin-Anhaltischen Maschinenbau AG, 2 December 1913 – operations problems
of the big calcium-nitrogen installation in Trostberg. With the BAMAG, Caro and
Frank, the Linde Company formed a utilization company for the Linde-Frank-Caro
method (Linde, *Aus meinem Leben*, p. 103), which took over the worldwide sale of
licences.

41. Linde built four nitrogen factories in Knapsack, Wittenberg (Piesteritz), Chorzow,
and Walshut with a total of 14,000 m³/h. GFLE, *50 Jahre Kältetechnik*, p. 117.

42. Two installations with a total of 10,000 m³/h. Linde-Archiv Höllriegelskreuth, Linde,
R., *Meine Tätigkeit*, p. 15; Wegener, *Kälte-Industrie*, p. 70.

43. Plumpe, *I.G. Farbenindustrie*, p. 212 f.

44. The gross profit of the department rose from an average of 3 million Marks between
1910 and 1914 to over 7 million Marks in 1917 and 1918. Additionally, amortization
increased from 0.5 to over 3 million. GFLE, *Geschäftsberichte* 1910–19.

45. For eye irritants (T-Stoff), lung poisons (Phosgen, Perstoff) and nose-throat irritants
(Clark, Dick, Adamsit) no gases, which could be obtained from gas separation, were
necessary. A pre-product of mustard gas was ethylene, which was however produced
only in 1921 and after (coke gas separation) through cryogenic separation. Sartori,
Chemie der Kampfstoffe, p.121. Hauslian/Bergendorf, *Der chemische Krieg*, 12. In 1915
chlorine was used to a limited extent as poison gas. Lepsius, Verdichtete und
verflüssige Gase in *Krieg*, 24 (1925), 1–7; 17–19; 57–61 and 69–72. Cryogenic
engineering possibly played a certain role in the Russian–German collaboration
on mustard gas reseach during the 1920s. Harris/Paxanem, *Form des Tötens*,
pp. 58–72.

46. Linde, R., *Meine Tätigkeit*, p. 20 f.

47. Linde, R., Die Trennung der Kohlengasbestandteile durch Kühlung und stufenweise
Verflüssigung [Typescript] 1931, Linde-Archiv Höllriegelskreuth, Akten Richard
Linde; Plank, *Geschichte der Kälteerzeugung*, p. 139.

48. Linde mainly co-operated with the Hoechst Company in Frankfurt. Bäumler, *Ein
Jahrhundert*, p. 145.

49. From 1900 to 1979, the Munich Division built some 2,800 systems to be put in
use throughout the world, registering during this time period 1,200 patents. Linde
AG, *1879–1979*, p. 23.

50. The personnel department shared the same space with advertising, the company
doctor and the works committee, which was located on the outskirts of the works
premises. Ibid., pp. 79 and 191.

51. An overview of these manufacturing plants is to be found in the memoirs of Franz Scholz, who began working at the operations office in 1918. Linde-Archiv Höllriegelskreuth, Scholz, *Erinnerungen*, pp. 3–12.
52. H. Hafner was the plant manager up until 1921; he was followed for a short period of time by Braumüller, who was then succeeded in 1922 until 1942 by Dr Alfred Heß, who organized production at Linde thoroughly for the first time.
53. In 1924 the department of operations had eight employees and an operations assistant. Jaeger, *Geschichte der Werksgruppe TVT*, p. 50.
54. The calculation of material and production costs were kept in the stockroom during the initial years. Scholz, *Erinnerungen*, p. 20. Linde-Archiv Höllriegelskreuth.
55. Linde brought in high-pressure compressors from G. A. Schütz, Luzern (PAHL, Kb 9, p. 204); Brotherhood & Co., London (Kb 9, p. 311); Whitehead, London (Kb 9, p. 371); CAIL, Paris (Kb 9, p. 384) and BAMAG (Schwartzkopff) Berlin (Kb 9, p. 301).
56. Jaeger, *Geschichte der Werksgruppe TVT*, p. 78.
57. Linde, R., *Meine Tätigkeit*, p. 12. Linde-Archiv Höllriegelskreuth.
58. This ban only led to their 'secret' use. Scholz, *Erinnerungen*, p. 16. Linde-Archiv Höllriegelskreuth.
59. Jaeger, *Geschichte der Werkgruppe TVT*, p. 78.
60. Vogel was once Linde's assistant at the Technical University Munich. Jaeger, *Geschichte der Werkgruppe TVT*, p. 49.
61. Jaeger, Personalbestand, Linde-Archiv Höllriegelskreuth. The first large increase in personnel came after the war; in 1950 the total employees in this department increased to over 50. By 1960 there were 124, by 1971 561 employees in engineering.
62. Chief engineer Ernst Mönch was Professor Moritz Schröter's assistant until 1890 and worked after that at the test lab under Carl Linde. Jaeger, Personalbestand, p. 118, Linde-Archiv Höllriegelskreuth.
63. In 1929 Mönch received two more employees and could therefore, in 1932 at the age of 70, finally retire. A separate department for purchasing was created in 1953. Jaeger, Entwicklung des Vertriebs. Linde-Archiv Höllriegelskreuth.
64. GFLE, *50 Jahre Kältetechnik*, p. 98.
65. Richard Linde writes that his father gave him crucial tips in 1911 on water-gas separation (Linde, R., *Meine Tätigkeit*, p. 7). Friedrich Linde and Rudolf Wucherer argued frequently with Richard Linde over questions of gas separation. Ibid., p. 2.
66. Scholz, Erinnerungen, p. 16. Linde-Archiv Höllriegelskreuth.
67. His employee was Dr Kölliger. In the 1920s, the chemists Heinrich Kahle, Ernst Karwat, and Paul Schuftan were added. Jaeger, *Geschichte der Werkgruppe TVT*, p. 49.
68. The chemistry department was led until 1938 by Dr Pollitzer, a former assistant to Walther Nernst, who was sent to Dachau in 1938 for being Jewish, and who was later released. Richard Linde arranged for him to be sent to Air Liquide in Paris, however, after the Germans invaded France, Pollitzer was captured by the GESTAPO and died in 1942 in Auschwitz. Linde-Archiv Höllriegelskreuth, Akte Pollitzer. Other Jewish employees in the chemistry department were Dr Paul Schuftan and Dr Lothar Meyer, both of whom succeeded in escaping further persecution in 1936 and 1938 respectively. Jaeger, Geschichte der Werkgruppe TVT, p. 116. Other (half-) Jewish employees in Höllriegelskreuth included Philipp Borchardt, who went to England in 1938 and Bruno Hippenmeyer. Lebenslauf Bruno Hippenmeyer, 18 December 1947. ATUM, PrAA, File Hippenmeyer.
69. Linde, R., *Meine Tätigkeit*, p. 19, Linde-Archiv Höllriegelskreuth.
70. GFLE, *50 Jahre Kältetechnik*, p. 126.

71. 'Now it appears completely incomprehensible that one did not recognize the simple relationships of heat transmission in both performances of the countercurrent [condensers].' Linde-Archiv Höllriegelskreuth, Linde, R., *Meine Tätigkeit*, p. 4.
72. Ibid., p. 3 f.
73. Ibid., p. 18.
74. Richard Linde began in 1907 to direct the most important erections; in 1908 he spent eight months travelling (ibid., p. 3). He began in 1909 to work on water-gas separation. Putting these systems into operation required months-long journeys for him from 1911 to 1914 (Holland, Gumpoldskirchen near Vienna, Budapest, Oppau, Berlin). Linde, R., *Meine Tätigkeit*, pp. 12 and 15, Linde-Archiv Höllriegelskreuth.
75. In her analysis of carrier engineer companies, Gail Cooper explores the varied policies of information transfer by machine manufacturers and engineering companies: 'In general, manufacturers encouraged the spread of engineering knowledge while engineering companies held that information more tightly.' G. Cooper, *Manufactured Weather*, p. 65. CUA, CC 2511, C 15, 18.
76. Linde was strict with his patent attorney F. C. Glaser in Berlin, repeatedly requiring a denial of access to his documented proof, 'because all of our achievements could be copied and used for the registration of a patent'. Letter from Linde to Glaser on 1 October 1905. PAHL, Kb 10, p. 445. For more on the history of patent disputes over rectification: Linde, *Aus meinem Leben*, pp. 122–33.
77. *ZgesKI* (1911), 132.
78. Hausen, *Der Thompson/Joule-Effect*. Prof. Oscar Knoblauch made the initial revisions. ATUM, PrAA, Akte Helmuth Hausen.
79. His dissertation was a continuation of the dissertations by E.Vogel (1909) and Fr. Noelle (1912) on the differentials of the Thomson-Joule Effect that were prompted by suggestions from Carl von Linde. The Linde Company supported Hausen financially; he was made a full-time permanent employee in February 1922 at Höllriegelskreuth. Many of his experiments were conducted in Höllriegelskreuth rather than Munich. Hausen, *Der Thompson/Joule-Effect*, p. 2; Hausen, Lebenslauf Helmuth Hausen, 20 June 1924. ATUM, PrAA, Akte Helmuth Hausen.
80. Hausen, *Theorie des Wärmeaustausches*.
81. Scientific publications from H. Hausen. Lebenslauf H. Hausen, 26 March 1935 and Korrespondenzakte Helmuth Hausen. ATUD, A 510. Linde-Archiv Höllriegelskreuth, Jaeger, Veröffentlichungen Hausens.
82. The calculations of condensation from a mixture from 8 December 1927 were changed on 16 December 1936, in: Hausen, *Rektifikationssäulen*, Heft 3, pp. 41–2. Linde-Archiv Höllriegelskreuth, Nachlaß Helmuth Hausen. I thank Dr Hans Kistenmacher at Linde AG for pointing me to Hausen's calculation records.
83. Hausen, Rektifikationssäulen H1 (1924–26), H2 (1926–27), H3 (1924–30), H4 (1930–42). Gegenstromapparate H1(1924–25), H2 (1925–26), H3 (1926–27), H4 (1928–29), H5 (1929–33), H6 (1933–37). Linde-Archiv Höllriegelskreuth, Nachlaß Helmuth Hausen.
84. Hausen, Austauschvorgänge bei der Zerlegung von Gasgemischen, in: *ZgesKI* 43 (1943), 248; Hausen/Schlatterer, *Einfluss der Rektifikation*.
85. According to Dr Hans Kistenmacher, director of the calculations department until 1990, this kind of open-book-technology remained unavailable to the majority of engineers for reasons of security. Verbal message from Dr Hans Kistenmacher. See: Kistenmacher, *Grundlegende Methoden*.
86. He was a founding member of the working group 'Destiller- and Rectification Apparatuses' in the Department of Process Engineering at the VDI and participated regularly at their conferences.
87. Hausen made drawings of the most interesting diploma work done, such as: Ludwigs Thiery, Bestimmung des Wirkungsgrades von Rektifikationsböden,

angefertigt 1932 bei E. Kirschbaum an dem Institut für Apparatebau in Karlsruhe. Berechnungsunterlagen Helmuth Hausen. Linde-Archiv Höllriegelskreuth, Nachlaß Helmuth Hausen.

88. According to Hermann Linde (verbal report), Hausen's execution of gas separation touched on the considerations of Fritz Linde in the Linde Company's festschrift of 1929. Hausen, *Die physikalischen Grundlagen*.

89. Hausen was asked by Richard Linde to examine the personal effects of Paulus Heylandt 1949/50 for usable new ideas in engineering and give his expert opinion. Linde-Archiv Höllriegelskreuth, Akte Heylandt.

90. Hausen/Linde, H., *Tieftemperaturtechnik*.

91. A biography of Walther Schottky from Reinhard Serchinger is to be published soon.

92. Jaeger, *Über Mathias Fränkl*. Linde-Archiv Höllriegelskreuth, Ordner Mathias Fränkl.

93. Fränkls father was a baker in Bad Reichenhall.

94. Autogen Gasaccumulator Krükl & Hansmann GmbH.

95. According to Hausen's recollections. Schreiben Helmuth Hausens an Erik Jaeger, 16 January 1967. Linde-Archiv Höllriegelskreuth, Ordner Mathias Fränkl.

96. The oldest document regarding a co-operation between Fränkl and Linde is Linde's added patent from 23 January 1927 to Fränkl's primary patent. A utilization contract was agreed upon on 18 June 1928 by Fränkl and the Linde Company. Hausen wrote his Habilitation two years later on the calculation of regenerators. Hausen, *Theorie des Wärmeaustausches*.

97. Fränkl-Patents regarding air separation. Linde-Archiv Höllriegelskreuth, Ordner Mathias Fränkl.

98. Description of a Linde-Fränkl installation, 1933. Linde-Archiv Höllriegelskreuth, Ordner Mathias Fränkl. Hochgesand, Das Linde-Fränkl-Verfahren, *Mitteilungen aus den Forschungsanstalten des Gutehoffnungshütte-Konzerns*, 4 (1935), 14–23.

99. Richard Linde, quoted in: Jaeger, Erik: Wer war eigentlich Herr Fränkl, *Werkzeitung* 7 (1967), 5–12, 12.

100. On Paulus Heylandt: Heylandt, Die Verwertung der Luft, *Chemikerzeitung* (1937), 10–11; Jaeger, Vorläufige Angaben zu einer Biographie Paul Heylandts; Prospekte der Heylandt Gesellschaft für Apparatebau Berlin-Britz; Linde-Archiv Höllriegelskreuth, Ordner Paulus Heylandt.

101. DRP 270383, 10 September 1908.

102. 1906: 'Flüssige Luft Maschinen und Apparate, System Paulus Heylandt GmbH' under the direction of the Hannover Bank Association with 1 million Marks in nominal capital. The 22-year-old 'physicist' Heylandt became the manager, in: *ZgesKI* 13(1906), 78; 1917: Gesellschaft für Apparatebau, Berlin-Britz. 1921: Gesellschaft für Industriegasverwertung, Berlin. According to his own records, Heylandt had over 400 patents.

103. Laschin, *Sauerstoff*, p. 31.

104. Report on experiments with carburetter tanks from the Gesellschaft für Industriegasverwertung mbH from 22 January to 26 February 1930. Linde-Archiv Höllriegelskreuth, Ordner Paulus Heylandt.

105. As Heylandt offered the Linde Company in 1929 to forego the sale of his patent in England for 2 million Marks, Fritz Linde wrote to the Linde director, Dr Brunn (Heylandt's negotiation partner): 'Dr Heylandt is repeating the same method he used two years ago by dealing with a third party to make certain agreements so as to leave himself a backdoor exit only to turn around and make an agreement with us. You know how distasteful we found this then and I find it even more difficult now to be concerned with these things.' Linde-Archiv Höllriegelskreuth, Akten Friedrich Linde, Korrespondenz mit Dr Brunn, 11 July 1929.

106. E.g. Prof. Rumpf at the TU Karlsruhe. Letter, Helmuth Hausen to Erik Jaeger, 16 January 1967. Linde-Archiv Höllriegelskreuth, Ordner Mathias Fränkl.

107. Knappich, Die autogene Schweißung, *Südwestdeutsche Industriezeitung* (1909), 248–52; Gräffkes, van: Zur Geschichte der Schweißbrenner, *Autogene Metallbearbeitung* 12 (1909), 91–3.

108. Because of the Fouchés invention, oxygen plants grew in France much earlier and faster than in England. Scurlock, *Matter of Degrees*, p. 489.

109. About the history of the Chemische Fabrik Griesheim: Pistor, *100 Jahre Griesheim*; Knapsack-Griesheim AG, *Unter uns*; Forstmann, *Chemische Fabrik Griesheim*, pp. 42–60.

110. Wyss had worked at the Bitterfeld works of the chemical factory Griesheim since 1896. In1899 he was given the task of compressing incidental hydrogen from chlor-alkali electrolysis in steel bottles, which were to be sold to balloonists. Wyss experimented with compressed hydrogen and designed the first hydrogen-oxygen welding torch for iron in 1903. Pistor, *100 Jahre Griesheim*, pp. 182–95.

111. Letter from Ernst Wyss to Theodor Plieninger dated 30 November 1906. Quoted as ibid., p. 184.

112. The chemist Dr Menne held the patent. DRP 137588 (26 May 1901).

113. Knapsack-Griesheim AG, *Unter uns*, p. 7 f.

114. Michaelis, Die autogene Schweißung, *ZfkflG* 11 (1908), 151–61.

115. The first Claude process oxygen plant was put into operation in 1908 in Bitterfeld by Griesheim. Wyss, Das autogene Schneiden, *ZkflG* 11 (1909), 104–11.

116. Schon seit 1897 hielt Linde in Sachen Chlorverflüssigung Kontakt zu Griesheim. Letter from Linde to Chemische Fabrik Griesheim, 18 February 1897. PAHL, Kb 10, p. 31.

117. Letter from Linde to director of VSW Michaelis on 14 January 1906. PAHL, Kb 10, 460. Außerdem: Absprache mit dem Griesheim Vorstand Theodor Plieninger, 13 December 1906. Linde-Archiv Höllriegelskreuth, Akten Dr Friedrich Linde, Korrespondenz mit Theodor Plieninger.

118. Contract dated 24 July 1908. Linde, *Aus meinem Leben*, p. 117.

119. Letter from the attorney Oppenheimer to the manager Andrea (Greisheim) Schreiben von Rechtanwalt Oppenheimer an Direktor Andrea (Griesheim) regarding developments in the purchase of Indusgas, which was eventually completed in 1920. Linde-Archiv Höllriegelskreuth, Akten Dr Friedrich Linde. Oppenheimer spoke of a 'Griesheim-Linde Group'. When Griesheim became part of IG Farben in 1925, Plieninger and Wyss became members of the IG Executive Board.

120. Contract, 27 May 1925. On 16 September 1925, Griesheim and Linde founded the 'new' United Oxygen Factories (after the first in 1904, see above) Linde-Archiv Höllriegelskreuth, Vereinigte Sauerstoffwerke-Geschichte, Chronologie.

121. The Nuremberg Light Association ostensibly wanted to test their luminary gas-oxygen lamp and the Linde Company had hoped this would lead to a new oxygen market. Linde-Archiv Wiesbaden, XII, Fusion DOAG.

122. The founding of the Süddeutschen Industriegasgesellschaft was initiated by Indusgas. The Linde Company made a quota deal with DIAG on 23 July 1914 and agreed on 17 July 1914 on shared patent rights against Hausmann & Co., a small independent oxygen manufacturer. Linde-Archiv Höllriegelskreuth, Akten Dr Friedrich Linde, Korrespondenz Indusgas 7 July and 27 July 1914.

123. Linde-Archiv Wiesbaden, XII, Merger DOAG.

124. Dräger, *Lebenserinnerungen*, p. 55.

125. Drägerwerk (ed.), *Der Retter Sauerstoff*. DMMS, FSA.

126. Drägerwerk (ed.), *Laboratorium für technische Physik, Chemie und Physiologie*. DMMS, FSA.

127. Geschichte der Firma Adolf Messer GmbH [Typescript], 1975, Linde-Archiv Wiesbaden, II.

128. In internal company histories, patent infringement was seen to have been the right of smaller companies (ibid.). Messer responded to these suits with a plethora of patent applications: Interessante Neuerungen bei der Herstellung von industriellem Sauerstoff, Autogene Metallbearbeitung (1923), 33–5. In 1922 Messer asked Linde to reduce patent infringement penalties. Linde-Archiv Höllriegelskreuth, Akten Friedrich Linde, Korrespondenz mit Adolf Messer, 20 September, 22 September and 22 November 1922.

129. Letter from Linde manager Philipp Borchardt to Dr Schück (IG Farben), 30 April 1929. 'I take the liberty of informing you that we have made an agreement with the Messer Company regarding the field, with which we previously had always competed with.' Linde-Archiv Höllriegelskreuth, Akten Philipp Borchardt.

130. Letter from Lindes to the Société des Applications de l'Acetylene, Paris, 19 February and 25 March 1906. PAHL, Kb 10, p. 478 and Kb 11, p. 6. Present: ISG received the AR: two representatives of the Linde Company, a banker, Mr Elkan and one representative of the Societé.

131. Together with the Spanish businessman Abello, Linde founded in 1906 a Spanish business that still exists today. Letter, Friedrich Linde to Abello, 2 July 1906; 15 April 1908; 19 February 1927, 30 September 1937. Linde-Archiv Höllriegelskreuth, Akten Dr Friedrich Linde, Korrespondenz 1898–1950.

132. The Lucerne Oxygen and Hydrogen Works was founded by Linde in 1909 together with Arnold Gmür. Carl Linde was chairman of the Supervisory Board of the most relevant Swiss oxygen company until 1932, Fritz Linde until 1956. Sauerstoff- & Wasserstoffwerke AG, *50 Jahre*, pp. 9–19.

133. Nordiska Syrgasverken A.B. with Linde as partial shareholder. Together with Syrgaswerken, Linde founded in 1910 the Dansk Ilt & Brintfabr. in Denmark and the Norsk Surstof & Vanstof Fabr. A.p. in Norway. Letter, Friedrich Linde to Nordiska Syrgasverken A.B., 14 July 1910. Linde-Archiv Höllriegelskreuth, Akten Dr Friedrich Linde, Korrespondenz 1898–1950.

134. Shareholder of Tentelewsche Werke St Petersburg.

135. Società Italiana Ossigeno ed altri Gas.

136. Österreichisch-Ungarische Sauerstoffwerke and in 1910 in Hungary the Hydroxygen AG. On 5 December 1910, Linde made an agreement with Oxygenunternehmung in Budapest to create an oxygen cartel for Hungary. Linde to Oxygenunternehmung, 5 December 1910. PAHL, Kb 11, p. 400.

137. The patent suit was won by Linde in both its second and third appeals (May 1908 and March 1909, respectively). Linde, *Aus meinem Leben*, p. 116. Agreement with Air Liquide in August 1908. Linde to Manager Delormé, 3 August 1908. PAHL, Kb 11, p. 245.

138. Letter, Linde to L'Air Liquide, 19 October 1908. PAHL, Kb 11, p. 264.

139. Linde, *Aus meinem Leben*, p. 115 f.

140. Air Liquide and the Linde Company even exchanged their own technical innovations with one another. Linde was able in the 1920s to start producing the cross-flow heat exchanger (Kreuzstromgegenströmer) first produced by Air Liquide (Linde, R., *Meine Tätigkeit*, p. 5) and Air Liquide received from Linde in 1930 the Fränkl process for oxygen production.

141. Agreement with Austria, Sweden, and Belgium (50/50) and discussions on Portugal, Greece, Turkey, and Egypt. Letter from Linde to ISG, 21 August 1908. PAHL, Kb 11, p. 254. On 30 September 1909 an agreement was reached on the quotas for Switzerland, Belgium, and Russia; on 19 August 1909 an agreement was reached for Italy. On 7 October 1909, Air Liquide was integrated within the International Oxygen Association, and Air Liquide introduced its pyro burner patent. On 23 July 1914, Air Liquide and the Linde Company together reached a quota agreement with Indusgas. Linde-Archiv Höllriegelskreuth, Akten Dr Friedrich Linde, Korrespondenz mit Air Liquide.

142. Letter, Linde to Ges. der Tentelewschen Fabrik in St Petersburg regarding the Air Liquide agreements of 8 March and 27 July 1909. PAHL. Kb 11, pp. 307 and 326. Letter Lindes to manager Murray of BOC regarding der Air Liquide agreements from 8 November 1909. PAHL, Kb 11, p. 332. Present at the meeting on market sharing in England: BOC 2/3, AL 1/3, later 50/50. Murray refused, only to lose out in the end.

143. Murray, *Methods of Liquefaction*, p. 207.

144. Davies, *William Hampson*, pp. 63–73. Hampson was an attorney, later a sociologist and found it therefore difficult to testify against either English or German cryogenicists.

145. Hampson was the arch-enemy of the English cryogenicist James Dewar; the refrigeration engineer James Ewing saw himself as Linde's loyal compatriot in England. *Engineering*, 65 (1898), 110, 337, 396, 474, 508, 541–2, 573, 607 and 795. Hampson was heavily criticized and made fun of by Dewar, Lennox, and Ewing.

146. Linde, *Aus meinem Leben*, p. 106. Letter Linde to manager Murray for BOX, 27 March and 5 August 1906. PAHL, Kb 11, pp. 14 and 75. Contract signing on 1 April 1906, meeting in Cologne on 20 September 1906.

147. Eisenmann, Brush. Brush's scientific writings are in Cleveland. Case, BC.

148. The laboratory in his home in Euclidstreet 'is one of the finest private laboratories in the world'. *Alliance Ohio Review* 10 January 1900. Case, BC, Box 17, 7.

149. Laboratory Notes. Case, BC, Box 6, 1. Brush tried for over 40 years to prove aether theory through experiments and financed as well experiments at the National Bureau of Standards and at the General Electric Labor in Schenectady from 1907 to 1920; none of these experiments, however, successfully proved the theories. Eisenmann, *Brush*, p. 153.

150. In October 1899, Brush wrote to Germany. On 8 November 1899, Friedrich, Carl von Linde's son, replied and sent Brush the English patent application written in poor English (which Linde without success had sent to the US in 1895). Brush corrected the documents. Case, BC, Box 10, 3.

151. Patents: 728178 (air liquefaction) and 727650 (fractionation), registered on 7 September 1895, patented on 12 May 1903; 815544 (gas separation), registered on 7 November 1902, patented on 20 March 1906; 795525 (rectification) and 815601 (nitrogen production), registered on 22 October 1903, patented on 20 March 1906. All patents are labelled with: Carl Linde of Munich, Germany, Assignor of one-third to Charles F. Brush, of Cleveland, Ohio.

152. Brush's argument: 'Dr Linde is a professor in the Institute of Technology at Munich, Bavaria, and is widely known as a scientific investigator' (p. 15). 'Tripler has no such scientific attainments as would make it all provable that he ever conceived the invention in issue or that he ever understood it and his witnesses are all lacking in scientific knowledge and were incapable of comprehending the invention' (p. 45). US-Patent Office, Interference Nr. 20751.

153. Brush's tactic was furthered by the bankruptcy of Tripler's Liquid Air Company in 1902, which was to have had a nominal capital of 10 million dollars. Scurlock, *Matter of Degree*, p. 488 f.

154. 'Now I would like to hear, whether you see a chance for sale or whether it would suit you, that other people occupy themselves with the question.' Carl von Linde to Charles Brush, 16 January 1903. PAHL, Kb 10, p. 277.

155. There is currently no thorough work done on the history of Linde Air Products. A series of stories about Linde Air Products are to be found at UCLDA. Chase, *History of Linde*. The Linde Story, *Focus* Heft 2(1982) 4–23, 3(1982)14–31, 4(1983) 20–7 and 5(1984) 22–8; *Story of the Linde Air Products Co*; Buckmann, *Notes*. From a German point of view: GFLE, *Lowest Temperatures*; Linde, *Aus meinem Leben*, pp.109–11.

156. In a letter to the Soc. des application de l'acetylene, Paris, 5 April 1906 Linde wrote that Brush 'after having conducted with extraordinary success our patent case

against Tripler', had done nothing further to develop the business. PAHL, Kb 11, Brief 20.

157. Letters 19 March, 20 April, 29 May, 20 June and 29 July 1906.
158. Linde to Cecil Lightfoot, 1 September 1906. PAHL, Kb 11, p. 80.
159. Letter Linde to Lightfoot, 3 December 1906. PAHL, Kb 11, p. 94.
160. Reisinger and his father-in-law Busch donated 250,000 dollars to Harvard University to build the Museum of Germanic Culture, which is now called the Busch-Reisinger-Museum. About Hugo Reisinger: Bergman, *Reisinger*.
161. Linde turned this down in a letter to Cecil Lightfoot, 5 November 1906. PAHL, Kb 11, p. 88.
162. Charles Brush wrote on 15 February 1907 and on 7 March 1907 letters to Wolf in an attempt to have him become a partner, yet he refused. Wolf died in 1915. Case, BC, Box 3.
163. Letter Carl v. Linde to Cecil Lightfoot, 3 December 1906. PAHL, Kb 11, p. 94.
164. Brush was one of the major figures in Cleveland's commercial circles; he was president of a number of his own companies and president of the Cleveland Chamber of Commerce as of 1909. See: Eisenmann, *Brush*, p. 168. The two most important new partners were G. A. Grasselli (Grasselli Chemical Co., later Du Pont) and John L. Severance (Standard Oil Co.). UCLDA, II, Linde Air Products.
165. Linde responded to Wolf's evaluation from 16 February 1907. PAHL, Kb 11, p. 109.
166. Linde considered this 'to be vital' as a counterweight to the Clevelanders. Carl v. Linde to Hugo Reisinger, 23 February 1907. PAHL, Kb 11, p. 109.
167. The initial capital stock totalled 250,000 dollars. Along with Severance, Brush and Grasselli, there were Calvary Morris (50 shares), J. H. Wade (100 shares) and Samuel Master (100 shares) as shareholders. The Cleveland faction totalled 145,000 dollars, Reisinger contributed 10,000 dollars and Linde 95,000 dollars. Letter, Brush to Linde, 7 August 1907. Case, BC, Box 3.
168. Linde insisted that the ISG have a stake. Legally, the corporation could not have been founded without this. Air Liquide thus also had shares in this Linde subsidiary, which allowed Air Liquide to gain equity for its Fouché welding torch licences. Letter, Brush to Linde, 21 January 1908 (response to Linde's letters of 20 November 1907, 29 November 1907 and 29 December 1907 and three letters from Air Liquide manager Javal to Brush). Case, BC, Box 3.
169. Buffalo November 1907. UCLDA, II, Linde Air Trade Catalogue.
170. Historical data on gas sales. UCLDA, II, Linde Air Trade Catologues. Wegener, *Kälte-Industrie*, p. 67.
171. Letter, Cecil Lightfoot and Friedrich Linde, 19 June 1914. Carl v. Linde to Hugo Reisinger, 13 June 1914. PAHL, Cb. 11, p. 468.
172. The U-boat *Germany* travelled twice to the USA in 1916 for trade purposes. Letter, Carl v. Linde to the manager of the Deutsche Bank, Berlin v. Stauß 20 December 1916. PAHL, Kb 11, 537 f. 'The Chicago Bank Hitchcock telegraphed: We stand ready to make payment for stock by wireless in Berlin if you can make delivery in America by submarine or otherwise.'
173. Linde-Archiv Höllriegelskreuth, Jaeger, *Linde Air Products*.
174. There may have been a partial sale. In 1917 the annual report listed a capital share of 811,000 Marks, in 1918 408,000 Marks (asset value 0) and in the years following an asset value of 0 was entered. Starting in 1911 (to 1917) the German Linde Company disclosed 546,000 Marks in dividends from Linde Air Products shares. Linde-Archiv Wiesbaden, XII, Bilanzunterlagen 1879–1930.
175. The Acetylene Apparatus Co., the largest gas works in Chigaco, to which Peoples Gas Co. belonged, bought the Colt Co. in 1911. The Chicago owners merged the company in 1912 with Oxweld Acetylene Co., which was founded by Union Carbide. UCLDA, BC, II, Chase, *History of Linde*.

176. Ibid., p. 47.
177. National Carbon Co. was founded in 1886 by Charles F. Brush. He was one of the primary backers of the merger. Ibid., p. 38.
178. Linde Air Products shareholders received 17.5 per cent of the shares of the new group with the merger. UCLDA, IV, Konsberg, *Union Carbide*, p. 1.
179. In a letter to the Reichsminister for Trade and Commerce, 4 March 1920, the Gesellschaft für Lindes Eismaschinen based their request for a capital increase on foreign competition which threatened to take the German company over. Linde-Archiv Wiesbaden, XII, Merger DOAG.
180. Richard Linde, Carl v. Linde's son, brought the first coke gas separators to the USA.
181. The first liquid gas tank trucks came from Germany. UCLDA, Chase, *History of Linde*.
182. Interview with Dr Leo I. Dana. Ibid.
183. Union Carbide Corp., *Cryogenic Science*, pp. 15–32.
184. According to Tim Coleman. UCLAD, HoL Round Table Discussion, p. 22.
185. A monograph on the history of the company has recently become available: Butrica, *Thin Air*.
186. Smith, Bethlehem, Private Archive, Chronology of Air Products Inc. 1934–45.
187. 'In March 1939, he finally hit upon an idea. Instead of distributing oxygen and acetylene in small and expensive and cumbersome cylinders, why not manufacture a relatively small and inexpensive oxygen producing unit which could be installed at the site?' Smith, Bethlehem, Private Archive, Air Products Then and Now.
188. Notes on the early History of APCI. Smith, Bethlehem, Private Archive, Aggressive Exploitation.
189. The English cryogenicist Scurlock stated that the USA lost interest in investment after Tripler's bankruptcy. Scurlock, *Matter of Degrees*, p. 488 f. Certainly, the strong position of the Europeans in the sector had its roots in the company's founding. Yet the leading positions of pioneers such as George Claude and Carl von Linde prove to be illustrative examples of the 'prime mover' theories of Alfred D. Chandler.
190. Quoted in GFLE (Ed.), *75 Jahre Linde*, p. 23.
191. All quotes from foreword by Friedrich and Richard Linde in GFLE (Ed.), *75 Jahre Linde*, p. 7.
192. Speech, Hermann Linde on Walther Ruckdeschel's 40-year anniversary with the company, 12 September 1966, p. 1. Linde-Archiv Höllriegelskreuth, Materialien Erik Jaegar.
193. Chamber music was a significant aspect of evening entertainment for Carl von Linde. Richard and Fritz together organized over 100 evenings of chamber music. All five of Richard Linde's sons were musically inclined and played together with their father regularly. Linde, H., *Eine Münchener Familie*, p. 41.
194. Speech, Hermann Linde on Walther Ruckdeschel's 40-year anniversary with the company, 12 September 1966, p. 2. Linde-Archiv Höllriegelskreuth, Materialien Erik Jaegar.
195. 'Direktor i.R. Rudolf Wucherer 90 Jahre alt', *Werkzeitung Linde*, 2 (1965), p. 7.
196. The name 'Ellira' was an anagram referring to 'el' for electric, 'li' for letter from Linde and 'ra' for rapid.
197. A thorough analysis of the German gas oligopoly has been submitted by Andrea Reiß as a master's paper in 2002. Reiß, *Die Entwicklung des Industriegase-Sektors*.
198. Rudolf Wucherer remained on the Management Board until he was 80 and on the Supervisory Board until he was 90. He was honoured with accolades from the TU Hannover in recognition of his achievements in the business and technology of gases.
199. The psychologist Bert Hellinger has developed the notion of 'family constellation' as a therapy tool.

200. The family and their closest friends in management were deeply affected by their deaths. Helmut died as a lieutenant outside of Leningrad in February of 1944. Werner Linde died in 1943 as a staff sergeant in Ukraine. 'It was especially tragic because he was first injured and then died during an air attack on his way home.' Richard Linde to Philipp Borchardt, 12 April 1946. Linde-Archiv Höllriegelskreuth, Ordner Dr. R. Linde, Privatakt, 1. Jan. 1942 bis 31. Dez. 1948, A- Hi.

201. All five sons were active members of 'Nordland', a nationalist youth organization in Germany. When Helmut told his parents of his loss in faith in God through his work with the Nordring, Richard stepped in. He informed the founder and director of Nordland (Rudolf Schäfer) in 1934 that 'he would forbid his two youngest sons from participating at Nordland events until he could be assured that the antichristian influence on his sons would be discontinued'. Richard Linde to Rudolf Schäfer, 27. December 1934. Linde, H., *Eine Münchener Familie*, p. 73.

202. Johannes's parents were Rudolf and Elisabeth Wucherer (Carl von Linde's youngest daughter). Following his studies in machine engineering at the Technical University Munich, Wucherer was an assistant for the thermodynamic scientist Mollier. His work focused on processes involving solutions of ammonia. In 1937 he went to Höllriegelskreuth and soon became a close co-worker of Richard Lindes, who was in charge of apparatus engineering. Johannes Wucherer's career was similar to that of his grandfather Carl von Linde, who also preferred work in thermodynamics, pursuing first work in temperatures reaching $-50°C$ and later in cryogenics. A number of calculation methods were developed by Wucherer for cryogenic processes in air and gas separation. When his father departed the Executive Board in 1954, Johannes Wucherer was entrusted with directing the technical department of apparatus engineering and called to join the board. *Werkzeitung Linde* 2 (1966), p. 7 f.

Chapter 3

1. Most recently: Ritschl, *Deutschlands Krise und Konjunktur*.
2. Jaegar, *Aus der Geschichte der Werksgruppe TVT*, p. 113.
3. Feldman/James, *Deutschland in der Weltwirtschaftskrise*; Feldman: *Vom Weltkrieg zur Weltwirtschaftskrise*.
4. Linde AG, *Geschäftsbericht und Bilanz 1931*, p. 3.
5. Linde AG, *Geschäftsbericht und Bilanz 1928*, p. 3.
6. Linde AG, *Geschäftsbericht und Bilanz 1931*, p. 4.
7. Letter from Richard Linde to his sisters, 26 November 1931 Linde, H., *Eine Münchener Familie*, p. 47.
8. Letter from Richard Linde to his sisters, 26 November 1931. Linde, H., *Eine Münchener Familie*, p. 47.
9. Linde AG, *Geschäftsbericht und Bilanz 1932*, p. 3.
10. Linde AG, *Geschäftsbericht und Bilanz 1930*, p. 4.
11. Linde AG, *Geschäftsbericht und Bilanz 1932*, p. 4.
12. Letter from Richard Linde to his sisters, 26 November 1931. Linde, H., *Eine Münchener Familie*, p. 47.
13. Petzina, *Die deutsche Wirtschaft in der Zwischenkriegszeit*, p. 109. Cf. also: Barkai, *Das Wirtschaftssystem des Nationalsozialismus*.
14. Linde AG, *Geschäftsbericht und Bilanz 1933*, p. 2.
15. As the most striking example here is to be mentioned the construction of the Reich autobahns, which did not make much sense in terms of traffic technology, given the backward state of motorization in Germany at the beginning of the 1930s. Despite the publicly announced significance of these large construction projects and despite the 'first spades', and opening of routes all over the German Reich, all celebrated with huge publicity whirlwinds – here all subcontractors were involved

- at no time were more than 140,000 labourers employed in the construction of the autobahns, and this maximum number was also achieved only for a short time. See: Schütz/Gruber, *Mythos Reichsautobahnen*.

16. In 1937/38 the real national product surpassed its pre-crisis level by almost one-fifth; as contrasted with its low point in 1932, it had more than doubled. The six million unemployed were in 1937 again incorporated into the working process.

17. That the real national income grew between 1933 and 1939 with an annual rate of 8.2 per cent is, however, until 1936 to be mainly traced back to the prolonged working hours. Cf. Hachtmann, *Industriearbeit im 'Dritten Reich'*.

18. Herbst, *Die nationalsozialistische Wirtschaftspolitik*.

19. Most recently a summary of the history of companies: Gall/Pohl (eds): *Unternehemen im Nationalsozialismus*.

20. Petzina, *Die deutsche Wirtschaft in der Zwischenkriegszeit*, p. 117.

21. At the start of 1939, Hitler ordered Schacht's resignation also as Reichsbankpräsident.

22. Wagner, *IG Auschwitz*, p. 30.

23. Letter from Richard Linde to his sisters, 12 February 1937. Linde, H., *Eine Münchener Familie*, p. 128 f.

24. Todt: Aufgaben der deutschen Technik, *Deutsche Technik* 4(1936), 478.

25. Sohns: Sind Heimstoffe zu teuer?, *Deutsche Technik* 4(1936), 587.

26. Hughes: *Das technologische Momentum*.

27. 'The path from coke to the finished car tyre is a long one though, as one can tell from the list of the intermediary products: coke – carbide – acetylene – acetylene hydrate – aldol – butylene glycol – butadien_– buna (rubber)' Gumz/Regul, *Die Kohle*, p. 102.

28. Welsch, *Geschichte der chemischen Industrie*, p. 155 f.

29. For that see Plumpe, *Die IG Farbenindustrie AG*, p. 265 ff.

30. Thomas, *Geschichte der deutschen Wehr- und Rüstungswirtschaft*. General Thomas, who authored this comprehensive work during the second half of the Second World War, was among the defence economy staff. He was arrested in connection with the failed revolt of 20 July 1944 and sent to the concentration camp in Dachau. The march of the American troops saved him from execution; he died on 29 December 1946 in an American prison. For Thomas' life see the introduction of the editor.

31. Beutler, 'Wesen, Aufgaben und Begriffe der Wehrwirtschaft', *Deutsche Technik*, 6 (1937), 301.

32. Mommsen, *Konnten Unternehmer im Nationalsozialismus apolitisch bleiben?*, p. 71. In Friedrich Linde's obituary in 1965: 'He rejected National Socialism decisively and without compromise', as the chairman of the Supervisory Board of the Linde AG, Dr Ing., Dr Ing. h.c. F. A. Oetken said in the memorial celebration for Dr Friedrich Linde in 1965. Linde-Archiv Höllriegelskreuth, Ordner Druckschriften vom 1. Januar 1967 bis 31. Dezember 1970.

33. Letter from Richard Linde to his sisters, 19 March 1932. Linde, H., *Eine Münchener Familie*, p. 22.

34. The only office, which he occupied in the service of the regime, was that of a 'Special Ringleader for Welding and Cutting Technology' (Sonderringleiters für Schweiß- und Schneidtechnik) within the Speer organization. Jaegar, *Aus der Geschichte der Werksgruppe TVT*, p. 114.

35. Linde, H., *Eine Münchener Familie*, p. 22.

36. Linde, H., *Eine Münchener Familie*, p. 88.

37. Letter from Richard Linde to his sisters, 16 December 1936. Linde, H., *Eine Münchener Familie*, p. 124.

38. To that testifies, among others, Franz Bäumler, who began as an apprentice in copper-smithy in 1943. He perceived the mood of the factory never as clearly National Socialistic, although he himself was until shortly before 'the eleventh hour'

a convinced Hitler youth. Linde VA, Zwangsarbeiter bei Linde in Höllriegelskreuth und Schalchen – Eine Bestandsaufnahme, Höllriegelskreuth o.J. [2001]. Linde-Archiv Höllriegelskreuth.

39. Dr Jakobsmeier in an interview with Franz Bäumler: 'Dr Hess was a party member, and there were others at Linde. What did it look like in the company?' Bäumler: 'I can say nothing precise about that. I was simply too young. We knew that the Linde family did not want to have anything to do with the politics of the Third Reich. Our workshop director, Herr Förs, belonged to the party, we did know that of course. He had to leave the company for a while after the war. That was by the way the case for several persons, including also our later director Dr Becker. He even had to work as an unskilled labourer in the workshop for a while. There were several people, who were forbidden to work, including my department head, Herr Egglsmann, who only in 1949 was allowed to start again, as the director of the welding school though. Or Meister Schlech, who disppeared immediately, he had been a real Nazi. But our top Nazi was Herr Zinkgraf. Right after the was war over, he disappeared and was never seen again.' Linde-Archiv Höllriegelskreuth.

40. 'Zum fünfundsechzigsten Geburtstag des Stellvertretenden Betriebsführers Betriebsdirektor Dr Alfred Heß', Werkzeitschrift für die Betriebe der Gesellschaft für Linde's Eismaschinen, 3 (1941), 15.

41. For more see: Linde VA: Geschichte der Lehrwerkstatt in Höllriegelskreuth 1924–1999, Höllriegelskreuth o.J. [1999]. Linde-Archiv Höllriegelskreuth.

42. Max Schäffer, trainer since 1943 in the educational workshop after injuries in the war and its director between 1966 and 1979 remembers Hess thus in an interview in the year 2000: 'Whether he was a strict Nazi, I cannot tell. [...] I don't have good memories of him. He was a typical spy. There were for instance always signs in the evening: the first said 'end of work, put tools away', the second one 'finishing time'. Usually people prepared to leave already with the first signal. After the first signal, Dr Hess would walk throughout the factory to catch someone who had finished. He was feared everywhere. Whoever got himself into hot water with him, could pack his stuff.'

43. For example: 'Der Parteitag des Friedens' by the press manager of the DAF in Düsseldorf and 'Hau-ruck – der Westwall steht' by the Gaupressewalter of the DAF in Hessen-Nassau and 'Englands Verbrechen an Europa', all in the factory magazine, 4th year. 1939, No. 18. During the war the magazine was dominated by reports from the fronts, military letters from Linde staff members – next to incessant slander of those opposed to the war and presentations glossing over the German raw material situation – then later with the appeals for persistence and announcements for the fallen in the war. Because of the shortage of paper the magazine was put out in 1944 with the combined numbers 4 to 6.

44. Esser, Die 'Nürnberger Gesetze'.

45. Feldman, Die Allianz und die deutsche Versicherungswirtschaft.

46. The large concentration camps set up by the National Socialists in 1930s Germany (Dachau, Theresienstadt, Neuengamme) must be distinguished from the work and extermination camps established during the war in the east. Release from a concentration camp within Germany was at least theoretically possible until the end of the 1930s. After the 'Wannsee Conference' where plans for the destruction of European Jews were made, concentration camps in Germany were used as transit points for Jews to be sent to extermination camps in the east.

47. Letter from Richard Linde to his sisters 7 December 1938. Linde, H., Eine Münchener Familie, p. 162.

48. Linde, H., Eine Münchener Familie, p. 160.

49. The Air Liquide is named as a leading French company for manufacturing oxygen in a letter from 1946, 'with which we have had very friendly relations since decades ago – also during the last war'. Letter from Linde AG, Gas Liquefaction Department

to the lawyer and notary Brumby, Berlin-Lichterfelde-West on 7 January 1947, gez. Wucherer und Eggendorfer, p. 8.

50. Thus Borchardt obtained on 7. 12.1938 an amount of 6,242.93 Reichsmarks 'as balance credit' paid in cash, and Pollitzer asked Direktor Eggendorfer in a letter written on 14 December 1938 about the transfer of 8,738 RM to the finance office in Munich and 10,000 RM to the account of his wife. Moreover he asked for 8,000 RM in a cross cheque and 8,000 RM in cash. Linde-Archiv Höllriegelskreuth.

51. Jaeger, *Aus der Geschichte der Werksgruppe TVT*, p. 116.

52. Philipp Borchardt wrote in his letter to Richard Linde from London on 31 March 1946 to Richard Linde: 'I know, that Pollitzer has died. What a loss for all of us! Exactly now, where I have major difficulties with the people there, his presence could have been priceless to me.' Linde-Archiv Höllriegelskreuth, Ordner Dr R. Linde, Privatakt, 1. Jan. 1942 bis 31. Dez. 1948, A- Hi.

53. Jaeger, *Aus der Geschichte der Werksgruppe TVT*, p. 116.

54. The Marx & Traube GmbH was founded in 1919 by Erich Marx and Dr James Traube with a capital of 50,000 RM. The aim of the company was commerce with machines and special tools for automobile repair.

55. Business co-leader Erich Kohlmeyer in his 'Kurzbericht über die Entwicklung der MATRA seit der Übernahme durch LINDE' to the Supervisory Board; Typescript from 31 March 1966, p. 1; Linde-Archiv Wiesbaden, File ZV 337.

56. Transcript of the 247th meeting of the Supervisory Board 10 January 1952, p. 5; Linde-Archiv Höllriegelskreuth.

57. The contracts from Matra made up about 7–8 per cent of the Güldner's total production in 1935.

58. Looking back, Erich Mrax evaluates the behaviour of the company in the process of Aryanization as following: 'At the contract negotiations Linde behaved decently in so far as it did not put any particular pressure, like with other companies, for the transfer of the shares, which went beyond the pressure to sell the non-Aryan shares. They did however defend the interests of their company widely with regard to the calculation of the sale price.' Letter to the Supervisory Board of the Gesellschaft für Lindes Eismaschinen AG on 8 June 1949, p. 2. Linde-Archiv Höllriegelskreuth.

59. GFLE, *75 Jahre Linde*, p. 23.

60. Aufstellung 'Bruttogewinne 1936' [Typescript], Linde-Archiv Höllriegelskreuth.

61. Linde AG, *Geschäftsbericht* 1940, p. 6.

62. Hellmann, *Höchst unauffällig*, p. 150.

63. Hellmann, *Künstliche Kälte*, p. 116. Hellmann adds: 'In civilian everyday life one could hardly trace anything of the frozen foods, and no one would have claimed in this time, that a refrigerator is a necessary component for existence. Not even one per cent of all households had it.' See for deep freeze storage also: Mosolff, *Der Aufbau der deutschen Gefrierindustrie*.

64. Linde AG, *Geschäftsbericht und Bilanz* 1934, p. 5.

65. Linde AG, *Geschäftsbericht und Bilanz* 1936, p. 3.

66. Linde AG, *Geschäftsbericht und Bilanz* 1937, p. 5.

67. Linde AG, *Geschäftsbericht und Bilanz* 1935, p. 3.

68. Letter from Richard Linde to his sisters, 4 March 1934. Linde, H., *Eine Münchener Familie*, p. 47.

69. Letter from Richard Linde to his siblings, 16 February 1935. Linde, H., *Eine Münchener Familie*, p. 81 f.

70. 'In Höllriegelskreuth operations are increasingly intense, almost everyone works for petrol production in Germany.' Letter from Richard Linde to his sisters, 4 March 1934. Linde, H., *Eine Münchener Familie*, p. 165.

71. Richard Linde: Report to the Supervisory Board on 30 November 38, p. 1, Linde-Archiv Höllriegelskreuth, Ordner Vorträge und Berichte von Dr R. Linde.

72. Richard Linde on 8 January 1938 in a family letter to his sisiters, Linde, H., *Eine Münchener Familie*, p. 147.
73. Jaeger, *Aus der Geschichte der Werksgruppe TVT*, p. 115.
74. Abelshauser (Ed.), *Die BASF*, p. 374.
75. Stokes: *Technology Transfer*.
76. Letter from Richard Linde to his sisters, 4 March 1934, Linde, H., *Eine Münchener Familie*, p. 47.
77. Linde AG, *Geschäftsbericht und Bilanz* 1934, p. 5.
78. Linde AG, *Geschäftsbericht und Bilanz* 1935, p. 3. These instructions of the Ministry of Economy stood in connection with the endeavour to contain price increases as a result of the armament business, so as not to hinder further growth in relevant branches.
79. Linde AG, *Geschäftsbericht und Bilanz* 1938, p. 5.
80. Cf. 'Aufstellung von Kartellen, Syndikaten und ähnlichen Abreden, beteiligt seit 1 January 1938'; Höllriegelskreuth, 29 April 1947; signed by Dr Wucherer and Dr R. Linde, Linde-Archiv Höllriegelskreuth.
81. Thus the document mentions in a total of 20 places: protection agreement, sales restriction, region agreement, abstention agreement, sales agreement, price and delivery agreement, patent exchange, licence contract with delivery restriction, market regulation agreement, sales regulation agreement and advertisement agreement.
82. This cartel was not, however, concealed. Thus the business reports of the 1930s regularly say: 'Like until now, we are a partner in the Vereinigten Sauerstoffwerken GmbH, Berlin, which regulates the prices and sales of technological gases for the included companies.' Linde AG, *Geschäftsbericht und Bilanz* 1938, p. 5.
83. Linde AG, *Geschäftsbericht und Bilanz* 1938, p. 6.
84. Petzina, *Die deutsche Wirtschaft in der Zwischenkriegszeit*, p. 123.
85. Büchmann, Steigende Weltgeltung der deutschen Industrie, *Deutsche Technik* 5 (1937), 590.
86. Beutler: Wesen, Aufgaben und Begriffe der Wehrwirtschaft, *Deutsche Technik*, 6 (1937), 303.
87. Büchmann, Steigende Weltgeltung der deutschen Industrie, *Deutsche Technik* 5 (1937), 590.
88. Linde AG, *Geschäftsbericht und Bilanz* 1933, p. 3.
89. Letter from Richard Linde to his sisters, 6 February 1936. Linde, H., *Eine Münchener Familie*, p. 105.
90. Letter from Richard Linde to his sisters, 12 February 1937. Linde, H., *Eine Münchener Familie*, p. 128 f.
91. Richard Linde reported in that year to the Supervisory Board: 'Unfortunately the share of the export contracts has gone down strongly as against the previous years, and indeed to 6 per cent against 40–50 per cent in the last years.' Richard Linde: Report for the Supervisory Board Meeting on 30 November 38, Linde-Archiv Höllriegelskreuth, Ordner Vorträge und Berichte von Dr R. Linde.
92. Richard Linde: Report for the Supervisory Meeting on 30 November 38, Linde-Archiv Höllriegelskreuth, Ordner Vorträge und Berichte von Dr R. Linde.
93. Richard Linde: 'Bericht an den Aufsichtsrat in der Sitzung am 21.5.1943', p. 1, in Linde-Archiv Höllriegelskreuth, File 'Vorträge und Berichte von Dr R. Linde'.
94. Richard Linde: 'Bericht an den Aufsichtsrat in der Sitzung am 3.2.41', p. 4, in Linde-Archiv Höllriegelskreuth, Ordner Vorträge und Berichte von Dr R. Linde.
95. Richard Linde: 'Bericht an den Aufsichtsrat in der Sitzung am 3.2.41', p. 4, in Linde-Archiv Höllriegelskreuth, Ordner Vorträge und Berichte von Dr R. Linde.
96. Richard Linde: 'Bericht an den Aufsichtsrat in der Sitzung am 3.2.41', p. 5, in Linde-Archiv Höllriegelskreuth, Ordner Vorträge und Berichte von Dr R. Linde.

97. 'Production at Fa. Linde (own factories and subcontracted suppliers) is devoted entirely to German contracts for the mineral oil programme and other installations crucial to war until the year 1945 (No. 4010 to 4019). Letter from Department B to central headquarters of the Reichsgruppe für Industrie, Berlin, on 20 January 1943, in which Linde suggested the founding of a branch operation in Italy, in order to 'take care of the contracts from Italy for the delivery of gas and air separation plants, which were designed to manufacture ammonia, explosives, Buna and fuel for the Italian factories crucial for war'. Linde-Archiv Höllriegelskreuth.

98. Of those Linde employees conscripted as soldiers, 102 died in the war by March 1943. *Werkszeitschrift für die Betriebe der Gesellschaft für Lindes Eismaschinen*, 8(1943), p. 1.

99. Report from Department B to the Supervisory Board on 17 September 1940 in Wiesbaden, Linde-Archiv Höllriegelskreuth.

100. Gas Liquefaction Department, Höllriegelskreuth, in a report to the Supervisory Board on 1 July 1946, p. 4, Linde-Archiv Höllriegelskreuth.

101. Linde AG, *Geschäftsbericht und Bilanz* 1943, p. 2.

102. Richard Linde: 'Bericht an den Aufsichtsrat in der Sitzung am 21.V.1943', p. 2, Linde-Archiv Höllriegelskreuth, Ordner Vorträge und Berichte von Dr R. Linde.

103. 'Already in the summer we had – and indeed on behest of the chief representative of chemical manufacturing – looked for an alterative workshop. [...] Then as the mentioned hurried contracts came about, the pressure to rapidly set up this alternative workshop increased greatly.' Richard Linde: 'Bericht an die Herren Aufsichtsratsmitglieder, Betr.: Liefergeschäft der Abt. B Apparatebau-Anstalt', 20 December 1943, p. 2 f., Linde-Archiv Höllriegelskreuth, Ordner Vorträge und Berichte von Dr R. Linde.

104. *Die Großunternehmen im Deutschen Reich*, p. 91.

105. The V 2 was the first production-line rocket to fly at high speeds with a parabolic flight path. Conventional weapon systems provided no defence; the rocket was capable of inflicting massive damages, for example, in the bombing of London. Bode/Kaiser, *Raketenspuren*.

106. For oxygen installation in Eperlecques see: http://www.v2rocket.com/start/deployment/watten.

107. 'At the beginning of 1940 it [the company] had 114 employees at the rocket centre working on that project alone.' Winter/Neufeld, *Heylandt's Rocket cars and the V2*, p. 66.

108. This emerges from a letter from Paulus Heylandt to Richard Linde in August 1943: 'The charge you made on the phone against our engineering office or the signatories, that they work too slowly and that the gentlemen in question are not properly employed, prompted me ... to investigate the matter with the result, that the charges are not correct. [...] The gentlemen were working at full capacity at the engineering office ... including for the A4 programme.' 'Betrifft: Arbeitseinsatz im A 4-Programm' Letter of Dr Paulus Heylandt, Berlin-Britz, 21 August 1943, to director Richard Linde. Richard Linde: 'Bericht an die Herren Aufsichtratsmitglieder für die Sitzung am 29 June 44', p. 3. Linde-Archiv Höllriegelskreuth, Ordner Vorträge und Berichte von Dr R. Linde.

109. Hermann Linde's speech in the 40th anniversary of the services of Walther Ruckdeschel, 12 September 1966, p. 3. Linde-Archiv Höllriegelskreuth, Materialien Erik Jaegar.

110. Richard Linde: 'Bericht an die Herren Aufsichtsratsmitglieder, Betr.: Liefergeschäft der Abt. B Apparatebau-Anstalt.' 20 December 1943, p. 1 f., in Linde-Archiv Höllriegelskreuth, Ordner Vorträge und Berichte von Dr R. Linde.

111. This expansion made further increase of the nominal capital necessary: on 20 January 1938 to 500,000 RM, on 12 January 1942 to 2 million RM, and on 16 June 1944 to 5 million RM.

112. 'Die Entwicklung der MATRA-WERKE GmbH, Frankfurt a. M., in der Zeit vom 31 December 1935 bis 31 December 1942', Report of the business co-leaders Kohlmeyer and Lauer to the Linde Supervisory Board on 2 March 1943 [Typescript], Linde-Archiv Wiesbaden, File ZV 337.

113. Untitled speech draft for the 50th anniversary of the Güldner-Motoren factories 1954, p. 8. Linde-Archiv Aschaffenburg. Fahr belonged to the buyers of the Güldner installation engines.

114. Linde AG, *Geschäftsbericht und Bilanz* 1939, p. 6.

115. Linde AG, *Geschäftsbericht und Bilanz* 1940, p. 7.

116. Linde AG, Niederlassung Höllriegelskreuth: Gelieferte Gross-Sauerstoffanlagen, Verzeichnis B, o.J. [before 1965], TR-Nr. 74–77.

117. H 288/89, according to the data sheet 78738 were delivered on 1 December 1943 and 1 January 1944. Linde-Archiv Höllriegelskreuth, Ordner H-Anlagen.

118. Wagner, *IG Auschwitz*, p. 10; Hayes, *IG Farben in the NAZI Era*; Stokes, *Von der IG Farbenindustrie*.

119. See for instance the memoirs of Primo Levi, who as inmate No. 174 517 of the concentration camp Auschwitz III worked in Buna factory in Monowitz and survived; the average life expectancy of a Jewish inmate there was three to four months. Levi, *Ist das ein Mensch?*

120. The installations were itemized in the list of delivered large oxygen installations, Index B, o.J. [before 1965], only as serial installation numbers; 'following contracts were withdrawn or dropped' from a list 'after the collapse', H'kreuth dates on the 1 July 1946 (Typescript, Linde-Archiv Höllriegelskreuth), that TR 78–80 in the contract amount of 1.5 million. RM were supposed to be delivered to Auschwitz, while TR 81 was to go to Christianstadt an der Bober (today Krzystkowice), where the 'GmbH was located for the utilization of chemical products'. Deviating from this, TR 78–81 are on the record sheet 78.752 named for set-up in Auschwitz, whereby the figure '80' is subsequently changed to '81'. Linde-Archiv Höllriegelskreuth, Ordner TR-Anlagen.

121. See: Herbert, *Fremdarbeiter*.

122. Spoerer, *Zwangsarbeit unter dem Hakenkreuz*, p. 16 f.

123. In an interview in 2001, Jakobsmeier asked Bäumler about a list of foreign labourers: 'Do you still have the mentioned list?' Bäumler: 'Unfortunately not. When leaving the company I looked in vain for someone who was interested in these documents. Nobody showed an interest, either in Wiesbaden nor in Höllriegelskreuth. I had no choice but to destroy it [...]. The same was true for the deputy director of our department, Herrn Barts. [...] He left Linde in 1972. On this occasion he left all his documents to me. All files concerning forced labourers went to his desk. And I never found the smallest indication of something which would have been incriminating. Personal records were also included in these documents in the form of diaries. Unfortunately I destroyed those also. I did not want them to fall into the hands of strangers. Herr Egglsmann and also Herr Barts handed down their personal documents to me, because they knew I was interested in historical questions. They were thought of as Herr Jaeger's work documents. Then when Herr Jaeger left, there was no longer anyone who could have found some use for them.' Linde VA [Dr Werner Jakobsmeier]: Zwangsarbeiter bei Linde in Höllriegelskreuth und Schalchen – Eine Bestandsaufnahme, Höllriegelskreuth o.J. [*ca.* 2001].

124. Records of the Executive Board meeting on 28 September 1998, Linde-Archiv Wiesbaden.

125. The data was compiled through employment and layoff records. The 'staff overview of the factories' (Die Belegschaftsübersicht der Werke) in Höllriegelskreuth and Schalchen started in 31 July 1944 and was updated monthly. Linde VA Höllriegelskreuth, Personalabteilung, Ein- und Ausstellungsbuch vom 1. 7. 1944 bis zum 31. 12. 1947.

126. Richard Linde: 'Bericht an den Aufsichtsrat in der Sitzung am 30. April 41, p. 1, Linde-Archiv Höllriegelskreuth, Ordner Vorträge und Berichte von Dr R. Linde.

127. Richard Linde: 'Bericht an den Aufsichtsrat in der Sitzung am 21.V.1943', p. 2, Linde-Archiv Höllriegelskreuth, Ordner Vorträge und Berichte von Dr R. Linde.

128. Richard Linde: 'Bericht an die Herren Aufsichtsratsmitglieder, Betr.: Liefergeschäft der Abt. B Apparatebau-Anstalt', 20 December 1943, p. 2. Linde-Archiv Höllriegelskreuth, Ordner Vorträge und Berichte von Dr R. Linde.

129. Richard Linde: 'Bericht an die Herren Aufsichtsratsmitglieder für die Sitzung am 29 June 44, Betr.: Liefergeschäft der Abt. B Apparatebau-Anstalt', 26 June 44, p. 2. Linde-Archiv Höllriegelskreuth, Ordner Vorträge und Berichte von Dr R. Linde.

130. 'The Russians – I can only speak of those, with whom I had something to do – were poorly dressed. They had no underwear to change and their hygienic facilities could not have been exactly comfortable. At any rate they smelled rather strongly. But no one looked starved. They were fed, as far as I know, in their barracks. They walked out together during the lunch break and came back together. Harassment – that simply did not fit. I cannot imagine that someone was intentionally mistreated. At Linde everyone knew everyone. And nobody was subject to suffering. ... if something had happened, I would have had to have known, since security was part of our department. Our department was so small, that I had to hear about every thing.' Interview Werner Jakobsmeier with Franz Bäumler, 2001. Linde VA [Dr Werner Jakobsmeier]: Zwangsarbeiter bei Linde in Höllriegelskreuth und Schalchen – Eine Bestandsaufnahme, Höllriegelskreuth o. J. [2001].

131. Petition of 54 foreign labourers to R. Linde on 3 January 1945. Linde-Archiv Höllriegelskreuth, Ordner Dr R. Linde, Privatakt vom 1.Jan. 1942 bis zum 31. Dez. 1948, A- Hi.

132. Report of Department D, Güldner-Motoren factories, Aschaffenburg to the Supervisory Board on 5 June 1946, p. 1. Linde-Archiv Höllriegelskreuth.

133. Letter from director Otto Hippenmeyer to general director Dr Friedrich Linde, Höllriegelskreuth: 'Kurzer Lagebericht über die benachbarten Werke vom 26 September 44', p. 2. Linde-Archiv Höllriegelskreuth, Ordner Dr R. Linde, Privatakt vom 1.Jan. 1942 bis zum 31. Dez. 1948, A- Hi.

134. Walther E. Ruckdeschel: Mein Lebenslauf., München 1989, p. 183. Linde-Archiv Wiesbaden.

135. The appropriation of less profitable iron ores, forced by the National Socialists' policy of autarchy, accelerated the efforts, 'which have already clearly proven, that under specific conditions, not only that one could save significantly on coke, if one works with the oxygen-enriched blow, but, what is at the moment still more important, the oven work can also be increased compared to operations without oxygen by 40 to 50 per cent'. Richard Linde: Report to the Supervisory Board Meeting on 30 November 38, Linde-Archiv Höllriegelskreuth, Ordner Vorträge und Berichte von Dr R. Linde.

136. Richard Linde: Report to the Supervisory Board in the meeting on 3.II.41, p. 3, Linde-Archiv Höllriegelskreuth, Ordner Vorträge und Berichte von Dr R. Linde.

137. Richard Linde: Report to the Supervisory Board in the meeting on 3. II. 41, p. 6 f, Linde-Archiv Höllriegelskreuth, Ordner Vorträge und Berichte von Dr R. Linde.

138. Jaeger, Aus der Geschichte der Werksgruppe TVT, p. 128.

139. Fighting fires in cork insulated special constructions was difficult. The fire was brought under control only after some days. 'With the cold store goods burned, a huge part of the winter ration of the population of Munich was destroyed. In sum this was 500 tons of butter, about 80 tons of butter oil, 3½ million eggs and about 18 cars of frozen goods. The cold-stored goods from the lower floors damaged by smoke, etc., mainly frozen meat and frozen conserves, were supplied for direct consumption or to conserve factories.' Bericht betr. Kühlhaus München/ Fliegerangriff 6./7. September 43. Confidential! Only for the gentlemen of the

chairmanship and the Supervisory Board! Signed by Hippenmeyer, Ombeck. Linde-Archiv Höllriegelskreuth, Ordner Dr R. Linde, Private File from 1. Jan. 1942 until 31. Dec. 1948, A- Hi.

140. Hermann Linde's speech in the 40th anniversary of the services of Walther Ruckdeschel, 12 September 1966, p. 3. Linde-Archiv Höllriegelskreuth, Materialien Erik Jaegar.

141. Linde AG, *Geschäftsbericht* 1945, p. 3.

142. Untitled draft speech for the 50th anniversary of the Güldner-Motoren factories 1954, p. 9. Linde-Archiv Aschaffenburg. Fahr was among the buyers of Güldner installation engines.

143. The relocation occurred apparently under the direction of the SS special inspection in Porta. On 24 February 1945 the Gas Liquefaction Department in Höllriegelskreuth sent a telex to the SS office: 'For Hammer-mills. Still waiting for your message, whether the tunnel's length is enough for setting up the ammonia compressor from Bielefeld. Dipl. Ing. Dr von Zur Mühlen will presumably arrive there at the end of the next week to oversee the relocation of the Bielefeld installation to Bielefeld.' Linde-Archiv Höllriegelskreuth, Durchschreibbuch Fernschreiben 1945, p. 21.

144. Department of Gas Liquefaction Höllriegelskreuth, Report to the Supervisory Board about the oxygen and acetylene factories on 1 July 1946 Linde-Archiv Wiesbaden.

145. This was an administrative unit established by the Germans in 1939 and consisted of those parts of Poland not incorporated into the Third Reich, including Warsaw, Krakow, Radom, Lublin, and Lvov.

146. Telex from 24 February 1945. Linde-Archiv Höllriegelskreuth, Durchschreibebuch Fernschreiben 1945, p. 25. The installation which extracted krypton from air, was set up in 1936 for the Vereinigte Glühlampen- und Elektrizitäts-AG in Ujpest/Hungary. Linde AG, Niederlassung Höllriegelskreuth, Gelieferte Gross-Sauerstoffanlagen, Verzeichnis B, o.J. [before 1965], No. 24. In 1949 Linde received from the same company a second contract for a krypton installation in Ajka/Miscolcz (No. 102). Whether this was to be a replacement construction or an expansion installation is not known.

147. Interview with Prof. Hermann Linde. Linde, H., *Aus der Geschichte der Tieftemperaturtechnik*. Cf. also Ruckdeschel, *Mein Lebenslauf*, p. 40.

Chapter 4

1. Erker/Pierenkemper (eds): *Deutsche Unternehmer zwischen Kriegswirtschaft und Wiederaufbau*, p. 36.

2. See Karlsch, *Allein bezahlt?* Karlsch estimates that the Soviet occupation zone lost ten times as much industrial capacity as the Western occupation zones (p. 233). Karlsch points out that although economic growth in the initial post-war years depended more on raw materials, energy, and transport capacity than on industrial capacity, over time the dismantling of East German industry resulted in significant barriers to growth, especially in the 1950s, which in some sectors could 'not be entirely overcome by the end of the GDR' (p. 234). On top of this came the ongoing costs of occupation, which were a stronger drag on the weak GDR economy than on that of the Federal Republic.

3. On the work of German researchers in the Soviet Union, see Albrecht/Heinemann-Grüder/Wellmann, *Die Spezialisten*.

4. The distribution of design worksheets and industrial property rights was handled very informally, 'upon request', so to speak. But there were also specific inquiries, mainly from large companies, that wanted to use particular important patents of competing German companies.

5. However, not only German companies were affected by this; in the Soviet satellite states that developed behind the 'Iron Curtain', all foreign-owned companies were put under state administration or expropriated altogether.
6. The population grew by more than 3 million due to the influx of refugees and emigrants from the east.
7. Hardach, *Wirtschaftsgeschichte Deutschlands*, pp. 115–19.
8. Between 1945 and 1948, the actual provisioning of the majority of the populace sometimes fell even below the level of 1,750 calories per day that the Western Allies deemed temporarily tolerable. But not all Germans went equally hungry: those who were self-supporting (about 14 per cent of the populace) generally experienced very few cut-backs, while 'John Doe' – that is, the unemployed, those disabled in the war, the elderly, and all white-collar employees whose jobs were classified as less physically strenuous (circa 30 per cent of the populace) – suffered real hunger.
9. In 1947, raw materials and semi-manufactured goods still accounted for 88 per cent of exports, while finished products amounted to only 11 per cent. See Jerchow, *Deutschland in der Weltwirtschaft*, p. 479.
10. Jerchow, *Deutschland in der Weltwirtschaft*, p. 474; Mausbach, *Zwischen Morgenthau und Marshall*.
11. A total of 1.5 billion US dollars flowed into Germany between 1948 and 1952.
12. In early 1949, the French occupation zone was merged into 'Bizonia'.
13. The Soviet Union reacted to the announced inclusion of Berlin in the Western currency system by completely blockading road and rail traffic between Berlin and the Western zones. The Western Allies responded with the spectacular and successful Berlin airlift and a trade embargo that was painful for the emergent German Democratic Republic (GDR).
14. The Federal Republic was able to end rationing of foodstuffs as of 31 March 1950, while certain products were still rationed until 1958 in the GDR.
15. Radkau, *Technik in Deutschland*, p. 313.
16. To regulate trade with the GDR, the Federal Republic signed the 'Interzone Trade Treaty' on 1 August 1952.
17. Theoretically, the centrally planned economy was supposed to guarantee optimal distribution and utilization of all resources through comprehensive governmental control; in fact, due to the dominance of political goals and the 'friction losses' from inflexibility that were inherent to the system, it was not able to outstrip the productivity of the market-oriented Western world as GDR leaders had hoped.
18. Linde AG: *Das ist Linde*, p. 12.
19. 'The difficulties of the post-war period, which are impeding the reconstruction and repair of our plants, have increased in business year 1947 with the severe currency erosion and economic disruption. ... Under these conditions, our production sites have not yet been able to reach normal production capacity.' Linde AG, *Geschäftsbericht 1947*, Executive Board Report, p. 2. At the end of 1949, damage from the war had been 'almost entirely eliminated'. Linde AG, *Bericht über die DM-Eröffnungsbilanz und Geschäftsbericht über das verlängerte Geschäftsjahr Juni 1948 bis Dezember 1949*, p. 12.
20. Letter from Charlotte Heylandt to the Linde Company on 23 May 1946: 'Unfortunately I still haven't heard anything from Russia, apart from a single postcard from my husband through the Red Cross at Christmas. It was undated and presumably written in September. He arrived in Moscow on the evening of 11 July 1945 after leaving Berlin that morning (by airplane). On the postcard he wrote that he had been very ill for an extended period since shortly after his arrival in Russia (I suspect a physical and emotional breakdown) but hoped with good medical care to be fit again in a few weeks. [...] My trip to visit my husband has apparently landed on a dead track, unfortunately. At the moment there is nothing I can do except school myself in patience and diligently continue learning Russian.'

Linde-Archiv Höllriegelskreuth, Ordner Dr R. Linde, Privatakt vom 1. Jan. 1942 bis 31. Dez. 1948, A-Hi. On German rocket scientists after the war, see Albring, *Gorodomlia*; Bode/Kaiser, *Raketenspuren*, pp. 146–54; Michels, *Peenemünde und seine Erben in Ost und West.*

21. Minutes of the Executive Board meeting of 27 October 1947. Linde-Archiv Wiesbaden.
22. Linde AG, *Geschäftsbericht* 1946, p. 2.
23. 'The German supplier is thus by no means the contractual partner. This is apparently intended to prevent German companies from accumulating a credit balance abroad.' From the collection 'Aus Tagebuchnotizen von Herrn Philipp Borchardt', prepared by Dr Erik Jaeger, p. 2, entry for 7 March 1946. Linde-Archiv Höllriegelskreuth.
24. Ibid.
25. Ibid., S. 3. Phoning was not yet an alterative, either; only in 1950, for example, did upper management in Höllriegelskreuth receive 'authorization for international long-distance business calls'. Those who received this authorization were Philipp Borchardt (technical director and authorized officer), Walter Ruckdeschel (technical authorized officer), Wilhelm Hartmann, Franz Frauscher (both managing agents), Rudolf Wucherer (member of the Executive Board), and Rudolf Becker (authorized officer). Formula 'Zulassung zum geschäftlichen Auslands-Fernsprechverkehr' der Oberpostdirektion München, 2 January 1950. Linde-Archiv Höllriegelskreuth.
26. Richard Linde: 'Bericht an den Aufsichtsrat, Abt. B, Gasverflüssigung, I.) Apparatebauanstalt Höllriegelskreuth' from 2 July 1946, p. 3. Linde-Archiv Höllriegelskreuth, Ordner: Vorträge und Berichte v. Dr R. Linde,.
27. Ibid., p. 5.
28. Linde AG, *Das ist Linde*, p. 101.
29. The annual report for 1949 estimated the impact of the currency reform 'along with the expropriation ('communization') of our property in the Eastern zone' at about 5.6 million DM in total. Linde AG, *Geschäftsbericht* 1949, p. 12.
30. Minutes of the Executive Board meeting of 17 January 1949. Linde-Archiv Wiesbaden.
31. Minutes of the Executive Board meeting of 17 January 1949. Linde-Archiv Wiesbaden.
32. Linde AG, *Geschäftsbericht* 1946, p. 2.
33. Linde AG, *Geschäftsbericht* 1949, p. 14.
34. Linde AG, *Geschäftsbericht über den Abschluss zum 20. Juni 1948*, p. 3.
35. Linde AG, *Geschäftsbericht 1945*, p. 2.
36. Linde AG, *Geschäftsbericht 1953*, p. 6.
37. Linde AG, *Geschäftsbericht 1945*, p. 3.
38. Linde AG, *Geschäftsbericht 1946*, p. 3.
39. Linde AG, *Geschäftsbericht über den Abschluss zum 20. Juni 1948*, p. 3.
40. Apparatebau-Anstalt Höllriegelskreuth, report to the Supervisory Board of 17 January 1949, p. 2. Linde-Archiv Höllriegelskreuth, Ordner Vorträge und Berichte von Dr R. Linde.
41. Apparatebau-Anstalt Höllriegelskreuth, report to the Supervisory Board of 17 January 1949, p. 3. Linde-Archiv Höllriegelskreuth, Ordner Vorträge und Berichte von Dr R. Linde.
42. Neebe, *Überseemärkte und Exportstrategien in der westdeutschen Wirtschaft*, p. 8.
43. 'Obviously LAPC has received some sort of reprimand from Washington recently due to our prewar contracts with LAPC because they conflict with the antitrust laws, which are coming to the fore again since the Democrats – contrary to expectations – have stayed at the helm.' Apparatebau-Anstalt Höllriegelskreuth, report to the Supervisory Board of 17 January 1949, p. 4. Linde-Archiv Höllriegelskreuth, Ordner Vorträge und Berichte von Dr R. Linde.

44. Apparatebau-Anstalt Höllriegelskreuth, report to the Supervisory Board of 17 January 1949, p. 5. Linde-Archiv Höllriegelskreuth, Ordner Vorträge und Berichte von Dr R. Linde.
45. Linde AG, *Geschäftsbericht* 1945, p. 3.
46. Linde AG, *Geschäftsbericht über den Abschluss zum 20. Juni 1948*, p. 3.
47. Linde AG, *Bericht über die DM-Eröffnungsbilanz und Geschäftsbericht über das verlängerte Geschäftsjahr Juni 1948 bis Dezember 1949*, p. 14.
48. 'To be sure, one can question whether it is advisable in the present circumstances to put plants into the Russian zone, even if they are built in one of the non-Russian sectors of Berlin. But after we learned that Messer plans to build a factory in Berlin and also enlarge their existing smaller plants, it seemed necessary for us to take action too, lest we allow ourselves to be driven out entirely from Berlin, where we had our largest plant.' Bericht über die Sauerstoffwerke an den Aufsichtsrat vom 28.11.1946, p. 2. Linde-Archiv Wiesbaden.
49. The request was approved. Linde AG, Verwaltungsstelle Berlin, letter of 8 March 1947. Telex from R. Wucherer on 4 February 1948. Linde-Archiv Höllriegelskreuth, Durchschreibebuch 12. 1947–8. 1948, p. 11.
50. Letter of 7 September 1949. Linde-Archiv Höllriegelskreuth, Durchschreibebuch Rudolf Wucherer 1949–1952, p. 1.
51. Linde AG, *Geschäftsbericht* 1952, p. 7.
52. On the other hand, prior to the currency reform there was a noticeable shortage of skilled workers, which was a consequence not only of the captivity of POWs but also of wages being paid in Reichsmarks.
53. Linde AG, *Geschäftsbericht über den Abschluss zum 20. Juni 1948*, p. 3.
54. Gas motors could be powered with cylinder or generator gas; since fuel was short and available only with ration cards, gas motors represented an important alternative for agriculture in the immediate post-war years. However, the fuel supply had already improved by 1948, and Güldner converted a substantial number of wood-gas tow tractors to diesel.
55. Linde AG, *Geschäftsbericht* 1945, p. 3. An in-house summary from 1953, 'Movement of machines and war damage', lists 94 machine tools that were transported to Gottmadingen in March 1945 in a chartered freight train. Of these, 20 machines were confiscated and removed 'by the First French Army on 30 June 1945' as well as – according to requisition slip no. 50003 – another 19 machines on 13 November 1945 by the Manufacture d'armes de Paris a' Augy (Yonne). Two more machines were commandeered in November 1945. In Aschaffenburg, the occupying powers removed a further 67 machine tools and 5 foundry machines. In all, 115 machines were dismantled. Report 'Maschinen- und Kriegsschaden' dated 21 April 1953. Linde-Archiv Aschaffenburg.
56. Statement in an untitled manuscript draft for the 50-year jubilee of Güldner-Motoren-Werke in 1954, p. 9 Linde-Archiv Aschaffenburg.
57. Linde AG, *Geschäftsbericht* 1946, p. 3.
58. Linde AG: *Das ist Linde*, p. 77.
59. Linde AG, *Bericht über die DM-Eröffnungsbilanz und Geschäftsbericht über das verlängerte Geschäftsjahr Juni 1948 bis Dezember 1949*, p. 14.
60. Richard Linde informed a job-seeking engineer in a letter of 28 March 1947: 'To be sure, my company has been scaled down, but since the reduction of our personnel was relatively small in scope, for the very reason that only a few were party members, we have no need for new men, so that I unfortunately see no possibility for our company to employ you.' Linde-Archiv Höllriegelskreuth, Ordner Dr R. Linde, Privatakt vom 1. Jan. 1942 bis 31. Dez. 1948, A-Hi.
61. 'Ansprache anlässlich des 50-jährigen Dienstjubiläums, gehalten am 7.I.46 von Dr R. Linde', p. 7. Linde-Archiv Höllriegelskreuth, Ordner Vorträge und Berichte von Dr R. Linde.

62. Director Otto Hippenmeyer on 25 October 1946 to Dr Richard Linde: 'During the return trip [from Sürth to Wiesbaden] in the unheated and unlit train with the cold setting in, I had hours of time to think about things that one just cannot understand, no matter how much one occupies oneself with them, and in particular your brother's fate. I keep asking myself if it isn't somehow possible for those people who are not tarnished to do something to help him. For example, I could corroborate under oath that your brother excoriated and rejected the Nazi regime with steely consistency, from A to Z, from the beginning to the end, and that in every way his actions drew the consequences of this, so that it is incomprehensible to me how he got into this position at all, in which he surely had hardly anything to say.' Linde-Archiv Höllriegelskreuth, Ordner Dr R. Linde, Privatakt vom 1. Jan. 1942 bis 31. Dez. 1948, A-Hi.

63. Durchschreibebuch 11.12.1947–23.8.1948, entry from 28 January 1948. Linde-Archiv Höllriegelskreuth.

64. Linde AG, *Geschäftsbericht* 1946, p. 7.

65. Abteilung Gasverflüssigung, Höllriegelskreuth, in a report to the Supervisory Board of 1 July 1946, p. 4. Linde-Archiv Höllriegelskreuth.

66. For example, the Einbeck branch operations of the Sürth division was reliant on the continued employment of former party members in 1946, in Friedrich Linde's opinion. Linde-Archiv Höllriegelskreuth, Korrespondenz Friedrich Linde.

67. On the theory and practice of denazification, see: Niethammer, *Die Mitläuferfabrik*.

68. The return of Borchardt, who had left Germany involuntarily, was obstructed by Allied regulations: 'The English minister for German affairs, Mr J.B. Hynd, let me know that the entire question of the return of refugees was just being processed and that he hoped to be able to make a statement about it in the House of Commons in the near future. I had submitted a petition to the Home Office in which I drew attention to the indefensibleness of the current situation, where all 'displaced persons' and even the prisoners of war are transported back to their homeland or sent home, while people like me are not allowed to return to their families for 'personal or familial reasons', as if I didn't have the same right as every prisoner of war to live in my homeland with my family. And yet there are surely not many who even have any desire to return.' Philipp Borchardt on 23 May 1946 to Richard Linde, Linde-Archiv Höllriegelskreuth, Ordner Dr R. Linde, Privatakt vom 1. Jan. 1942 bis 31. Dez. 1948, A- Hi.

69. The tribute 'On Dr Richard Linde's Retirement from the Executive Board' in the company newsletter was written by Philipp Borchardt. In it, he expressed his friendly admiration: 'It is not only his successful work as an engineer or as a member of our company's Executive Board – it is his entire human personage that we accompany into retirement with gratitude and admiration.' Borchardt seemed almost apprehensive about the future: 'His colleagues and the entire workforce now face the challenge of continuing their work without their previous intellectual leader, to whom they and the company owe their rank and reputation – though they also hope that for a long time to come Dr Linde will not withhold his advice and his help if decisions must be made in which his expertise and experience would be invaluable.' Linde AG, *Werkszeitschrift* 2 (1950), p. 3 f.

70. 'Bericht betreffend Rückerstattung von Matra-GmbH.-Anteilen'. Enclosure to a letter from Director Ombeck (Wiesbaden) to Richard Linde in Höllriegelskreuth on 23 March 1948, p. 1. Linde-Archiv Wiesbaden.

71. For example, in the Supervisory Board meeting of 1 April 1948 and the Executive Board meeting of 29 August 1949, where agreement was unanimous 'that a settlement and an amicable agreement with Mr Marx is impossible due to his demands (50 per cent voting rights), and a judicial decision should be effected'. Linde-Archiv Wiesbaden.

72. Telex from R. Wucherer on 4 March 1948 to the head office of the Linde Company in Wiesbaden, Linde-Archiv Höllriegelskreuth, Durchschreibebuch 12.1947–8.1948, p. 11.

73. 'Bericht betreffend Rückerstattung von Matra-GmbH-Anteilen'. Enclosure to a letter from Director Ombeck (Wiesbaden) to Richard Linde in Höllriegelskreuth on 23 March 1948), p. 1. Linde-Archiv Wiesbaden.

74. Although the law was drawn up by German jurists such as Otto Küster, Adolf Arndt, and Walter Roemer, it was then enacted as a military government law, not as a German law. For its wording, see: Weißstein/Riedel, *Kommentar zum Militärregierungsgesetz*, or *Amtsblatt der US-Militärregierung* vom 10.11.1947. Law 59 regulated the return of plots of land, factories, and securities.

75. For the wording of the law, see: Gesetz Nr. 951 zur Wiedergutmachung nationalsozialistischen Unrechts vom 16.8.1949, in *Regierungsblatt der Regierung Württemberg-Baden*, 20 (1949), 187. On this topic, see also: Pross, *Wiedergutmachung*, p. 51.

76. 'In order to have a free hand in the negotiations with the lawyers and to reach a settlement if need be, the Supervisory Board authorizes the Executive Board to agree to 26 per cent voting rights (qualified minority) in regard to Mr Marx's claims. It is to be feared that the chamber might render a judgement awarding Herr Marx 50 per cent and relegating the Linde Company to the role of a limited partner despite its larger share of the capital.' Minutes of the 241st meeting of the Supervisory Board on 6 December 1949 in Höllriegelskreuth, p. 8 f. Linde-Archiv Höllriegelskreuth.

77. From: 'Kurze Übersicht über die Entwicklung der MATRA-WERKE GmbH' [Typescript]. Linde-Archiv Höllriegelskreuth.

78. The current historical debate on the restitution of Jewish property after 1945 can be found in: Goschler/ Ther (eds), *Raub und Restitution*.

79. Linde AG, *Geschäftsbericht 1945*, p. 2.

80. Linde AG, *Geschäftsbericht 1945*, p. 3.

81. Linde AG, *Bericht zur D-Mark-Eröffnungsbilanz und Geschäftsbericht 1949*, p. 14.

82. 'For your further orientation, it should be said here that the steel cylinder apparatus for each factory requires significant financial investments and practically nothing can be done with the industrial gases produced if there are not enough steel cylinders available for filling and delivery to the customer.' Letter from Linde AG, Abt. Gasverflüssigung, to the lawyer and notary Brumby, Berlin-Lichterfelde-West, dated 7 January 1947, signed Wucherer and Eggendorfer, p. 3. Linde-Archiv Höllriegelskreuth.

83. The operating company Wasserstoff-Sauerstoff-Werke GmbH Schwarzenberg/ Erzgebirge was maintained to hold open Linde's legal claim to restitution in case Germany was reunified. It had to pay taxes (corporate and wealth taxes, and until 1957 the 'Notopfer Berlin', a special levy to help the war-ravaged city) on its deposit capital (1961 = 20,000 DM). Aktennotiz betr. Steuerzahlungen für Wasserstoff-Sauerstoff-Werke GmbH Schwarzenberg Erzgebirge vom 16.1.1961, gez. R/Hf. Linde-Archiv Wiesbaden.

84. This announcement was a response to a statement to that effect by the Süddeutsche Bank. Letter of 19 December 1955. Linde-Archiv Höllriegelskreuth.

85. Linde AG, *Das ist Linde*, p. 100. On the negotiations over Linde's leasing and later takeover of the oxygen and acetylene plants in Alsace-Lorraine, see: 'Bericht der Abteilung B für die Aufsichtsrats-Sitzung am 17. September 1940 in Wiesbaden'. Linde-Archiv Höllriegelskreuth.

86. Linde AG, *Geschäftsbericht und Bilanz 1945*, p. 3.

87. Letter from Linde AG, Abt. Gasverflüssigung, to the lawyer and notary Brumby, Berlin-Lichterfelde-West, dated 7 January 1947, signed Wucherer and Eggendorfer, p. 2. Aufstellung von Kartellen, Syndikaten und ähnlichen Abreden, beteiligt seit

1.1.1938, compiled on 29 April 1947, signed Wucherer, Dr R. Linde. Linde-Archiv Wiesbaden, Ordner ZV 456.

88. Linde AG, *Geschäfts-Bericht und Bilanz* 1931, p. 4.
89. On the history of IG Farben, see: Plumpe, *Die I.G. Farbenindustrie AG.*
90. The Allied Control Council law No. 9 of 30 November 1945 laid the foundation for severing the cartel relationships between IG Farben's factories, redistributing the ownership rights to its industrial facilities and individual assets, completely controlling its research and production activities, seizing certain industrial facilities and individual assets as reparations, and destroying those facilities that were used exclusively for purposes of warfare. Bäumler, *Ein Jahrhundert Chemie*, p. 11. The liquidation of IG Farben's property has not been completed to this day, nearly a half century after the enactment of the Allied Control Council Law. The company's shares continued to be traded on the stock market, though it had been in trusteeship since 1952. However, in November 2003 IG Farben declared bankruptcy after the collapse of a deal to sell most of its remaining holdings.
91. Stokes, *Divide and Prosper*. In 1951, independent joint-stock companies were founded.
92. Bäumler, *Ein Jahrhundert Chemie*, p. 18.
93. Letter from Linde AG, Abt. Gasverflüssigung, to the lawyer and notary Brumby, Berlin-Lichterfelde-West, dated 7 January 1947, signed Wucherer and Eggendorfer, p. 4. Linde-Archiv Höllriegelskreuth.
94. Ibid., p. 4 ff. Similarly chaotic and peculiar events occurred in 1945 in the Borsigwalde oxygen plant 'under the trusteeship of a Mr Knör appointed by the French military government, who was previously an engineer and assistant to the now-retired head of this plant, and who behaved very ruthlessly toward us, his former parent company, during the Russian military occupation.' Ibid., p. 7.
95. Linde AG, Zentralverwaltung, letter of 30 August 1946 to Dr Friedrich Linde, signed Richard Linde and W. Flossel. Linde-Archiv Höllriegelskreuth.
96. World market prices for raw materials and industrial products climbed by 20 to 30 per cent within a few months, and by 60 per cent in the course of the year. 'The German export industry, which is oriented toward peacetime needs, can gain market share due to the use of foreign production capacity for military contracts.' Winkel, *Die Wirtschaft im geteilten Deutschland*, p. 70.
97. Radkau, *Technik in Deutschland*, p. 320 f.
98. Radkau, *Technik in Deutschland*, p. 318.
99. Schildt/Sywottek (eds): *Modernisierung im Wiederaufbau.*
100. Dienel, *Das Bild kleiner und mittlerer Unternehmen.*
101. 'In the 1970s, capital expenditures per job were higher in West German agriculture than in industry.' Radkau, *Technik in Deutschland*, p. 317.
102. 'France, Italy, the Federal Republic of Germany, and the three Benelux countries agreed to far-reaching economic co-operation in March 1957: with the Treaty of Rome, which went into effect on 1 January 1958, they founded the European Economic Community (EEC) and the European Atomic Energy Community (Euratom).' Hardach, *Wirtschaftsgeschichte Deutschlands im 20. Jahrhundert*, p. 224.
103. Linde AG, *Geschäftsbericht* 1950, p. 6.
104. In the 1950 annual report, it was already noted that Linde's earnings were 'hurt by rising production costs, especially the increase in material and energy prices as well as wages'. Linde AG, *Geschäftsbericht* 1950, p. 6.
105. In 1952, for example: 'Several groups of our products have come under perceptible price pressure, which we have been able to counter successfully so far with further rationalization of production.' Linde AG, *Geschäftsbericht* 1952, p. 7; or 1956: 'In a few divisions, however, successful rationalization achieved through previous investments is no longer sufficient to compensate for cost increases that have

resulted from higher material prices and especially from the jump in wages and salaries – partly due to the reduction of working hours [to 45 hours per week].' Linde AG, *Geschäftsbericht* 1956, p. 4.

106. Linde AG, *Das ist Linde*, p. 101.

107. Annual report 1956: 'The increase in inventories compared to last year was around 35 per cent for raw materials and operating supplies, approximately 12 per cent for semimanufactured goods, and about 52 per cent for finished products [...].' Linde AG, *Geschäftsbericht* 1956, p. 5.

108. Linde AG, *Geschäftsbericht* 1955, p. 5; 1956: p. 6.

109. Linde AG, *Das ist Linde*, p. 101.

110. Linde AG, *Geschäftsbericht* 1956, p. 7.

111. Linde AG, *Geschäftsbericht* 1957, p. 4.

112. 'Even if we basically welcome it, the common market poses conundrums for us.' Speech by CEO Dr Hugo Ombeck at the company's annual general meeting in Munich on 18 July 1957. Linde-Archiv Höllriegelskreuth, Ordner Ansprachen Hauptversammlung 1957–72.

113. Linde AG, minutes of the Executive Board meeting on 15 March 1950. Richard Linde retired from the Supervisory Board in 1955; he died in 1961.

114. Philipp Borchardt to Richard Linde on 23 May 1946: 'I can assure you that I feel the loss of two such promising young people as a personal sorrow, as I looked forward to experiencing how the younger generation would have carried on when it became too much for us one day. Hermann and Gert are so young that I can no longer entertain such a hope – at least not for myself.' Linde-Archiv Höllriegelskreuth, Ordner Dr R. Linde, Privatakt vom 1. Jan. 1942 bis 31. Dez. 1948, A- Hi.

115. Rudolf Wucherer retired from the Supervisory Board in 1965 and died just a year later.

116. Linde AG, *Geschäftsbericht* 1954, p. 1.

117. Linde AG, *Geschäftsbericht* 1961, p. 1.

118. Linde AG, Schreiben des Vorstandes an die Aktionäre als Anlage zum Geschäftsbericht 1960, June 1961, zu Punkt 8: Neuwahl des Aufsichtsrates.

119. However, noise from the gates (before the tunnel was built) interfered with rehearsals, so that the orchestra moved on after two decades.

120. Linde AG, minutes of the Executive Board meeting of 9 January 1952.

121. Linde AG, *Geschäftsbericht* 1953, p. 9.

122. This was in line with a trend at the time. Under the auspices of the first Housing Subsidies Act (Wohnungsbauförderungsgesetz), 3 million housing units were built between 1951 and 1956, and an additional 500,000 to 600,000 new units followed annually in the years thereafter. A total of 9 million new housing units were built in West Germany between 1949 and 1966.

123. The Maschinenfabrik Sürth already had its own company health insurance fund before it was taken over by Linde in 1920, which was consolidated with the BKK Linde in 1956. Linde AG, *Geschäftsbericht* 1953, Sozialbericht, p. 9.

124. The recreation home in Schalchen was established in 1951 in 'connection with the further expansion' of the Schalchen by Traunstein/Obb branch factory. Linde AG, *Geschäftsbericht* 1951, p. 9.

125. Linde AG, minutes of the Executive Board meeting of 9 January 1952.

126. Already in 1950, for example, over 4.1 million Deutschmarks were spent on voluntary social benefits (pensions, appropriations for relief funds, and 'Christmas and year-end payments'). The company's social report did not neglect to mention that voluntary social benefits amounted to 'about 19 per cent of salaries and wages and *three* times [emphasis in the original] the dividends budgeted to be distributed to shareholders during this period'. Linde AG, *Geschäftsbericht* 1950, p. 14. By the mid-1950s, voluntary social benefits had shrunk to 15.6 per cent of the payroll. Linde AG, *Geschäftsbericht* 1955.

127. Linde AG, *Geschäftsbericht* 1950, Executive Board Report, p. 8, and 1951, p. 6.

128. In 1956, the previous year's sales were 'achieved again'; new orders were 'satisfactory overall despite some fluctuations' Linde AG, *Geschäftsbericht* 1956, p. 6. In 1957 the 'previous year's sales were achieved. Thanks to some large export orders, the volume of orders has ... risen.' Domestic business was only 'satisfactory'. Linde AG, *Geschäftsbericht* 1957, p. 6. In 1958 large export orders led to 'especially high sales revenue'. New orders were 'satisfactory'. Linde AG, *Geschäftsbericht* 1958, p. 6.

129. The Executive Board had decided already in 1951 to expand the Sürth facilities in such a way as to enable manufacture of commercial refrigeration systems in the future, which till then had been delivered by the Maschinenfabrik Esslingen. Linde AG, minutes of the Executive Board meeting of 12 December 1951, Linde-Archiv Wiesbaden.

130. Linde AG, *Arbeit bei Linde*, p. 14 f.

131. Linde AG, *Geschäftsberichte*, 1950, p. 8; 1954, p. 2; 1956, p. 6 and 1958, p. 6.

132. 'The leasing contract was cancelled by the Süddeutschen Kalkstickstoff-Werken and we considered closing the factory there. We actually concluded at that time already that a reservoir of manpower was available there, which was lacking here [in Höllriegelskreuth], close to the city that had been 80 per cent destroyed, so that we decided to build a new factory there in the open countryside. The rightness of this decision became apparent much later.' Hermann Linde in a speech celebrating the 40-year service anniversary of Walther Ruckdeschel, 12 September 1966, p. 4. Linde-Archiv Höllriegelskreuth, Materialien Erik Jaeger.

133. Linde AG, *Geschäftsbericht* 1952, p. 7.

134. 'Linde also uses this separation process, which is based on the different boiling points of the gases, in separating other gas mixtures in order to eliminate technically important components from for example coke oven gas, blast furnace gas, cracked gas, and so on. This extracts highly pure hydrogen, methane, ethylene, benzene, and other gases, some of which are essential to the plastics industry.' Linde AG, *Das ist Linde*, p. 24.

135. Statement in: Linde AG, *Geschäftsbericht* 1955, p. 6. The plant went on stream in 1958; Linde later received another order from abroad. 'Power consumption for the refrigeration is about an order of magnitude greater than in air separation; the avoidance of cold losses is thus an unusually important problem ... Still, thanks to the collaboration of theory and practice, and after an astonishingly short development period, we succeeded in bringing an industrial-scale plant on stream almost at the first go, and fulfilling the guarantee.' Linde AG, *Das ist Linde*, p. 25.

136. Linde AG, *Geschäftsbericht* 1957, p. 7.

137. 'Business in the area of gas separation plants, which has been very brisk up to now, has quieted down.' Linde AG, *Geschäftsbericht* 1958, p. 7.

138. In 1953: 'The economic recovery in the Federal Republic of Germany, especially the lively activity in the construction market, has also impacted the gas business and kept our factories quite busy ...' Linde AG, *Geschäftsbericht* 1953, p. 8. In 1954: The gas business 'has grown very satisfactorily'. Linde AG, *Geschäftsbericht* 1954, p. 8. Sales in 1955 'continued to be high'. Linde AG, *Geschäftsbericht* 1955, p. 6. In 1956: sales 'climbed continuously'. Linde AG, *Geschäftsbericht* 1956, p. 7. In 1957 sales again 'climbed continuously'. Linde AG, *Geschäftsbericht* 1957, p. 7. Sales in 1958 were 'good'. Linde AG, *Geschäftsbericht* 1958, p. 7. The gas business in 1959 'expanded again'. Linde AG, *Geschäftsbericht* 1959, p. 8.

139. Ruckdeschel, *Mein Lebenslauf*, Linde-Archiv Wiesbaden.

140. 'Unfortunately, various factories in the chemical industry, which either generate oxygen as a by-product of nitrogen production or do not consume all that they produce in their own oxygen plant, are flirting with the idea of marketing this [excess] oxygen.' Minutes of the 247th Supervisory Board meeting of 10 January

1952, p. 8. Linde-Archiv Höllriegelskreuth. Linde tried to buy up at least a portion of this overproduction in order to stifle competition. Oxygen was purchased in liquid form by the chemical factories Trostberg, Griesheim, and Hüls.

141. Minutes of the 247th Supervisory Board meeting of 10 January 1952, p. 8. Linde-Archiv Höllriegelskreuth.

142. The AM department produced compressors, tension-release turbines, carbonic acid production plants, dry ice systems, gas cylinder and tank valves, pneumatic tools and their fittings.

143. A typical statement can be found in the 1956 annual report: 'Price setbacks like salary increases and material price increases in small-scale refrigeration have negated the benefits of rationalization despite higher sales.' Linde AG, *Annual Report* 1956, p. 7. Rationalization measures in the manufacturing sector were also mentioned in the annual reports for the following two reports; in 1957: 'The measures taken towards rationalization have improved our market position.' Linde AG, *Geschäftsbericht* 1957, p. 7. 1958: 'Further measures towards rationalization have been adopted.' Linde AG, *Geschäftsbericht* 1958, p. 7.

144. For instance, storage capacity had to be increased in order to store products ready for delivery since turnover of refrigerators and typical commercial refrigerator units were 'only seasonally possible' at that time. Linde AG, *Geschäftsbericht* 1955, p. 5.

145. Hellmann, *Höchst unauffällig*; Centrum für Industriekultur Nürnberg/Münchner Stadtmuseum (ed.), *Unter Null*, p. 151.

146. Hellmann, *Künstliche Kälte*, p. 240 f.

147. A large 110 l-Model from Bosch still cost 770 DM in 1952; in 1954, it cost 598 DM and, in 1958, the refrigerator, which had subsequently become more compact, cost 478 DM. And at the Cologne Autumn Fair in 1959, Bosch, Bauknecht, Linde, and other firms were offering their 110 l-refrigerators at the price of 380 DM. Hellmann, *Künstliche Kälte*, p. 240.

148. Dienel, *Eis mit Stiel*; Centrum für Industriekultur Nürnberg/Münchner Stadtmuseum (ed.), *Unter Null*, p. 103.

149. Linde AG, *Geschäftsbericht* 1960, p. 14.

150. Linde AG, *Geschäftsbericht* 1950, p. 9.

151. During the first half of 1950, Güldner's tractor A 28 was at the top of its vehicle class for newly licensed vehicles in the Federal Republic. Herrmann, *Traktoren in Deutschland*, p. 73.

152. Linde AG, *Geschäftsbericht* 1953, p. 8.

153. 1955 represents a typical year's sales: 'The spring months brought a strong demand for field tractors. During the second half of the year, demand receded.' Linde AG, *Geschäftsbericht* 1955, p. 6

154. 'Via the creation of a sales organization, we were able to increase our share of the market, particularly in the area of field tractors.' Linde AG, *Geschäftsbericht* 1954, p. 8. During the next year, Güldner reached the highest number of newly licensed vehicles with 5,986 tractors. Hermann, *Traktoren in Deutschland*, p. 73.

155. *Landmaschinen-Rundschau*, 1 (1955), 12.

156. Ibid., 5 (1955), 120.

157. Stuhr, *Fahrzeuge mit hydrostatischem Fahrantrieb*. Linde-Archiv Aschaffenburg.

158. Linde AG, *Das ist Linde*, p. 80.

159. Linde AG, *Geschäftsbericht* 1956.

160. Linde AG, *Geschäftsbericht* 1958, p. 7.

161. *Landmaschinen-Rundschau*, 6 (1958), 148.

162. 'Wer geht mit wem zusammen?', *Landmaschinen-Markt* 14 (1958), 587.

163. 'The jubilee-motor was manufactured in an original way in front of a large number of invited guests. The last pieces were put together by the guests. For example, the Bavarian Finance Minister attached the exhaust pipe and an editor put in the air

filter. All of those present were then able to confirm, to their astonishment, that the motor actually worked after their "construction".' *Landmaschinen-Rundschau*, 6 (1958), 148.

164. Notice 'Zusammenarbeit FAHR-GÜLDNER', ibid., 14 (1958), S. 587.
165. Winkel, *Die Wirtschaft im geteilten Deutschland*, p. 76.
166. 'In fact the actual gross social product in 1966 rose only 2.9 per cent, industrial production only 1.2 per cent. In both cases, this was the smallest level of growth since the currency reform.' Hardach, *Wirtschaftsgeschichte*, p. 238.
167. Gerald Ambrosius and Hartmut Kaeble denote the actual end of the post-war boom starting in 1974: 'If one disregards the European-wide collapse of the economic situation in 1966/67, 1973 was the last year in which the social product reached high growth rates. In 1974 a worldwide crisis set in and ended a period in which periodic downturns were not experienced or recognized as recessions but rather as disruptions in growth. Ambrosius/Kaelble, 'Einleitung: Gesellschaftliche und wirtschaftliche Folgen des Booms der 1950er und 1960er Jahre', Kaelble (ed.), *Der Boom*, p. 10 f.
168. Winkel, *Die Wirtschaft im geteilten Deutschland*, p. 76.
169. Linde AG, *Das ist Linde*, p. 101.
170. 'A survey of the shareholders' sociological classes accounting for 87 per cent of the capital showed that out of 7,760 shareholders: 72 per cent owned up to 3,000 DM nominally, 21 per cent 3,100 to 10,000 DM nominally, 6 per cent 10,100 to 50,000 DM nominally, and 1 per cent of shareholders 50,100 DM nominally and more. Interestingly, about 13 per cent of the recorded capital was found in investment groups. Investment groups are known for being extremely dispersed. Thus, it can probably be assumed that this 13 per cent of the capital was also in the hands of small-scale savers. The biggest shareholding for one person did not account for more than 5 per cent of the capital.' Linde AG, *Das ist Linde*, p. 101.
171. Linde AG, *Geschäftsbericht 1961*.
172. Dr J. Wucherer: 'Gedanken zu Punkt 21c) der Tagesordnung zur Vorstandssitzung am 6.12.1961' from December 5, 1961. Linde-Archiv Wiesbaden.
173. There were no concrete sales goals; only in the meeting of the Board of Management on August 1, 1965 were the first sales prognoses for the factories discussed. Until this point, sales development had only been retrospectively noted. Even more crucial for the long-term development of the company was the policy to plan investments according to depreciation as opposed to strategic investments in areas in which the highest profit yields were to be expected! Linde AG, Board of Management Meeting Protocol from January 14, 1966, Linde-Archiv Wiesbaden.
174. On the spatial conditions in the context of further barriers: Dienel: 'Räumliche Bedingungen heterogener Forschungskooperationen', Jochen Gläser et al. (eds), *Zusammenarbeit von Wissenschaft und Technik*.
175. Linde AG, Protocol of the Board of Management, December 9, 1965. Linde-Archiv Wiesbaden.
176. Cryogenics, process technology, industrial gases, welding technology, cold stores, compressors, turbines, pneumatic tools, refrigeration and air-conditioning technology, hydraulics, diesel motors, tractors, floor conveyors.
177. Dienel, *Techniktüftler?*; Frieß/Steiner (eds), *Forschung und Technik in Deutschland*, pp. 170–85; ibid., *Das Bild kleiner und mittlerer Unternehmen*, Reith/Schmidt (eds), *Kleine Betriebe*, pp. 100–23.
178. In France there was a similar mood. Especially influential (Franz-Josef Strauß wrote the foreword for the German translation): Servan-Schreiber, *Die amerikanische Herausforderung*. See also: Blauhorn, *Erdteil zweiter Klasse?*; Erik Hoffmeyer, *Die amerikanische Herausforderung*.

179. This applied not only to the Board of Management at that time, but also to the department head and area leader level. April 15, 1965: Wucherer became 90; December 6, 1965: Friedrich Linde became 95.
180. In 1971, Sürth celebrated its 100-year anniversary. For the history of Sürth, see: Dokumentation der Chronik, Linde-Archiv Wiesbaden.
181. Linde AG, *Das ist Linde*, p. 63.
182. Linde AG, *Das ist Linde*, p. 49.
183. Linde offered reciprocating compressors for air and industrially used gases, in particular dry-run compressors for oil-free media, sliding vane compressor, turbo ventilators, expansion turbines for air and gas splitting, carbon dioxide – and dry ice systems, gas bottling plants for oxygen, hydrogen, and acetylene, gas cylinder valves, rotating and striking pneumatic tools, including drills, turbo and rotor grinding machines, saws and screwdrivers, riveting hammers, chiselling and scaling hammers, mashers, and pit removers as well as compressed air motors, vibrators, and mountings for air pipes. Linde AG, *Das ist Linde*, pp. 54 and 62.
184. 'Bericht über die Besprechung bei Maschinenfabrik Sürth am 5. und 6. Februar 1963'. Linde-Archiv Wiesbaden.
185. Criticized were the too high administrative costs in the construction office as well as in highly bureaucratic deliveries of low worth. In general, the competition did not draw up extensive rough-drafts of plans, making installation simpler and cheaper. 'Bericht über die Besprechung bei Maschinenfabrik Sürth am 5. und 6. Februar 1963' Linde-Archiv Wiesbaden. Nevertheless, the typical Linde sector-oriented thinking remained a problem, as Hermann Linde wrote to Megerlin, his colleague in Mainz-Kostheim, on January 20, 1964 in reply to his letter from January 17th: 'You are perhaps correct to say that it is very difficult for us here in Höllriegelskreuth to judge marketing problems for domestic appliances, domestic refrigerators, and so on.' Linde-Archiv Höllriegelskreuth, Collection H. Linde.
186. Thoughts on the new organizing of the marketing department for refrigeration on May 7, 1963. Linde-Archiv Wiesbaden; Strategiepapier 'Vertrieb Kälte vom 25.11.1963 von Dr J. Wucherer. Linde-Archiv Wiesbaden.
187. Dr Klein left Linde in 1966 due to a conflict and became marketing director at Rank Xerox. Linde AG, Protocol of the Board of Management meeting on 25 October 25 1966, Linde-Archiv Wiesbaden.
188. In the same strategy paper on January 3, 1964, Dr Klein writes rhetorically, playfully providing question and answer: 'Is it not more correct for the Linde Society to forego certain sectors and thereby put an end to these losses? – Answer: We have always emphasized the production and control of all refrigeration areas as a signature feature at Linde. The elimination of one single sector must be carefully examined. Reason: The entire refrigeration needs of the Western world will increase in the next decades according to the prognoses of market researchers. The refrigeration market thus has no structural weaknesses, rather, on the contrary, expansion and growth possibilities.' 'Neuorganization der Kältesparte', 3 January, 1964, Author Dr Klein, Mainz-Kostheim, Linde-Archiv Höllriegelskreuth.
189. 'Neuorganisation der Kältesparte', January 3, 1964, Author Dr Klein, Mainz-Kostheim, p. 2. Linde-Archiv Höllriegelskreuth.
190. Linde AG, Protocol of the Board of Management Meeting on November 19, 1965, Linde-Archiv Wiesbaden.
191. Linde AG, *Geschäftsbericht* 1960, p. 19.
192. Linde AG, *Geschäftsbericht* 1961, p. 18.
193. Hellmann, *Höchst unauffällig* ; Centrum für Industriekultur Nürnberg/Münchner Stadtmuseum (ed.) *Unter Null*, p. 151.
194. Hellmann, *Künstliche Kälte*, p. 244 f.
195. 'Over 80 per cent of all domestic refrigerators are now table models built according to kitchen measurements.' Hellmann, *Künstliche Kälte*, p. 245 f.

196. Linde AG; *Geschäftsbericht* 1964, p. 24.
197. Linde AG; *Geschäftsbericht* 1965, p. 26.
198. Delivery costs for the expanded household appliance programme, June 23, 1965; delivery costs for the expanded household appliance programme, June 29, 1965, both in the Linde-Archiv Wiesbaden. It was only later that drying machines, microwave ovens, and so on came to this segment. Fittingly, there are 'white goods' as well as 'brown goods'. Brown goods are composed of home electronic goods such as televisions, record players, and tape players.
199. Protocol of the Board of Management Meeting on July 20, 1965 and July 29, 1965. Linde-ArchivWiesbaden.
200. Polonius had studied economics and started working for Linde in central administration in 1953. He became manager of the tax and audits department, later took over sales management for the domestic refrigerator factory in Mainz-Kostheim, then in 1964 went to Sürth to become manager of the sales administration. He became the single business manager of the Linde Household Appliance AG. And after AEG held the majority of shares in this company, he was named director of the business leadership in 1971. After Linde had decided to give up its shares in the Linde Household Appliance AG, Polonius was appointed to the business management committee for the Refrigeration and Installation Systems division on October 1, 1977. *Linde heute* 1 (1978), 13.
201. The present and future of the divisions in refrigeration, sales and results analysis with an eye on future development possibilities, April 1965, p. 42. Linde-Archiv Wiesbaden.
202. Hermann Linde and Walther Ruckdeschel did not otherwise have a problem-free relationship with each other, as Linde retrospectively describes in his address for his 40-year service jubilee in 1966: 'We both had several common problems over the years, in particular during the period when we split the management of the apparatus assembly department. We did not always share the same opinion, but our differences were always objective and differences of factual information; sometimes, I had to see after the fact, that you were correct. You, however, never failed to be fair, you respected divergent views, and you never allowed differences in opinion to spoil personal relationships.'
203. The confrontation with the Italian partner was not only marked by translation problems: 'The negotiations had until this point suffered from the mistake that the negotiations had been delayed by the change of negotiators and the too strongly emphasized juridical problems ...' File notes from Oetken: Re.: Co-operation Linde – Zanussi, July 13, 1966. Linde-Archiv Wiesbaden.
204. ' ... that, as you well know, the takeover of the Linde trademark on the part of the joint venture is the main basis of our agreement.' Translation letter from Zanussi to Linde on April 22, 1966, p. 11. Linde-Archiv Wiesbaden.
205. 'I say this because in this moment the important problem in 2 words is: [THE] TRUST.' Private letter from Pusco to Simon, May 6, 1966 (capitalized in original). Linde-Archiv Wiesbaden.
206. Protocol of the Board of Management meeting on August 30, 1965 Linde-Archiv Wiesbaden.
207. Linde AG, *Geschäftsbericht* 1967, p. 14.
208. In the USA, self-service shops in retail grocery stores had already started to show up all over the country before the Second World War. In contrast, the Federal Republic took somewhat longer. In West Berlin, a city which always had to fulfil a special symbolic role as the 'shop window of the West' in the middle of the GDR, the local Meyer branch opened the first large self-service shop in 1953. Twelve years later, the overwhelming majority of its branches, 92 out of 120 Meyer-branches, were turned into self-service shops. Bertz, *Keine Feier ohne Meyer*, p. 101.

209. This happened slowly, because AEG continued to sell refrigerators under the Linde brand name.
210. According to a survey in April 2003 that was published in the *Financial Times*, Linde's name recognition was at about 50 per cent in Germany. See also the speech from VV Wolfgang Reitle during the Leadership Conference in 2003 in Berlin, 'Linde – oh, the chocolate company!?' In a market psychological examination from the beginning of the 1980s, the handwritten Linde insignia and the Linde hexagon used in the 1970s and 1980s were compared and found to be 'still popular' (Marktpsychologische Untersuchung zum Image und zur Imagewerbung von Linde, February 1981, p. 24).
211. Sack, *Güldner Traktoren und Motoren*, p. 13.
212. Linde AG, *Geschäftsbericht* 1964, p. 25.
213. 'Qualität und technischer Stand der GÜLDNER-Erzeugnisse', Documents on the Extra Minutes of the Board of Management Meeting on November 20, 1964, Manager of the Division Aschaffenburg Pöhlein, November 13, 1964. Linde-Archiv Höllriegelskreuth.
214. Herrmann, *Traktoren in Deutschland 1907 bis heute*, p. 74. In 1964, Güldner's production programme included air-cooled diesel motors, tractors, hydrostatic motors, hydro-motors and pumps, diesel-hydrostatic engines, transportation vehicles (Güldner-Hydrocar), platform vehicles with variable hydrostatic transmissions, and the forklift 'Güldner-Hubtrac'. Linde AG, *Das ist Linde*, p. 78.
215. 'In the field of hydrostatic vehicle engines, Güldner was the pioneer and, for many years, it was the single manufacturer of floor conveyor machinery with variable, adjustable, hydrostatic engines. Many customers from all areas of technology as well as manufacturers of floor conveyors still use the Güldner "Hydro-Stabil" – transmission in their products today.' Linde AG, *Das ist Linde*, p. 85.
216. 'Umsatzausweitung bei Güldner – Neue Versuchs- und Entwicklungsanlagen vorgestellt. Press release from Linde AG from May 21, 1964.
217. Internal Document, February 12, 1969. Linde-Archiv Höllriegelskreuth.
218. Linde AG: Protocol of the Board of Management Meeting on April 2, 1965. Linde-Archive Wiesbaden.
219. Linde traced the decisive drop in tractor licences in the Federal Republic to the 'cancellation of interest subsidies in the Green Plan and the German Central Bank's restrictive measures' in particular. Linde AG, *Geschäftsbericht* 1966, p. 27.
220. Linde AG, Division Güldner Aschaffenburg, Factory Manager Pöhlein, Presentation for the Linde Board of Management on February 12, 1969. Linde-Archiv Höllriegelskreuth.
221. Linde Press Release from March 19, 1969.
222. Sales in 1965 from the individual sectors according to the Board of Management Protocol from January 14, 1965: Munich: 329.6 million DM, Sürth 241 million DM, Aschaffenburg 112 million DM, cool stores 7.2 million DM.
223. Dienel (ed.), *Der Optimismus der Ingenieure*, pp. 9–25.

Chapter 5

1. Radkau, *Technik in Deutschland*, p. 319 ff.
2. Schildt and others locate the roots of this upheaval to some degree in the 1950s: cf. Schildt/Sywottek, *Modernisierung im Wiederaufbau*; Schildt, *Moderne Zeiten*. Others argue for a later beginning point (1968): Schissler (ed.), *The Miracle Years*.
3. Herbert (ed.): *Wandlungsprozesse in Westdeutschland*. From the perspective of gender history, the 1970s rather than the 1960s are considered to be the crucial period of change. (Debates over § 218, the women's autonomy movement, reforms to marriage and family law.) Oertzen, *Teilzeitarbeit und die Lust am Zuverdienen*.

4. Weimer, *Deutsche Wirtschaftsgeschichte*, p. 229 f.
5. Dienel, *Die deutsch-ungarische Zusammenarbeit*; Fischer (ed.), *Deutsch-ungarische Beziehungen*, pp. 481–521.
6. The working group on industrial refrigeration received contracts to deliver 12 expander turbines and 10 liquid O_2 pumps to the People's Republic of China from German and French manufacturers. *Linde heute* 1 (1973), 5.
7. A few examples: 'Internationale Ausstellung Chemie 70 in Moskau', *Werkzeitung Linde* 6 (1970), 8 ff.; 'Turboverdichter für Russland', *Linde heute* 4 (1973), 3; 'Linde in Budapest', *Linde heute* 5 (1972), 1, 'Tiefkühlspezialgeschäfte für Polen', Linde heute 4 (1974), 5; 'Linde baut für Forschungsinstitut in UDSSR', *Linde heute* 6 (1974) 2 f.; 'Internationale Messe in Budapest', *Linde heute* 4 (1974), 2; 'Linde auf der INCHEBA', *Linde heute* 4 (1973), 4 f.; 'Hähnchenschlachtereien für Polen', *Linde heute* 4 (1973), 6. The success of their Eastern Europe business is mirrored in the Linde annual reports from 1970–79.
8. Record of interview with Hermann Linde, 21 February 2001, p. 24 f. included in the TV series *Zeitzeugen*, Bayerischer Rundfunk.
9. Interview Müller, p. 15.
10. The Linde engineers working in the GDR to erect plants and put them into operation were under strict orders not to establish private contacts or relations there. Should relationships start and eventually lead to wedding plans or a request to exit the GDR, this would threaten the co-operation itself. It was made clear to the Linde management that they were to prevent such developments from taking place amongst their employees.
11. Karlsch/Stokes, *Faktor Öl*, p. 377.
12. OPEC's tremendous price increase was a result of the war of October between Syria, Egypt, and Israel; however, it would probably have occurred without the war, cf. zur Ölkrise Hohensee 1996. Hohensee refers to an oil price rather than an oil crisis because of the fact that the crisis was a consequence not of the finiteness of oil, but rather it was the increase in prices due to political issues which resulted in a crisis.
13. Pfister (ed.), *Das 1950er Syndrom*; Bergmeier, *Umweltgeschichte*.
14. Carson, *Der stumme Frühling/Silent Spring*, Meadows (ed.), *Die Grenzen des Wachstums/The Limits to Growth*.
15. Raschke (ed.), *Die Grünen*.
16. Mesarovic/Pestel (ed.), *Menschheit am Wendepunkt*.
17. Environmental policy had already become an issue for the established parties. A Department of Environmental Protection had been established within the Ministry of the Interior since 1969 and in 1971 the Federal government declared the first environmental programme. Further developments followed: the first world environment conference in Stockholm, the establishment of a United Nations environmental programme (UNEP), the declaration of the International Day of the Environment for 5 June, the opening of the Federal Environmental Agency in Berlin, the enactment of the Federal Emissions Control Act.
18. Aside from its gas works and cold stores, Linde's own factories did not consume particularly high levels of energy. Energy costs made up only 2.4 per cent of their total operating performance. Nonetheless, in 1975, nearly 8,000,000 DM more were spent on electricity and fuel oils for essentially the same total consumption in 1972. (Speech, Hermann Linde at the Hauptversammlung der Linde AG, 13.5.1976.)
19. The system could save up to 84 per cent of the normal heat and ventilation costs. *Linde heute* 6 (1974), 8.
20. König, *Geschichte der Konsumgesellschaft*, p. 60.
21. *Linde heute* 6 (1973), 7; Linde AG, *Geschäftsbericht 1973*, p. 29 f.

22. In 1991, the single high temperature reactor at Hamm was taken offline after only three years of operation. For a summary of the history of this technology, see: Kirchner, *Der Hochtemperaturreaktor.*

23. Linde AG, *Geschäftsbericht,* 1971, p. 21. This is where a treatment plant for radioactive gases was designed and built. During the same year, Linde built a purge gas plant for the nuclear power plant Unterweser and Neckar-Westheim. Linde AG, *Geschäftsbericht,* 1974, p. 28.

24. Linde AG, *Geschäftsbericht,* 1977, p. 29.

25. Linde AG, *1879–1979,* p. 73.

26. Factories with more than 110 employees were to have Economic Committees with a simple right to information. Milert/Tschirbs: *Von den Arbeiterausschüssen,* p. 73 ff.

27. Linde AG, *Geschäftsbericht* 1953, p. 1.

28. Linde AG, *Geschäftsbericht* 1961, p. 5.

29. Milert/Tschirbs, *Von den Arbeiterausschüssen,* p. 81.

30. Sentencing by the First Division from 1 March 1979 over the verbal negotiations from 28, 29, 20 November and 1 December 1978. (1 BvR 532, 533/77, 419/78 und BvL 21/78.)

31. Jaeger, *Der Auslandsbezug;* Lerche, *Der Europäische Betriebsrat.*

32. Coincidentally, the year of 1976 was also the year in which the contract boom for ethylene plants, which were essential to plant construction turnover, came to a halt. This was presumably due to the increase in the price of synthetic materials as a consequence of the oil crisis contracts, which in turn affected the demand for ethylene as a source material.

33. Müller, *Entwicklung TVT,* p. 2 ff.

34. Müller, *Entwicklung TVT,* p. 2.

35. *Linde heute* 1/2 (1971), 5; *Linde heute* 4 (1973), 4 and *Linde heute* 1 (1972), 1.

36. Linde AG, *Geschäftsbericht* 1972, p. 21; *Linde heute* 5 (1972), 1.

37. Linde AG, *Geschäftsberichte* 1970–80.

38. Commercial plant construction is a business prone to wild oscillations even today. Hundreds of the client's engineers often work on these projects during the planning phase in Höllriegelskreuth.

39. Linde AG, *Geschäftsberichte* , 1973, p. 28; 1975, p. 31.

40. In 1983, 80 per cent of the Selas-Kirchner GmbH's shares were held by the Linde AG (Linde AG, *Geschäftsbericht* 1983, p. 9), in 1985 100 per cent (Linde AG, *Geschäftsbericht* 1983, p. 9). For more information see: Kreuter, *Beiträge zur Geschichte,* p. 8 f.

41. Cf. 'Messer Griesheim Schweißtechnik von Linde', Linde-Papier, 23.11.1971; 'Linde/Messer Griesheim Tausch Schweißtechnik gegen Tieftemperaturtechnik', *Industriemagazin* 4 (1972), 92–3.

42. Linde AG, *1879–1979,* p. 83.

43. Neue Werksgruppen-Bezeichnungen, Linde-intern, Rundschreiben Nr. 2, 10.01.1972; Linde: Zellteilung für gezieltes Marketing. *Absatzwirtschaft* 3 (1973), 6–8.

44. The weakness of Herman Linde's role on the Executive Board perhaps played a role.

45. Linde AG, *Geschäftsbericht* 1974, p. 32. In 1978, a marketing organization for gas was established in Brazil, the Linde do Brasil Ltda, and in 1979 a new gas transfiller in Melbourne, Australia was put into operation.

46. Cf. 'Neugliederung der Werksgruppe Sürth der Linde AG', Linde-Paper, 27.10.1971, Linde-Archiv Wiesbaden and 'Neugliederung der Werksgruppe Sürth der Linde AG', Rundschreiben Nr.66, 4.11.1971, Linde-Archiv Wiesbaden.

47. Linde AG, *Geschäftsbericht* 1971, p. 38.

48. Cf. press attention at the time, for example: 'Linde AG strukturiert in Kostheim um', *Wiesbadener Tagblatt*, 21.3.1975; 'Unruhe bei Linde', *Kölner Stadtanzeiger*, 3.4.1975; 'Linde-Kundendienst zieht nach Sürth', *Wiesbadener Tagblatt*, 21.8.1975.

49. Cf.: 'Linde AG takeover of Tyler Refrigeration International GmbH', confidential, 23 October 1975. Linde-Archiv Höllriegelskreuth.

50. *Linde heute* 3 (1973), 4.

51. Linde AG, *Geschäftsberichte* 1972, p. 25, 1973, p. 34, 1974, p. 34, 1976 p. 35.

52. Industrial and Commercial Refrigeration in Tunisia, *Linde heute* 1/2 (1971), 1, England, *Linde heute* 1 (1972), 11, Near East, *Linde heute* 2 (1972), 2, and South Africa from 1890 up to the 1970s, *Linde heute* 3 (1973), 9–13.

53. Twenty Linde snow canons at a ski resort on Fujiyama to be used throughout the year. *Linde heute* 5 (1973), 13/14. For more on Linde in Japan generally: *Linde heute* 4(1973), 18.

54. Including Clark, Jungheinrich, Still, Lansing, Atlas, and others.

55. In the Federal Republic of Germany these were: Aschaffenburg, Augsburg, Berlin, Bielefeld, Dillenburg, Essen, Freiburg, Hamburg, Heilbronn, Kaiserslautern, Kassel, Munich, Nuremberg, Regensburg and Salzgitter; abroad there were the regions Western Europe, North America, Canada, Middle East, Africa, Japan. There were also Linde-Güldner licensed manufacturers in Great Britain, Spain, Austria, Australia and Japan. Internal Linde document. Linde AG, Werksgruppe Güldner-Aschaffenburg, 1969. Linde-Archiv Wiesbaden.

56. Abels, Hydrostatisches Gabelstaplergetriebe mit Primär und Sekundärverstellung, *Ölhydraulik und Pneumatik* 15 (1971), 51–5.

57. Abels/Caspari, 'Die neuen elektrischen Linde-Gabelstapler', *HD* 2723 (1971).

58. Mutual prejudices towards both North and South were felt on the Executive Board. They were perhaps most present in development decisions made against the North. The rural values of diligence, proficiency, and honesty were seen to be less pronounced in North Germany. Similar judgements were made however about the Southern population as the Güldner division continued in the red. In the 1960s, Max Pöhlein made a reference to this by blaming the Prince Bishop of Würzburg for settling communities of criminals in the area 200 years before and from where many Aschaffenburg workers came. This 'Australian argument' has not reappeared since the 1970s.

59. Linde AG, *Geschäftsbericht* 1973. Looking back: Hans Meinhardt: Mit eiserner Hand, *Industriemagazin*, 2 (1984), 60.

60. Cf. Interview with the director of Werksgruppe Güldner-Aschaffenburg Reinhard Lohse, who joined the Executive Board in 1979. Lohse stated: the Japanese often use a trojan horse. *Deutsche Hebe- und Fördertechnik Dfh* 12 (1984), 44–7.

61. In 1988, Hans Meinhardt said of his move from Aschaffenburg to Wiesbaden: 'It was of course a rather large adjustment for me, to move to a relatively quiet headquarters that concentrated primarily on administration', *Capital* 3 (1988), 157.

62. This is the opinion of the interview partner. Interview IÜ-111.

63. Meinhardt was decisive in the establishment of staff units for organization, marketing, and planning.

64. Herzog, Strukturwandel der Linde AG als Lernprozess. Herzog identified two phases: in the first phase (1954–67), Linde identified performance gaps, acquired new knowledge and developed new strategies for structural change. During the second phase, (1968–84) the strategies were put into action and realized.

65. For industrial firms this was not unusual. For more on the history of advertising and rise of marketing, see: Reinhardt, *Von der Reklame zum Marketing*; Gries/ Ilgen/Schindelbeck, *Kursorische Überlegungen zu einer Werbegeschichte als Mentalitätsgeschichte*; ibid., *Ins Gehirn der Masse kriechen!*, pp. 1–28.

66. Organization plan Group Headquarters, 1961: an advertising and press office was set up directly below the Executive Board.

67. Linde VA [Werner Jakobsmeier]: *Werbung und Public Relation*, p. 2 ff. Linde-Archiv Höllriegelskreuth.
68. 'Die Werbetätigkeit der Gesellschaft Linde in den Jahren 1887 bis 1937. Bemerkungen zu einer Ausstellung im Dr-Friedrich-Linde Haus, März 1965'. Material Erik Jaeger, Linde Archiv Höllriegelskreuth.
69. Linde paper on the duties of the staff unit, 16.6.1965, Linde-Archiv Wiesbaden.
70. An early marketing employee remembers his early attempts to explain the advantages of modern marketing falling upon deaf ears. Back then, a division head who was also a member of the Executive Board took the marketing employee to a horizontal driller being manufactured and said 'Do you see this? Every morning I walk by this and I know exactly how far things are.' This was what he considered market and product research or marketing to be. He knew everything, he [. . .] needed only to go through the factory and he could see how far things were, that was how things were seen then.' (Interview: I29L-21 f.)
71. Directory of positions, marketing 13 May 1965, Linde-Archiv Wiesbaden.
72. This was also the case for the staff units of marketing, legal issues and insurance, finance, and social and personnel issues. Corporate restructuring, *ca*. 1967, Linde-Archiv Wiesbaden.
73. 'Press relations, which are in essence business activities and of advertising import, will be cultivated in the future by the marketing research department.' Those responsible for this were then put in charge of the two Linde publications, 'Werkzeitung' and 'Linde Berichte aus Technik und Wissenschaft'. Comments on Headquarter's organization plan, Wiesbaden 6 January 1969, Linde-Archiv Wiesbaden.
74. Interviews: IO-6, 21; IA-11; IS-69, 77, 79, 83, 90. The organizational structures of Hoesch AG, Mannesmann and steel companies in the Rheinland were carefully followed.
75. The situation varied amongst the divisions. Refrigeration had a tradition of long-standing documentation and standardization, Plant Engineering did not. Richard Linde's attitued was characteristic of the situation at the time (CEO 1924–45). His son Hermann Linde later said: 'For him, organization was not a major issue – he had that in his head.' Hermann Linde's address celebrating Walther Ruckdeschel's 40 years of service, 12 September 1966, p. 1. Linde-Archiv Höllriegelskreuth, Materialien Erik Jaeger.
76. In 1968 Plötz became the engineering director of Industrial Gases, in 1970 a deputy member and in 1971 a regular member of the Executive Board of the Linde AG (for industrial gases).
77. Günther Kammholz (Scholven Chemie, later VEBA), an important Linde client and member of the Linde Supervisory Board from 1978–92. Cf. Kreuter, *Beiträge zur Geschichte*, p. 4.
78. Still underdeveloped in this respect, the department ran into capacity and transactional problems, lacking as it was in experience. Hermann Linde himself was reluctant to accept the idea that such documentation should be placed elsewhere. Young women from Great Britain were hired to do the job. They 'brought a new idiom and mini-skirts to the company, but not much in the way of relief'. Ibid., p. 5.
79. 'Organisationsplan für die Einkaufsabteilung der G.L.' with plant organization plan, Wiesbaden, 23. 3. 1920. Linde-Archiv Wiesbaden.
80. Jaeger, *Aus der Geschichte der Werksgruppe TVT*, p. 78.
81. Group Headquarters organization plan 1961; organization plan for Linde Güldner Nr. 6660315, March 1966. Archiv Linde AG, Wiesbaden. In Cologne there was at first resistance to the establishment of the department. (Interview: I28M-102 ff.)
82. At Höllriegelskreuth this included control over factory-internal measures to be taken as well as the determination of capital expenditure requirements and the allocation of commission numbers, plan codes, illustration numbers, and other

number issues as well as formulas and compiling of reports for factory management. 'Aufgabengebiet der Organisationsabteilung Höllriegelskreuth', p. 2, 7.1.1964. Linde-Archiv Höllriegelskreuth. The same was true for the O-Departments at Aschaffenburg, Kostheim, and Sürth.

83. The regional organization departments had limited free movement regarding regulations stipulated by organization headquarters. Unusual cases were clarified with organization Group Headquarters. Personnel exchanges between organization headquarters and satellite organization departments were planned, which was to create a common structure of organization duties. The tasks of organization headquarters were to be implemented by such personnel who were then sent to the individual organization departments. 'Zusammenarbeit zwischen Zentraler Organisationsabteilung und den örtlichen Organisationsstellen', ZV Wiesbaden, 7.5.1965. Linde-Archiv Wiesbaden.

84. The unit responsible for plant engineering placed 'informing the organization headquarters in Wiesbaden of organization plans in need of approval from Wiesbaden (such as changes to organization plan issues present at numerous branches such as wage systems, the implementation of organizational means, and so on)' last on its list of duties. 'Aufgabengebiet der Organisationsabteilung Höllriegelskreuth', p. 2, 7 January 1964. Linde-Archiv Höllriegelskreuth.

85. Simon was of the opinion 'that the increasing complexity of the technology at hand limits the elasticity of the company and its leadership more than has been necessary. The decision to build or expand production facilities requires planning for larger spans of time. Furthermore, corporate planning must no longer be tied to the domestic economy, but be focused rather on larger markets.' Protokoll der Vorstandssitzung vom 2.4.1964. Linde-Archiv Wiesbaden.

86. Agthe, *Unternehmensplanung*; Horn, *Zukunftsgestaltung durch Unternehmensplanung*; Albach, *Beiträge zur Unternehmensplanung*.

87. Steiner, *Die DDR-Wirtschaftsreform*, Hoffmann, *Aufbau und Krise*.

88. This herculean task was to be accomplished by 'two men' at the 'Headquarters Planning Staff'. Neugliederung der Unternehmensplanung, Anlage 3 e Funktionsbeschreibung der Stabsstelle Planung undatiert, ~1966/67. Linde-Archiv Wiesbaden.

89. The department comprised of two positions. Organization plan 'Organisation der Hauptabteilung Organisation und Planung zum Organisationsplan 1/69', Stand 15. April 1970. Linde-Archiv Wiesbaden.

90. Organisationsplan Zentralverwaltung 23 June 1971. Linde-Archiv Wiesbaden.

91. Marius Herzog writes of Phase 2 of Linde's corporate restructuring. Herzog *Strukturwandel*, p.73 ff., 106 ff.

92. A Linde Group Headquarters employee at the time remembers: 'It was not discussed much, but this probably had to do with the fact that they didn't spend much time together – good old Hermann Linde, who didn't think enough in terms of business and focused instead on technology and Meinhardt just made sure that enough cash was coming in.' Interviews IA-72 cf. also Interview: IÜ-44.

93. Linde AG, *Arbeit bei Linde*, p. 22 f.

94. He switched to the Linde Supervisory Board in the same year. Ombeck came from steam engine and turbine engineering and had worked for Linde since 1913. He became a member of the Executive Board in 1928 and CEO in 1954. Ombeck was responsible for the expansion of the department of commercial refrigeration systems and was often involved with the head offices in Wiesbaden. *Werkzeitung Linde* 2 (1964), 3.

95. Simon went to Hoechst dye works in 1930 and was an assistant to the top management when he left in 1947. He built up the administrative position in Wiesbaden for accounting and worked on legal and tax issues as well as the Linde corporation's shares, taking over bookkeeping for the branches. He quickly became

syndic in 1948, received limited commercial authority in 1949 and then power of procuration in 1950. As Ombeck, who was CEO at the time, retired in 1961, Simon became deputy chairman and was also in charge of the Güldner- Motorworks Aschaffenburg and the Matra Works in addition to his management of headquarters. *Werkzeitung Linde* 10 (1968), 7 and 2 (1971), 2.

96. Interview I28M–29.

97. Cf. (Interview Förg p. 2) and Interview I28F-30 ff.

98. Dr Wolfgang Baldus, director of Development at Höllriegelskreuth and Dr Herberg Baldus, director of Technology, sons of a mathematics professor at the Technical University Munich. W. Baldus married one of J. Wucherer's daughters. The three Hailer brothers were the sons of an earlier director of the Höllriegelskeuth factories, who died in an industrial accident in 1950. The Linde family consequently took on the responsibility of the boys' education (cf. Interview Wolfgang Baldus p. 2 f.; Interview Gottfried Hailer p.16 and Werner Hailer, p. 2).

99. Walter Linde (1928–) worked at the level just below this and later managed business in the USA until the 1980s. He was the son of Hildegard Wucherer (Rudolf Wucherer's daughter) and Selmar Linde as well as Carl von Linde's great grandson and Johannes Wucherer's nephew.

100. Ruckdeschel worked in the Erecting Department, was frequently abroad for his work and possessed excellent language skills as well as profound technological knowledge. During the Second World War he was CEO of the Linde subsidiary Heylandt Apparatus Engineering. After the war he focused on re-establishing Linde's success in the international market. *Werkzeitung Linde* 8 (1966), 10; Interview I28M-21. Walther Ruckdeschel, *Mein Lebenslauf.* Linde-Archiv Wiesbaden.

101. Hermann Brandi was on the Executive Board at August Thyssen and Hermann Holzrichter on Bayer's Executive Board (Interview I28M-46). Brandi became chairman of the Supervisory Board in 1968, Holzrichter was his successor. *Werkzeitung Linde* 1 (1969), 6.

102. Karl Beichert, Executive Board member (1943–51) from Wiesbaden wanted to see Wiesbaden gain more influence in the 1940s. He was an accountant for Linde and the first businessman to become a member of the board; cf. Interview I28M-28 and *Werkzeitung Linde* 10 (1968), 7.

103. Wucherer in his 'Thoughts on point 21c) Agenda for Executive Board meeting 6 December 1961', 5 December 1961, Linde-Archiv Wiesbaden.

104. Excerpt from the protocol from the Executive Board meetings on 8 and 15 January 1962: 'Wucherer gave all members of the Board a copy of his file notes from 8 January on suggestions for corporate organization …Wucherer emphasized this was not intended to establish a Directorate General, but rather to develop a principle of collegiality as a base for the organization of the entire corporation.' Linde-Archiv Wiesbaden.

105. Wucherer in his 'Thoughts on point 21c) Agenda for Executive Board meeting 6 December 1961', 5 December 1961, Linde-Archiv Wiesbaden.

106. Cf. Meinhardt's report on his visits to IBM, WTC New York, 25 September 1946; the American Institute of Management, New York, 1 October 1964; the UNION CARBIDE Corp., New York, 2 October 1964; the EMHART Corp., Connecticut, 8 October 1964; the Eaton Manufacturing Co., Cleveland, 29 September 1964 and Thompson RAMO WOOLDRIDGE Inc., Cleveland, 29 September 1964. Zentralverwaltung Linde, Wiesbaden..

107. 'The plan was to have an Executive Organ made up of two CEOs. In addition to already existing duties, the Executive Organ […] was obliged to prepare Executive Board decisions, oversee the realization of decided measures to be carried out, deal with overall engineering, economic and financial planning and represent the corporation publicly in the press or at associations. […] Staff unit positions should be made available to help the Executive Organ achieve its mission.' Excerpt from

the protocol for the Executive Board meeting from 25 October 1966, Linde-Archiv Wiesbaden.

108. 'The Executive Board carries responsibility for the entire corporation ... business policies for individual divisions and their regular business should be carried out by each division within the frameworks laid out by the Executive and Supervisory Boards.' Excerpt from the protocol for the Executive Board Meeting on 25 October 1966, Linde-Archiv Wiesbaden.

109. Internal document at Linde: 'Standort Unternehmensleitung, Zentrale Verwaltung', Wiesbaden 9 August 1966, cf. Point 5, Corporate Organization, protocol for the Executive Board meeting on 2 February 1967. Linde-Archiv Wiesbaden.

110. Executive Board meetings took place every few months, since the end of 1965 the Executive Board has met interchangeably at Wiesbaden and München-Höllriegelskreuth.

111. Cf. Letter of 27 April 1966 from the Linde Executive Board to the chair and deputy chair of the Supervisory Board Dr F. A. Oetken, Dr H. Th. Brandi. Enclosure: proposal for an integrated corporation plan at Linde, 1966; internal Linde document: 'Aufgabengebiete des Exekutivorgans', 5 December 1966; internal Linde document: 'Aufgaben des Exekutivorgans des Vorstands', undated (probably 1966); 'Neugliederung der Unternehmensorganisation', probably 24 January 1967; Protocol from 2 February 1967 Point 5, Corporate Organization. Linde-Archiv Wiesbaden.

112. Point 5, Corporation Organization, protocol for the Executive Board meeting 2 February 1967. Linde-Archiv Wiesbaden.

113. 'Corporate Organization Restructuring' undated (probably 1966/67). Linde-Archiv Wiesbaden.

114. Cf. the enormous variety of Linde activities: reports on corporate clients, Linde AG, 1969, Linde-Archiv Wiesbaden.

115. 'Corporate Philosophy at Linde'. Internal Linde document, *ca.* 1978; 'Referat von Herrn Dr Meinhardt vor Leitenden und AT-Angestellten der ZV am 13.5.1985 zum Thema 'Geschäfts- und Führungsgrundsätze der Linde AG sowie Aufgaben und Stellung der Zentralverwaltung.'' Linde-Archiv Wiesbaden. Cf. also. Meinhardt, *Optimierung des Portfolios*, pp. 135–46; Henzler (ed.), *Handbuch strategische Führung*.

116. The mobile symbolized this balance at Linde Aus. Cf. also Peter von Zahn's documentary: 'Corporate Group in Equilibrium', 1984, which was made at the end of restructuring.

117. Meinhardt, *Optimierung des Portfolios*, pp. 135–46; Henzler (ed.), *Handbuch strategische Führung*.

118. Cf. 'Referat über den Aufenthalt bei der Firma Raytheon Company in Lexington, Massachusetts, USA in der Zeit vom 9. September bis 5. November 1968, mit Vorschlägen für Linde', 29 January 1969, Linde-Archiv Wiesbaden.

119. Cf. Herzog on restructuring: Herzog, *Strukturwandel*, p. 68.

120. As of 1965, Meinhardt participated as protocol director on the Executive Board meetings. In 1966, Meinhardt was taken into the Executive Organ as an assistant to Wucherer and Simon. Organization plan 'Unternehmensorganisation', 1966, Linde-Archiv Wiesbaden.

121. Organization plan 1968. Linde-Archiv Wiesbaden.

122. Cf. 'Führungsanweisung für die zentralen Stabs- und Verwaltungsstellen', Stand 10/76, Linde-Archiv Wiesbaden; Allgemeine Führungsanweisung Stand 9/78). Together with the Marketing Management Institute's Dr R. Matheis, Linde led in 1977 the seminar 'Mitarbeiterführung als Teilaufgabe optimalen Managements' for top-level Group Headquarters employees. Seminar documents: Linde-Archiv Wiesbaden. Cf. Kieser, *Organisationstheorien*, p. 96 ff.

123. Since its founding in 1953 to 2000, nearly 600,000 managers have attended the Harzburger Academy. Hachmeister, Die Rolle des SD-Personals, *Mittelweg 36* 2 (2002), 17–35.
124. Interview K 112.
125. Interview: K 123–8.
126. Interviews 1337, 161, 164, M 200, 201.
127. Interview L 60.
128. In the mid-Seventies, there were two directorship positions in management at Plant Engineering in Munich, whereas there were four such positions in the Wiesbaden administration.
129. Cf. Linde AG, *Geschäftsberichte* 1960–76.
130. 'Referat über den Aufenthalt bei der Firma Raytheon Company in Lexington, Massachusetts, USA in der Zeit vom 9. September bis 5. November 1968, mit Vorschlägen für Linde', 29 January 1969, p. 23.
131. Protocol of the Executive Board meeting on 20 July 1972 and 21 November 1972. Linde agreed to the solution after a long, hard-fought battle in the Supervisory Board.
132. Looking back in 1976, Meinhardt refers cautiously to Hermann Linde's spokesmanship: 'He was a kind of spokesman for the Executive Board.' *Der Spiegel* 38 (1976), 72 f.
133. Cf. 'Spezielle Führungsanweisung für Betreuung', Stand 1/72; Stand 10/76; 'Stellenbeschreibungen 'Betreuung', 'Stellenbeschreibung Gesamtvorstand', Stand 8/76; 'Stellenbeschreibung Leitung der Zentralverwaltung', Stand 10/76. Linde-Archiv Wiesbaden.
134. Cf. special management directives for individual Divisional Board members, Stand 9/78. Linde-Archiv Wiesbaden. The competencies came from job descriptions for individual Divisional Board members and the description of the Divsional Board itself.
135. Cf. for example 'General Management Directive', Stand 9/78. Linde-Archiv Wiesbaden.
136. In 1974 Linde had 14 directors, 10 of which were for the divisions, 4 at Group Headquarters. (Status 1 January 1974.) Company Report 1973, p. 8. The number of directors depended on the organization structure, that is the number of divisions and staff units.
137. Interview IÜ-72.
138. Interview Linde on 21 July 1999.
139. Hermann Linde's speech at the shareholders' meeting for the Linde AG, 13 May 1976. Linde compared the number of orders received in millions of DM. He differentiated between plant engineering and mechanical or vehicle engineering/construction. The number of orders received increased afterwards in plant engineering:1973 = 291 million DM, 1974 = 597 million DM, 1975 = 819 million DM, while there was no increase in mechanical and vehicle engineering/construction during the same time periods: 1973 = 534 million DM, 1974 = 584 million DM, 1975 = 525 million DM).
140. Vgl. 'Linde AG entwickelt Tankschiff-System für den Transport von verflüssigtem Erdgas', Linde-Papier, 3 April 1973. Cf. also Linde AG, *Geschäftsbericht* 1973, p. 29.
141. Interview IÜ-71.
142. Interview Linde, 21 July 1999; Interview IÜ-31.
143. A former Group Headquarters employee remembers: 'there was only one voice in opposition to this plan, and this was that of Dr Linde (pause), who wanted to lead the business for his family like it had been done so previously, with a headquarters in Munich as well. It was an intense discussion with lots of arguing over where the

future administrative headquarters should be located. There were major differences of opinion between Meinhardt and Linde.' Interview IÜ-32.

144. Cf. 'Linde zeigt Wiesbaden nicht die kalte Schulter', *Allgemeine Zeitung*, 22 July 1971; 'Stadt stellt Linde AG weiteres Gelände zur Verfügung', *Wiesbadener Kurier*, 22 July 1971.

145. Interview IÜ-32.

146. Works Council and Employee Resolution, Linde AG, 1 September 1976; 85.5 per cent of 2,250 employees in Höllriegelskreuth and Schalchen voted on the resolution.

147. 'Greuel im Vorstand', *Der Spiegel* 38 (1976), 72 f. Other press responses: 'Auseinandersetzungen im Linde-Vorstand', *Hannoversche Allgemeine*, 4 September 1976; 'Streit an der Linde-Spitze', *Wiesbadener Kurier*, 3 September 1976; 'Krach bei Linde', *Mannheimer Morgen*, 3 September 1976; 'Dr Hermann Linde verläßt Linde AG', *Main Spitze*, 8 September 1976; 'Vereist', *Frankfurter Rundschau*, 9 September 1976; 'Hermann Linde scheidet aus', *Frankfurter Allgemeine Zeitung*, 8 September 1976; 'Linde ohne Linde', *Wiesbadener Kurier*, 8 September 1976.

148. Executive Board, Linde AG. Press notes, Linde AG, 6 September 1979.

149. Linde worked at the Technical University Munich.

150. 'Stimmrechtsbeschränkung genehmigt', *Frankfurter Allgemeine Zeitung*, 8 June 1973; 'Stimmrechtsbeschränkung spaltet Linde-HV', *Ebersberger Zeitung*, 8 June 1973.

151. Cf. Linde *Geschäftsberichte* 1980 ff.

152. In 1982 cold stores on average had an occupancy of barely 68 per cent, just under that of 1981. Utilization improved starting in 1983, but the average occupancy rate, at 80 per cent, remained below that of the golden years of 1976–80.

153. Linde *Geschäftsberichte* 1983, p. 41 and 1984, p. 19.

154. In 1982, the company's turnover was at 2.7 million DM and employed nearly 18,400 people.

155. Installation engineering (in the refrigerator and freezer trade as well as refrigeration and air-conditioning systems) and central plant engineering were at issue here. Linde-Pressemitteilung Werksgruppe Kälte- und Einrichtungstechnik vom 12 January 1984. Linde-Archiv Wiesbaden.

156. Meinhardt, *Optimierung des Portfolios*, Henzler (ed.), *Handbuch strategische Führung*, p. 142.

157. In 1977 nobody spoke of a fourth division any more. The three divisions were given the titles Plant Engineering, Industrial Gases and Vehicle and Mechanical Engineering by Meinhardt. 'Drei große Zweige machen Linde krisenfest', *Wiesbadener Tagblatt*, 1 April 1977.

158. 'The company advocates freedom, democracy and social commitment in our society as it also recognizes an economic environment based on performance, competition and private property. The performance of a company in a social market economy is influenced primarily by the creativity, commitment and sense of responsibility of its employees. Lecture given by Dr Meinhardt to AT managers and employees of the ZV on 13 May 1985 on 'Business and Leadership Basics at the Linde AG and the Duties and Position of its Headquarters'. Linde-Archiv Wiesbaden.

159. 'I expect performance and a high level of commitment. I cannot stand average, weak people.' 'Mit eiserner Hand', *Industriemagazin*, 2 (1984), 59; Linde AG, *Dr Hans Meinhardt*, p. 19.

Chapter 6

1. In addition to the price of oil, increased spending by German tourists in foreign countries and funds transfers by foreigners living in Germany also contributed to the deficit.

2. According to OECD statistics, the three countries with the highest debts in 1983 were Brazil, with a debt of 97 billion dollars; Mexico with a debt of 83 billion dollars; and Argentina, with foreign debts of 42 billion dollars. In 1983 the Latin American continent was de facto incapable of debt payment.

3. Weimer, *Deutsche Wirtschaftsgeschichte*, p. 304.

4. Lindner, *Den Faden verloren*.

5. A candid revelation about the Social-Liberal government came from the incumbent finance minister Otto Graf Lambsdorff (FDP). Lambsdorff's memorandum of 9 September 1982 entitled 'Concept for a Political Plan to Overcome Weak Growth and to Fight Unemployment' became a 'divorce decree' for the Social-Liberal coalition. Weimer, *Deutsche Wirtschaftsgeschichte*, p. 308.

6. On 3 December 1984 one of the greatest catastrophes in the history of the chemical industry occurred in Indian Bhopal, in a factory owned by Union Carbide of India Limited. Between 2,000 and 10,000 people died and several hundred thousand were injured. http://www.krisennavigator.de/rifa4-d.htm.

7. Two large projects (Böhlen) were set up directly by Franz Josef Strauss, Hans Meinhardt and the GDR foreign currency procurer Alexander Schalck-Golodkowsky.

8. In his speech to the stockholders at the 1991 annual meeting, Hans Meinhardt called attention to the positive economic effect caused by the high demand in the new states: 'Without this demand, the Federal Republic would not have had an economic development any different from the other Western European countries', and added that one should also consider this fact 'when discussing the costs of German reunification'. Speech to the executive committee of the Executive Board of Linde AG on 22 May 1991 in Munich.

9. Richter/Furubotn, *Neue Institutionenökonomik*, p. 371 f.

10. The archival materials for 1980 to the present are not as publicly accessible as the materials for previous decades. Therefore, the character of this business history has to change for this later period and orient itself more towards the description of external facts.

11. Board of directors of Linde AG, minutes of the meeting on 10 July 1984. Linde-Archiv Wiesbaden.

12. Linde AG, *Geschäftsbericht* 1988, p. 16.

13. Hans Meinhardt, who had been with the business since 1955, commented on the success of the previous years in a 1990 interview: 'I had not experienced anything like it at Linde.' *Welt am Sonntag*, 9.12.1990, 39.

14. Meeting of the Executive Board on 29.5.1985. Linde-Archiv Wiesbaden.

15. Schultz-Gambard, Maßnahmen deutscher Wirtschaftsunternehmen, *Zeitschrift für Frauenforschung*, 4 (1993), 20.

16. Molvaer/Stein, *Ingenieurin, warum nicht?*, p. 37.

17. Dienel, *Fauen in Führungspositionen*, p. 97.

18. Linde AG, *Geschäftsbericht* 1990, p. 16.

19. The company had already passed the 6 billion mark in 1990, with sales amounting to 6.1 billion DM. Linde AG, *Geschäftsbericht* 1990, p. 23.

20. Linde AG, *Geschäftsbericht* 1996, 11–13.

21. As late as 1985 Linde still had to defend itself from a potential takeover by BOC. Minutes of the meeting of the Executive Board on 29.5.1985, Linde-Archiv Wiesbaden.

22. 'Linde-Vorstand mit Lob übergossen', *Süddeutsche Zeitung*, 27. 5.1998, 33.

23. 'Linde strebt immer höher hinaus', *Süddeutsche Zeitung*, 13. 11.1998, 25.

24. 'Der bayerische Finanzminister in der ZV', *Linde heute* 3 (1989), 4.

25. Klaus-Peter Schmid in *DIE ZEIT*, 26.6.2003.

26. For a discussion of the compensation for forced labourers and the foundation initiative, see: Eizenstat, *Unvollkommene Gerechtigkeit*, and Spiliotis, *Verantwortung und Rechtsfrieden*.

27. Linde AG, *Geschäftsbericht* 1997, p. 10.
28. *Linde heute spezial*, 9 (2002), p. 4.
29. Information from *Linde heute*, 6 (1996), p. 4.
30. 'Linde-Chef hinterlässt ein wohlbestelltes Haus', *Süddeutsche Zeitung*, 14. 5. 1997, 31.
31. Hans Meinhardt was also the chair of the Supervisory Board of Karstadt, Beiersdorf and other companies.
32. 'Sie wollen ewig herrschen', *DIE ZEIT,* 15.5.2003, 20.
33. *Wirtschaftswoche*, 31.1.1992, 92.
34. *Wiesbadener Kurier*, 28.3.1996.
35. Linde AG, *Geschäftsbericht* 1997, pp. 24, 30, 36 and 42.
36. Linde AG, *Geschäftsbericht* 1990, p. 20.
37. Müller, *Entwicklung der Werksgruppe TVT*, p. 21.
38. Müller, *Entwicklung der Werksgruppe TVT*, p. 22.
39. Univation Technologies was founded in 1997 as a joint venture of Union Carbide and Exxon Chemical. In 1999 Exxon fused with Mobil to become ExxonMobil Chemical. Union Carbide has been a subsidiary of Dow Chemical Company since 2001.
40. *Linde heute* 5 (2002), 15.
41. In the late 1980s, Linde developed its own method of producing propylene through propane hydration independent of steam separation processes. It produced a large-scale demonstration and experiment plant at BASF in Ludwigshafen. Müller, *Entwicklung der Werksgruppe TVT*, p. 14. In 1990 there were successful pilot attempts of the dehydration of propane.
42. Until 1945, the acetylene plant in Bützow belonged to the Hanseatische Acetylen-Gasindustrie AG, the subsidiary of the German AGA. In 1945 a provisional start-up was successful; 'a few days later, however, the Russians took over'. Memo of the Hanseatische Acetylen-Gasindustrie AG to the head of the Svenska A/B gas accumulator Lidingö on 11.6.1945, in file 'AGA in Oest'. AGA collection in Stockholms Företagsminnen/Centre for Business History in Stockholm.
43. The market shares according to sales in 1991 were assessed as follows: Linde 110 million DM, MGI (Messer Griesheim) 90 million DM, AL (Air Liquide) 80 million DM und AGA 7 million DM. 'Einschätzung Marktanteile in Mio DM/Neue Bundesländer/September 1991' from 5.9.1991, in file: 'LEUNA Allgemein' (Dr Eggendorfer), Linde-Archiv Höllriegelskreuth.
44. Linde AG, *Geschäftsbericht* 1990, p. 13.
45. Dr Eggendorfer to Dr Meinhardt: 'Informationen zu LEUNA / Brandenburg / CSFR' from 18.07.1991, in Ordner LEUNA Allgemein (Dr Eggendorfer), Linde-Archiv Höllriegelskreuth.
46. *Linde heute* 6 (1996), 5.
47. 'Genehmigungsverlauf neue Bundesländer/Generelles', in Ordner LEUNA Allgemein (Dr Eggendorfer), Linde-Archiv Höllriegelskreuth.
48. *Linde Today* 4 (2003), 9.
49. Linde AG, *Geschäftsbericht* 1990, p. 15.
50. Linde AG, *Geschäftsbericht* 1992, p. 31.
51. *Luft- und Kältetechnik* 2 (1992), 82.
52. In general Schalchen was the most important manufacturing site for heat exchangers, tank systems, and the construction of coolers for air and gas separation plants. In the following years a part of the heat-exchanger production went to Schalchen too.
53. In the previous year Linde was awarded the 'Engineering Excellence Award 1984' by the 'Australian Institution of Engineers' for exceptional achievements in the planning and construction of the petrochemical plant of the ICI-Australia Engineering Pty. Ltd in Botany. Linde AG, *Geschäftsbericht* 1984, p. 29.

54. For these projects, Strauß negotiated directly with the South African prime minister Botha.
55. Linde AG, *Geschäftsbericht* 1990, p. 22.
56. Linde AG, *Geschäftsbericht* 1999, p. 28.
57. Minutes of the chairman/group leader meeting of TVT about the methanol production plant for Georgia Pacific Group, on 7.2.1983.
58. Minutes of the board meeting 7.2.1983. Linde-Archiv Wiesbaden.
59. Linde AG, *Geschäftsbericht* 1990, p. 16.
60. Minutes of the Executive Board 5.6.1984. Linde-Archiv Wiesbaden.
61. F. J. Strauß lobbied heavily for the Böhlen contract.
62. Interview of Werner Jakobmeier with Gerhard Full (2001): 'Before the opening of the borders, the business situation for plant design and construction was much better than today. In the central planned economy, financial resources were deployed as a unit, so that the most necessary projects could be realized. This completely disappeared with the end of the Soviet Union.' Linde-Archiv Höllriegelskreuth.
63. Hydrogen technology held the promise of a positive influence on all of the important social questions of the 1980s: 'OPEC cartel, scarce reserves of natural resources, climate damage through increasing carbon dioxide in the atmosphere, smog alarms due to auto and industry exhaust, varying oil prices as well as risks in nuclear energy plants are catchphrases that make the solar-hydrogen-engineering interesting as the possible solution to many problems. [...] It is already clear now that growth industries increasingly need this natural resource and energy source.' Dr Walter Klauser in *Linde heute* 1 (1987), 5.
64. Ludwig Bölkow was the driving force behind these efforts. After leaving the MBB-board, he promoted hydrogen technology.
65. Partners in this project were Linde (10 per cent), Bayernwerk AG (70 per cent), Messerschmidt-Bölkow-Blohm (10 per cent) and KWE (10 per cent).
66. The first industrial hydrogen liquefier in the FRG was in Linde-Werk Lohhof near Munich. *Linde heute* 1 (1987), 16.
67. Linde AG, *Geschäftsbericht* 1984, p. 31.
68. In 1980 Linde took over the Unox process for waste water treatment through the introduction of oxygen from Union Carbide Corp., USA, with all patents, rights, and business activities.
69. The 'Laran' process is used to handle water heavily contaminated with biological waste. The process uses exclusion of air and allows the reduction of 50–90 per cent of the contaminants with only a small amount of energy. A methane-rich burnable gas is a byproduct. The 'DS' process allows the storage and industrial use of sludge. The sludge is oxidized by an oxygen-rich gas at around 60°C and then putrefies through air exclusion. Later two other processes were added: the 'Metex' process for the separation of heavy metals from waste waters through biosorption and a process for eliminating phosphates.
70. Interestingly, the plant was at first rejected by the neighbouring inhabitants; 'the unfounded fear of an oxygen tank played a major role' in this rejection. *Linde heute*, 1 (1987), 15.
71. Brüggemeier/Rommelspacher, *Blauer Himmel über der Ruhr*.
72. Müller, *Entwicklung der Werksgruppe TVT*, p. 18.
73. Minutes of the board meeting of 22.6.1999, Linde-Archiv Wiesbaden.
74. Minutes of the board meeting of 5.11.1998, Linde-Archiv Wiesbaden.
75. In 1990 Linde took over all units of SOGAS in Portugal. In 1990–95 it took over the Gas and Equipment Group Ltd in England, and in 1991 it acquired Electrochem Ltd in England, which focused on special gases.
76. See *Linde heute* 5 (1997), 12 f and 6 (1997),16 f.

77. In 1937 Linde had installed a helium purification plant and a helium compressor in the airport in Frankfurt/M, which would have been used for the Airship LZ 130. LZ 130's predecessor, the Zeppelin LZ 129, was filled with highly combustible hydrogen and exploded in 1837 in Lakehurst/USA, probably due to electrostatic charging.
78. Minutes of the board meeting of 6.5.1998, Linde-Archiv Wiesbaden.
79. Proceeds worldwide: 1981 = 922, 1982 = 864, 1983 = 923, 1984 = 1.123 million DM. Linde AG, *Geschäftsbericht 1984*.
80. Lecture by Dr Reinhard Lohse, the head of the Material Handling Division, at the conference for leading employees. Quoted from *Linde heute* 3 (1987), 7.
81. *Welt am Sonntag*, 9.12.1990, pp. 39 and 49.
82. In 1998 Linde considered suing Toyoto over the optical similarities between their forklifts and Linde's: 'The board decided to look into potential measures Toyota could take against Linde imports in China before making a decision about the lawsuit.' Linde-Archiv Wiesbaden.
83. CAD = Computer Assisted Design.
84. The German Lansing GmbH in Roxheim could only be taken over by Linde in 1992, after a positive decision of the German federal supreme court overturned an injunction order by the German antitrust division, which had been upheld by the supreme court of Berlin. In 1992 German Lansing GmbH became part of the working group FH. At the end of 1998 the location, which had 225 employees, was closed. Today Lansing has fewer than 1,000 employees in the Linde Company.
85. Linde AG, *Geschäftsbericht 1992*, p. 28.
86. Linde AG, *Geschäftsbericht 1990*, p. 26.
87. Linde AG, *Geschäftsbericht 1999*, p. 33.
88. In 1995 Linde founded Linde Gas Xiamen Ltd as a production centre for liquid oxygen and nitrogen, argon, carbon dioxide, and hydrogen at the site of the forklift plant.
89. 'Full made clear in strong words that reneging on the agreement (on the increase in capital) would endanger the trust in a positive continuation of co-operation in the Linde Xiamen project, and that we reserve the right to take necessary measures.' Minutes of the board meeting on 9.6.1998, Linde-Archiv Wiesbaden.
90. Minutes of the board meeting on 14.1.1999, Linde-Archiv Wiesbaden.
91. Linde AG, *Geschäftsbericht 1999*, p. 37.
92. Minutes of the meeting of the Executive Board on 11.5.1981, Linde-Archiv Weisbaden.
93. Minutes of the board meeting from 6.7.1999, Linde-Archiv Wiesbaden.
94. In the same year Linde developed a new type of refrigeration equipment, the salad bar, for another manufacturer. The salad bar, which is a self-service counter where customers can put together their own salad, had already been a standard piece of equipment in US supermarkets for years.
95. Minutes of the board meeting on 16.1.1985, Linde-Archiv Wiesbaden.
96. Stickel/Tröster (eds), *49,98 Tante Emma – Megastore*, p. 39.
97. Linde AG, *Geschäftsbericht 1996*, p. 39.
98. Linde AG, *Geschäftsbericht 1997*, p. 38.
99. For the Criosbanc takeover, the formula was as follows: 1/3 assets value (=equity and hidden assets) x 1.24 and 2/3 income value (remaining yield x 6.5) x 1.24 (minutes of the board meeting of 16.1.1998). Other examples of takeovers with the Linde formula include Jetschke Industriefahrzeuge. Minutes of the board meeting on 9.6.1998, Linde-Archiv Wiesbaden.
100. Linde AG, *Geschäftsbericht 1998*, p. 38.
101. Linde AG, *Geschäftsbericht 1999*, p. 37.

Chapter 7

1. Gerhard Full in *Linde heute* 6 (2002), 9.
2. Acetylene gas had been used for lighting since the 1890s; around the turn of the century, gas streetlights were installed in Stockholm and other large Swedish cities.
3. Here AGA worked closely with the appropriate state agencies: 'The production of the system was no coincidence, since also the Swedish Lighthouse Department had a long experience of developing products for beacon lighting; parallel to AGA's experiments with acetylene gas the Lighthouse Department had conducted similar experiments.' Westberg, *Var optimist!*, p. 144 f.
4. Almqvist, *Technological Changes*, p. 8.
5. He received an honorary doctorate in 1918 from the University of Lund, and was a member of both the Academy of Engineering Sciences and the Royal Swedish Academy of Sciences.
6. Almqvist, *Technological Changes*, p. 9.
7. The number of countries in which AGA maintained subsidiaries grew strongly up until 1928: 'Following the war-time demand for acetylene for beacons, lighting and welding, by 1928 subsidiary companies had been formed in a further 17 European and Latin-American countries for the manufacture and sale of acetylene.' Almqvist, *Technological Changes*, p. 13.
8. 'As an effect of the trade restrictions of the First World War, the company came to focus on the domestic market.' Westberg, *Var optimist!* p. 144 f.
9. AGA bought the rights to the union melt electric welding method for Sweden; in Germany Linde introduced the process and marketed it under the name 'Ellira'. In 1940, AGA attempted to buy the $CaSiO_3$ powder needed for this process via its German subsidiary Autogen Gasaccumulator. AGA director Dalén to Autogen Gasaccumulator AG, Berlin, on 20 April 1940. AGA Historisches Archiv, Direktionen Gunnar Dalen, Korrespondenz Autogen Gasaccumulator Berlin 1938–1945.
10. In 1919, with LM Ericsson and ASEA as shareholders, AGA founded its own company, Svenska Radio AB, which produced radio parts and tubes. After parting with this company, AGA introduced in 1927 the first radio device with a built-in loudspeaker that could be connected to the electric grid – that is, the prototype of the modern radio receiver. In the following year, AGA took over the radio division of Baltic AB as the basis for its own production. Almqvist, *Technological Changes*, p. 15.
11. From 1919 onward, the AGA car was produced by a German company, Autogen-Gas-Accumulator AG (AGA), in the Lichtenberg district of Berlin. At the end of the First World War, this subsidiary had about 1,000 workers, who had been involved in German armament production. A small and relatively simple car intended for mass production was designed by FN in Belgium as a peacetime conversion product. 'Despite treacherous handling and lousy brakes, the AGA was something like a Volkswagen [literally, a 'people's car']' (Ostwald, *Deutsche Autos*, p. 38). In Berlin, Breslau, and other large cities, the AGA car was one of the most common models used as a taxi at that time. But the AGA was not blessed with long-term success. 'The idea was good, but economic reality and the competition proved too much.' Almqvist, *Technological Changes*, p. 13. The Aktiengesellschaft für Automobilbau (AGA), spun off in 1920, was thus sold to the Stinnes group in 1922. When this company went bankrupt, the short era of car manufacturing at AGA came to an end in 1925; about 8,000 cars had been made.
12. AGA had already produced mobile flashing beacon systems for military airstrips during the First World War. In 1920 the company received the contract for supplying leading lights and airfield lighting for the important Paris–London air route. In 1921/22 AGA managed to win a major contract: 300 lighted buoys for the Chicago–

Cheyenne night airmail route, one leg of the New York–San Francisco airmail connection. *AGA-Journal, House-organ for Gasaccumulator*, 3 (1925).

13. 'It was far from easy for Dalén to convince his Board to invest in such an odd product, but by 1934 the cooker was being sold in all four corners of the world, and was later to be manufactured in at least ten countries.' Almqvis, *Technological Changes*, p. 21. By 1948, 100,000 AGA ranges had been produced. AGA itself stopped production in 1957; under licence, an English company, Glynwood International, makes about 2,000 ranges per year under the name of AGA.

14. A 1933 attempt to form national monopoly agreements with other European lighthouse makers (in order to dictate prices as the sole provider in each country) failed – due in part to Julius Pintsch AG in Berlin. Julius Pintsch AG on 16 May 1933 to AGA director von Heidenstam, AGA Historisches Archiv, Ordner Direktionen R. v. Heidenstam, Korrespondenz Specforetag 1933–1935.

15. Westberg, *Var optimist!*, p. 145.

16. Submarine periscopes under licence from the Italian Galileo plant were produced, as well, in a bomb-proof underground workshop built on Lidingö in 1942. Almqvist, *Technological Changes*, p. 23.

17. The co-operation led to AGA-Baltic's concentrating all its inventive resources to comply with the demand of the military. Westberg, *Var optimist!*, p. 144.

18. The 1965 annual report announced that the radio and television division had come under strong pressure due to a sharp drop in sales in the wake of increasing imports of low-priced appliances, and that earnings were not satisfactory. AGA, *Annual Report* 1965, p. 4. In 1971, AGA sold its shares in RTM Marknadsaktiebolag, which had been jointly operated with Philips Sweden; after nearly a half century, there were now no more radios and televisions being made under the AGA name.

19. Almqvist, *Technological Changes*, p. 30.

20. See the AGA *Annual Report* 1967. AGA Innovation was later integrated into the gas division, since the most important jobs came from this area. AGA, *Annual Report* 1976.

21. For example, as early as 1967 there were thoughts of using oxygen 'for reconditioning lakes' (AGA, *Annual Report* 1967, p. 18). In 1969 AGA Innovation and the battery division Tudor AB began developing new storage cells for electric vehicles to decrease air pollution in congested urban areas. AGA, *Annual Report* 1968, p. 32. In 1977 AGA brought to market a system for utilizing geothermal energy; in 1978 the heating division experimented with solar collectors. AGA, *Annual Report* 1977, p. 23.

22. Almqvist, *Technological Changes*, p. 31.

23. AGA, *Annual Report* 1970, p. 22 f.

24. In 1961, AGA was actually the world's largest manufacturer of flat radiators. AGA, *Annual Report* 1961. A broad economic recession in Western Europe caused a decline in the construction sector and thus also in radiator sales; radiator manufacturing was sold in 1979.

25. AGA, *Annual Report* 1983, p. 2

26. *Neue Zürcher Zeitung*, 17 September 1985. The NZZ added: 'AGA is behaving anticyclically here; after all, steel companies such as Sandvik and Fagersta in the Swedish region of Bergslagen are selling off their energy generation plants to insurance companies and municipalities.'

27. AGA, *Annual Report* 1994, p. 5.

28. All in all, welding technology did not yet play an important economic role in the inter-war years: 'Welding was to play a small economical role in company activities into the 1920s.' Almqvist, *Technological Changes*, p. 12.

29. Westberg, *Var optimist!*, p. 146.

30. 'A limiting factor in the large scale production of atmospheric gases (oxygen, nitrogen and argon) has hitherto been the high transportation costs. This problem

is greatly reduced by the switchover to the transportation of gases in their liquid state.' AGA, *Annual Report* 1968, p. 32.

31. 'The main applications for AGA's gases have traditionally been welding, cutting and brazing with the oxy-acetylene flame.' But in the long run, acetylene welding would be superseded by electrical processes, and fierce competition developed with propane and butane for lighting and heating applications. The future had already started to become apparent in 1967: 'However, development work now under way is aimed at finding new applications, particularly for the so-called atmospheric gases – oxygen, nitrogen and argon.' AGA, *Annual Report* 1967, p. 18.

32. AGA, *Annual Report* 1980, p. 4.

33. AGA, *Annual Report* 1982, p. 11.

34. 'AGA has a decentralized organization with substantial responsibility and authority residing in the board and management of the local company. This form of organization has proven very effective for correctly judging market opportunities, building up long-term customer relationships and keeping the competition at bay.' AGA, *Annual Report* 1980, p. 18.

35. In 1996, the regional orientation was abandoned in favour of an organization with three business divisions: Manufacturing Industry, Process Industry, and Health Care. AGA, *Annual Report* 1996, p. 2 f.

36. 'AGA was an acetylene company for a long time, and this is to some extent explainable by the fact that AGA's growth embraced many more products besides gas than was the case with the other companies.' AGA, *Annual Report* 1980, p. 18.

37. In 1964, the AGA group consisted of six main divisions: Gas, Signal Equipment, Electronics, Radio/TV, Radiators, and Batteries. Gas accounted for about 50 per cent of revenues, Electronics for 8 per cent, and all other divisions about 10 per cent each. The AGA group's net profit increased by 32.5 per cent from 1963 to 1964; consolidated sales actually rose by 72 per cent between 1959 and 1964 (from 426.1 million Kr. to 733.1 million Kr.). Foreign activities consisted mainly of gas production for industrial applications. AGA, *Annual Report* 1964.

38. AGA, *Annual Report* 1968, p. 27; AGA, *Annual Report* 1969, p. 25.

39. 'AGA has always had a strong position in gases in small and medium-sized volumes.' AGA, *Annual Report* 1982, p. 11.

40. 'AGA decided not to enter the tonnage market at that time and was to some extent outdistanced by the competition in pipeline gases.' AGA, *Annual Report* 1980, p. 23.

41. AGA, *Annual Report* 1971. AGA dissolved this alliance in 1987.

42. 'Group operations, it was decided, should be concentrated on business areas in which AGA's technical and marketing skills were sufficient to enable the Group to expand in international markets.' AGA, *Annual Report* 1985, p. 4.

43. Gas deliveries by volume grew 5 to 10 per cent annually between 1967 and 1977. AGA, *Annual Report* 1977, p. 6.

44. AGA had about an 8 per cent share of the global market in 1977. Combustible gases (acetylene and others) made up about 30 per cent of AGA's sales, while atmospheric gases (oxygen, nitrogen, and argon) constituted 70 per cent. AGA, *Annual Report* 1977, p. 6.

45. AGA, *Annual Report* 1996, p. 9.

46. 'The gas companies operate to a high degree on local markets, since the technology and the economics of transportation require that production be located near the customer. ... Demand and growth in demand are based solely on the conditions on the local markets. What the supplier must therefore do is to estimate the development for the products on each geographic market and in each region and optimally dimension production capacity, product mix, and distribution systems.' AGA, *Annual Report* 1980, p. 11.

47. At the start of its expansion, AGA already had branches in 23 countries in Europe, the USA, and Latin America. AGA, *Annual Report* 1977, pp. 10–12.

48. 'For a modern air separation plant, costs of capital account for roughly 50 per cent of total costs, and energy for about 40 per cent, whereas wages amount to less than 10 per cent.' AGA, *Annual Report* 1982, p. 8 f.

49. Until about 1967 nitrogen was still regarded mostly as a waste product of gas production; starting in 1970, demand began to rise by about 10 per cent annually, with an upward trend. AGA, *Annual Report* 1977, p. 6.

50. 'Market share has been rapidly enlarged in stiff competition with German and American companies, and today AGA is a serious competitor in atmospheric gases as well.' AGA, *Annual Report* 1981, p. 14.

51. 'Given the present nature of the Group's business, it was a natural choice for the Board of Directors to appoint the head of the Gas Division, Marcus Storch, to my successor as President of the AGA.' AGA, *Annual Report* 1980, Letter from the President Sven Agrup, p. 4.

52. 'We kept our share of the world market in 1997 but the operating margin was lower than expected. This meant that we failed to reach the objective of increased earnings per share, which in turn reduced the return on equity.' AGA, *Annual Report* 1997, p. 3.

53. 'We were not satisfied with the 1998 result. We were affected by the year's financial crises and were hit particularly hard in Russia and Ukraine, as well as in South America. This result is also due to the fact that it has taken longer than anticipated for the positive rationalization effects to have an impact.' AGA *Annual Report* 1998, p. 2.

54. 'The efficiency improvement programmes, started in 1998 ... involve a radical restructuring of the entire AGA Group. Small, outmoded filling stations will be closed, transport will be outsourced to independent companies, and production and distribution will be coordinated across borders in several regions. Local product programmes will be replaced by international ones. The effect will be a strengthened market position [...] .' AGA, *Annual Report* 1998, p. 4.

55. 'In order to grow successfully long term we are investing aggressively ...' Group president Marcus Storch and chairman of the Investment Committee Lennart Selander in a circular to the directors and managers on 25 January 1995. AGA Historisches Archiv, Ordner Gas and Welding Division.

56. AGA, *Annual Report* 1998, p. 5.

57. 'In West Germany, AGA used to be solely an acetylene company, but during the last few years a distribution network for atmospheric gases has been built up. A decision has now been made to construct a large air-separation plant in Southern Germany.' AGA, *Annual Report* 1978, p. 10.

58. AGA Gas GmbH, *AGA – 75 Jahre Fortschritt*.

59. 'The equipment for these plants is purchased from the supplier who can offer the most suitable equipment on any given occasion.' AGA, *Annual Report* 1977, p. 12. In the inter-war years, AGA received a 10 per cent across-the-board discount on machines bought from Linde. Letter, Gunnar Dalen to GFLE, 22 July 1948, AGA Historisches Archiv, Direktionen Gunnar Dalén, Korrespondenz Tyskland 1944–50.

60. AGA director Gunnar Dalén to Friedrich Linde on 22 October 1947, in: AGA Historisches Archiv, Direktionen Gunnar Dalén, Korrespondenz Tyskland 1944–50. But at that time Linde had to reply that 'due to a total shortage of the raw materials we need to make our machines – copper, tin, zinc' the Gas Liquidation Department of Linde's Ice Machine Company was at present only able to take on so-called finishing contracts 'that consist of asking our customers to supply the materials required for the contract in question'.

61. Visit to the oxygen station of the Linde Corporation in Herrenhausen-Hanover, on 5 May 1951, report by J. Aversten, AGA Historisches Archiv, Direktionen Gunnar Dalén, Korrespondenz Tyskland 1951–60.

62. An example of this is in a letter of 15 October 1954. AGA Historisches Archiv, Direktionen Gunnar Dalén, Korrespondenz Tyskland 1951–60.

63. For example, Gunnar Dalén informed the director of Linde's Ice Machine Company AG in a letter of 30 June 1960: 'Considering the longstanding and friendly partnership between you and us, we would be very sorry to see you make the El- & Gassvetsning company an offer of a 400 m³ oxygen installation.' AGA Historisches Archiv, Direktionen Gunnar Dalén, Korrespondenz Tyskland 1951–60.

64. Response to Dalén's letter of 29 October 1947, ibid.

65. Letter from director Gunnar Dalén to the Executive Board of Linde's Ice Machine Company AG in Höllriegelskreuth by Munich on 8 November 1954, ibid.

66. AGA, *Annual Report* 1974, p. 29.

67. *Linde heute* 2 (1976), 1.

68. Ebner first entered the private equity business in 1991 with Pharma Vision, BK Vision, Gas Vision (later Spezialitaeten Vision), and Stillhalter Vision. Through these four associated companies he attracted money from private investors in order to purchase large blocks of shares. The funds rapidly and systematically built up strategic holdings in the companies he focused on. In early 2000, at the height of the stock market boom, his private fortune was estimated at more than 3 billion Euros. But his luck did not hold: in August 2002, Ebner had to sell substantial shares of his funds to the Zuercher Kantonalbank because his BZ Bank had major financial problems. On Ebner's person and business activities, see: Sedlmaier, *Firmenjäger*, pp. 173–201.

69. The BZ Bank played a central role in the merger of the Schweizer Bankgesellschaft and the Schweizerisch Bankverein into UBS. By means of the BZ Bank, Ebner is also said to have pulled the decisive strings in the sale of the Winterthur insurance company to Credit Suisse in 1997. Sedlmayer, *Firmenjäger* p. 189 f.; *Financial Times*, 1 August 2002.

70. 'Apart from Switzerland, Ebner also invested heavily in Italy and – at the end – primarily in Sweden. In his fund "Spezialitaeten Visionen", for example, he held stakes in only four companies: Olivetti, Investor AB, Hero, and Industrivärden.' Sedlmayer, *Firmenjäger* p. 199. In addition, Ebner's colleague on the board of directors of the BZ Bank, Johann Björkmann (1944–), is a Swede who sits on the board of Skanditek Industrieforwaltnins.

71. Here, intensive negotiations had been under way since mid-1998 with Hoechst and stockholders in the Messer family, which however failed due to the high purchase price Hoechst demanded and the threat of merger control proceedings from Brussels. Minutes of the Executive Board meetings on 23 October 1998, 26 January 1999, and 14 August 1999, Linde-Archiv Wiesbaden.

72. www.aga.com, no. 990908, Statement by the Board of Directors of AGA.

73. 'Given the circumstances in the gas industry today, we can all understand the advantages of joining forces with Linde. They have a culture that is well suited to AGA, and I think most of you will find Linde's strategy very positive. There will be continuity, and the AGA name and company will live on in the eyes of the customer.' AGA, *AGA Now, Internal Information*, 9 (1999).

74. The 1999 year-end report noted that AGA had 669 fewer employees than in the previous year, 103 of whom were accounted for by the sale of its English subsidiary. AGA, *AGA News*, 6 (2000).

75. Linde AG, *Geschäftsbericht* 2001, p. 44.

76. The EU competition department authorized the acquisition of the industrial gas division of Hydrogas-Messer, a joint subsidiary of Hydrogas and Messer Griesheim.

But this takeover concerned only a filling plant in Köping with 15 employees, while the carbon dioxide business of Hydrogas-Messer was not affected.

77. Negotiations with the shareholders of Messer Griesheim GmbH, Hoechst AG (66.7 per cent) and Messer Industrie GmbH (33.3 per cent) were thus terminated, and Linde Technische Gase GmbH, which had been converted into a joint stock company, Linde Gas AG, in anticipation of a takeover in 2000, was reabsorbed into its parent company Linde AG in October 2001.

78. 'AGA inte längre ett 'publikt' bolag.'*AGA Nyheter* 30 (2000).

79. Linde AG, *Geschäftsbericht* 2000, report of CEO Gerhard Full, p. 6.

80. Linde AG, *Geschäftsbericht* 2000, p. 48.

81. The introduction of INOmax led to a press scandal in Germany when the ZDF television newsmagazine Frontal21 reported on 19 March 2002 that the price of NO therapy for babies had increased fifty-fold due to the Linde product and its monopoly. http://zdf.de/ZDFde/druckansicht/0,1986,1016949,00.html.

82. *Linde heute spezial* 9 (2002), 4.

83. Linde AG, *Geschäftsbericht* 2000, p. 9. To refinance the bank loans taken out to buy AGA, Linde placed a seven-year, €1 billion Euro-bond issue and additional notes totalling €366. Ibid., p. 16.

84. Ibid., p. 6.

85. *Linde heute spezial* 9 (2002), 6.

86. *Linde heute*, 6 (2002), 15.

87. Linde AG, *Geschäftsbericht* 2000, p. 8.

88. Ibid., p. 4.

89. Linde AG, *Geschäftsbericht* 2001, p. 10.

90. Ibid., p. 36.

91. Linde thus made an effort after 2001 to pay off its large debts from the AGA acquisition on an accelerated schedule. In 2001, Linde repaid 300 million, and was able to reduce its financial debt by a half billion Euros (€501 billion) in 2002.

92. In 2000, Linde made a determination of which institutional investors held Linde stock. In addition to the three known major shareholders Allianz (12.5 per cent), Commerzbank (10.7 per cent), and Deutsche Bank (10 per cent) that together accounted for 33.2 per cent of Linde's stock, other institutional investors held 35 per cent (a third of which was in foreign possession). About 32 per cent of share capital was in the hands of private investors. Linde AG, *Geschäftsbericht* 2000, p. 24.

93. Wolfgang Reitzle at Linde's annual general meeting in Munich on 27 May 2003.

94. Seminal works on this topic: Lorange/Roos, *Strategic Alliances*; Gilroy, *Networking in Multinational Enterprises*; Zentes (ed.), *Kooperationen, Allianzen und Netzwerke*.

95. Grafoner had studied electrical engineering at the University of Dortmund, and then went to Brown Boveri and Company (BBC) in Ladenburg in 1978. Grafoner worked for AEG from 1985 to 1996 and then VDO in Schwalbach, until Mannesmann acquired VDO. Information according to Linde AG press release of 27 July 2000.

96. Meinhardt emphasized in an interview on 28 August 2001: 'this is a basic principle that I brought back from over there [from the USA]: make no compromises at the top. He who is at the top – the higher up, the more dangerous and the more endangered he is.' Hans Meinhardt in an interview with Hans-Liudger Dienel on 28 August 2001.

97. 'Offen für Veränderungen – Gerhard Full und Dr Wolfgang Reitzle im Gespräch', *Linde heute spezial* 9 (2002), 7.

98. Reitzle thus also became chairman of Aston Martin, Land Rover, Jaguar, and Volvo.

99. *Linde Today*, 3 (2003), 9.

100. Linde AG, *Geschäftsbericht* 2000, p. 20.

101. Large quantities of hydrogen are needed to refine heavy petroleum residue, which was previously burned as a waste product, into valuable products; the more stringent environmental regulations for the desulphurization of fuels will also create a higher demand for hydrogen in the future.

102. Linde AG, *Geschäftsbericht* 2001, p. 50.

103. Linde AG, *Geschäftsbericht* 2000, p. 29. Linde undertook its initial experiments with liquefied natural gas as a fuel for commercial vehicles as early as 1972 in co-operation with Daimler-Benz AG. Linde delivered both the fuel system for the experimental bus and the plant for preparation, liquefaction, storage, and regasification of the natural gas. See: 'Hat der Erdgasmotor eine Zukunft?' *Linde heute* 4 (1972), 16 f.

104. Linde AG, *Geschäftsbericht* 2001, p. 46.

105. Linde AG, *Geschäftsbericht* 2001, p. 18.

106. Linde AG, *Geschäftsbericht* 2000, p. 30.

107. The transport of large quantities of liquefied gas by ship was a topic at Linde for the first time in March 1944, when the Linde Corporation received an inquiry from the Reichswerke Hermann Göring regarding a plant for extracting liquid methane to be delivered to Berlin via the Mitelland Canal on barges. A contract never came about for war-related reasons. Jaeger, *Aus der Geschichte der Werksgruppe TVT*, p. 128.

108. Linde AG, *Geschäftsbericht* 2000, p. 29.

109. In 1998 BOC had proposed the spin-off of its plant construction segment to Linde. At that time, Linde turned down this offer and attempted to take over BOC altogether. Minutes of the Executive Board meeting of 28 September 1998, Linde-Archiv Wiesbaden.

110. The impulse for closer co-operation came from (among others) Lurgi (Dr Matthes). Minutes of the Executive Board meeting of 26 January 1999, Linde-Archiv Wiesbaden.

111. The outgoing CEO Gerhard Full in a 2003 interview with Werner Jakobsmeier.

112. *Linde Today* 2 (2003), 26–9.

113. Linde AG, *Geschäftsbericht* 2000, p. 10.

114. Warehousing equipment is generally somewhat less vulnerable to recession, but in hydraulics the economic downturn made itself felt already in 2001 in multiple purchasing sectors – especially in harvesters and construction machinery – and led to a decline in orders. Linde AG, *Geschäftsbericht* 2000, p. 53.

115. Linde AG, *Geschäftsbericht* 2000, p. 35.

116. *Linde heute spezial* 9 (2002), 3.

117. According to Wolfgang Reitzle in his speech at the annual general meeting on 27 May 2003.

118. Wolfgang Reitzle at Linde's 2003 balance-sheet press conference.

119. Linde's share of the South American forklift market was 8.5 per cent in 1997. Minutes of the Executive Board meeting of 23 June 1998, Linde-Archiv Wiesbaden.

120. *Linde heute spezial* 9 (2002), 4.

121. Linde AG, *Geschäftsbericht* 2000, p. 40.

122. Ibid., p. 41.

123. Wolfgang Reitzle at Linde's 2003 balance-sheet press conference.

124. Hervé, *Ende des Familienkapitalismus*, Berghan/Unger/Ziegler (ed.): *Die deutsche Wirtschaftselite im 20. Jahrhundert*, pp. 75–93.

Appendix

Abbreviation listing

AB	plc., public limited company
ACHEMA	exhibition of chemical machinery
AEG	Allgemeine Elektricitätsgesellschaft AEG
AGA	Aktiebolaget för Gasaccumulator, Swedish Gasaccumulator plc.
AM	general mechanical engineering
AR	supervisory board
ARAL	Company name of the 'Benzol-Verband' (B.V. for short) since 1952; composed of aromatics (benzole) and aliphatics (benzene)
ASAG	Sulzer AG Archives
ATE	The Alfred Tewes Machine and Armature Factory
ATUD	Technical University Dresden Archives
ATUM	Technical University Munich Archives
ATUM, PAA	Technical University Munich Personnel Records Archives
ATUM, PrAA	Technical University Munich Doctoral Records Archives
ATUM, RA	Technical University Munich Rectoral Archives
B.V.	limited liability company
BASF	Badische Anilin- und Sodafabriken
BayHStA	Bavarian Public Records Office
BBC	Brown Boveri Corporation
BG	*Badische Gewerbezeitung* (business newspaper from Baden)
BIGB	*Bayerisches Industrie- und Gewerbeblatt* (Bavarian industry and business newspaper)
BKK	company health insurance
BMW	Bayerische Motorenwerke AG, Bavarian Motor Works.
BOC	British Oxygen Company
BOX	British Oxygen Works
Br. Ref. Co.	(Linde) British Refrigeration Company
Buna	Synthetic rubber
CAD	Computer Aided Design
CAIL	CAIL, S.A. DES ANCIENS ETABLISSEMENTS CAIL
CASE	Case Western Reserve University Archives, Cleveland, Ohio, USA
Cb.	Copybook
CDAX	Composite DAX
CDU	Christian Democratic Union, German political party
CEO	chief executive officer
CNG	Compressed Natural Gas
COMECON	Council for Mutual Economic Assistance
COSIGUA	A Brazilian steel group
ČSSR	Czech and Slovak Socialist Republic (CSSR)
CSU	Christian Social Union (German political party)
CUA	Cornell University Archives, Ithaca, NY, USA

DAF	Deutsche Arbeitsfront (NS workers' and employers' union)
DAX	German stock exchange
DESY	German Electron Sychrotron
DIAG	German Industrial Gas Association
DKV	German Refrigeration Association
DKW	abbreviation for various products of the Rasmussen Group, here: Deutsche Kühlschrank-Werke
DMMS APV	German Museum Munich, Polytechnical Society Archives
DMMS FSA	German Museum Munich, Company Documents Archives
DOAG	Deutsche Oxydric AG
DPJ	Dinglers Polytechnical Journal
DRP	German Reich's patent
DTMA	German Technical Museum Berlin Archives
DÜWAG	Düsseldorfer Waggonbau AG (railway carriage construction plc. in Düsseldorf)
EG	Einkaufsgenossenschaft, central purchasing co-operative
EKI	Development policy commentary and information
EKO	Eisenhüttenkombinat Ost (steel producer in Eisenhüttenstadt)
ETH	Eidgenössische Technische Hochschule, Swiss Technical University
EU	European Community
Euratom	European Atomic Energy Community (Euratom)
EWG	European Economic Community (EEC)
F&E	Forschung und Entwicklung, Research and Development
FCKW	CFC chloroflourocarbons
FDP	Free Democratic Party of Germany
FH	Department of Material Handling and Hydraulics
Fiat OM	automobile company in Turin, Italy (FIAT)
FN	Belgian armament manufacturers (FN)
FSA	company documents archives
GDR	German Democratic Republic
GFLE	Gesellschaft für Linde's Eismaschinen, Company for Linde's Ice Machines
GfMK	Gesellschaft für Markt- und Kühlhallen, Company for Markethalls and Warehouses
GHH	Gute-Hoffnungs-Hütte
GStAPKM	Geheimes Staatsarchiv Preußischer Kulturbesitz, Secret Central Archives of the former Prussia
GuG	Geschichte und Gesellschaft, Journal for History and Society
HANOMAG	Hannoversche Maschinenbau AG, Hannover Mechanical Engineering Group
HD-Polyethylen	High-Density Polyethylene
HERA	Hadron Electron Ring Accelerator
H-FCKW	HCFC halogenated chloroflourocarbons
	German refrigerator factory in Scharfenstein, Saxonia
HStA	Hauptstaatsarchiv, Public Records Office
IAG	International Oxygen Association
IAS	International Accountings Standards
IG Farben	IG Farben
IGA	Industriegas GmbH (industrial gas group)
IIR	International Institute of Refrigeration
IKB	German Industrial Bank plc.
INO	Inhaled Nitric Oxide
IWF	International Monetary Fund
JIPO	Domoradice spol. s.r.o. in Ceský Krumilov Czechia, now Linde Pohony s.r.o.

JULI	Jungheinrich-Linde
k.s.	limited partnership
KA	Kältetechnischer Anzeiger
KCA	Komplexe Chemieanlagen
KFL	Komatsu Forklift Company
KG	limited partnership
KHD	Klöckner-Humbold-Deutz plc.
KHDA	Klöckner-Humbold-Deutz Archives
KM	refrigeration machines and systems department
KPdSU	Soviet Communist Party
LAB	Berlin State Archives
LLC	Linde BOC Process Plants
LNG	liquified natural gas
LPP	Linde Process Plants
LuK	Zeitschrift für Luft- und Kältetechnik, journal for air and refrigeration technology
MA	Augsburg Engineering Works
MAN	Maschinenfabrik Augsburg – Nürnberg, Augsburg-Nurnberg Engineering Works
MAPAG	Augsburg-Plattling Engineering Works plc.
Matra	Matra
MBB	Messerschmidt, Bölkow & Blohm
Mefo	Metallurgic Research Institute
MF	engineering works
Min. d. Inn.	Ministry of the Interior
MuK	Markt- und Kühlhallen AG
MWM	Motorenwerke Mannheim
NMAH-Archives	National Museum of American History Archives, Washington D.C., USA
NMR	Nachlass Maschinenfabrik Riedinger
NO	nitrogen monoxide
NS (-Zeit etc.)	National Socialism
NSDAP	German National Socialist Workers' Party
N.V.	plc., public limited company
OKW	German Armed Forces High Command
OPEC	Organization of Petroleum Exporting Countries
PAA	Personnel records archives
PG	NS party member
PrAA	doctoral records archives
PS	horsepower (h.p.)
PSA	phthalic anhydride
Pte. Ltd.	Private Limited
PTR	Physikalisch-Technische Reichsanstalt
Pty.	property
R.O.W.	Rhinland Olefin works Wesseling
R.W.M.	Ministry of the National Economy (NS government)
RA	rectoral archives
RDS	raw materials and foreign exchange department
RGW	Rat für gegenseitige Wirtschaftshilfe (= COMECON)
RM	Reichsmark
ROCE	Return on Capital Employed
S.A.	plc., public limited company
S.A.R.L.	Limited liability company
S.A.T. Linde	Société d'Application des Techniques Linde S.A.R.L.

s.r.l.	Limited liability company
SA	National Socialist paramilitary organization (SA)
SD	National Socialist secret service
SIAC	Société Industrielle de l'Anhydride Carbonique S.A.
SOGAS	Sociedade Gases e Prod. Químicos Sa.
SPD	German Social Democratic Party
spol. s.r.o	Limited liability company
TEGA	Technical Gases and Gas technic G.m.b.H.
TG	Technical Gases Department of Linde
TH	technical university
TUI	Touristik Union International
TVT	Department of Low Temperature and Process Technology
UCLDA	Union Carbide, Linde Division, Archives, Danbury, Connecticut, USA
UdSSR	Union of Soviet Socialist Republics (= USSR)
UNEP	United Nations Environmental Programme
UV	ultraviolet
V2, V-Waffe	V2 rocket, 'retaliatory weapon'
VA	Verfahrenstechnik und Anlagenbau, Engineering Department of Linde
VDI	Verein deutscher Ingenieure, The Association of German Engineers
VDI-Z	*VDI* (journal)
VDK	Verband deutscher Kältemaschinenfabriken, Society of German Refrigeration Machine Manufacturers
VDO	Vereinigte Deuta OTA (VDO), a company which manufactures driver information systems
VEB	State-owned enterprise in the GDR
VEBA	Vereinigte Elektrizitäts- und Bergwerks AG (VEBA)
VR	(Socialist) Peoples' Republic
VSW	Vereinigte Sauerstoffwerke (VSW)
WGA	Werksgruppe Güldner Aschaffenburg (WGA)
ZA	newspaper archive
ZEHAG	A refrigeration company in Zürich, now a part of the Linde Group
ZEKI	*Zeitschrift für Eis- und Kältetechnik* (magazine for ice and refrigeration technology)
ZfU	*Zeitschrift für Unternehmensgeschichte* (magazine about company history)
ZgesKI	*Zeitschrift für die gesamte Eis- und Kälteindustrie* (magazine serving the entire ice and refrigeration technology industry)
ZkflG	*Zeitschrift für komprimierte und flüssige Gase* (Journal about compressed and liquid gases)
ZV	Zentralverwaltung, central administration

Linde AG Supervisory Board

Ackermann, Josef (1997–)
Baumann, Karl-Hermann (1998–)
Beiten, Gerhard (1979–)
Bouillon,* Rüdiger (1985–)
Brandi, Hermann Th. (1964–73, deputy chairman 1965–68, chairman 1968–73)
Bude,* Hans Gerhard (2003–)
Buhl, Bernhard (1930–40)
Burgard, Horst J. (1978–93)
Buz, Heinrich von (1879–18)
Buz, Richard (1918–32)
Derkum,* Paul (1953–61)
Dhom, Robert (1978–83)
Diekmann, Michael (2003–, weiterer deputy chairman 2003–)
Dittmar, Gerhard (1978–86)
Dorn,* Anselm (1965–73/1976–83, deputy chairman 1978–83)
Eßing,* Günter (1988–90)
Finck, August von (1969–78)
Flügge,* Heinz (1973–85)
Forchel,* Otto (1990–2002)
Förg, Wolfgang (1993–98)
Frommel, Richard (1907–11)
Full, Gerhard (2003–)
Götte, Klaus (1978–83)
Hahl,* Gernot (1998–)
Hartig,* Joachim (1993–)
Hawreliuk,* Heinz (1983–88)
Hess, Johannes (1930–1946, chairman 1943–46)
Hofmann,* Klaus Heinrich (2000–03)
Holzrichter, Hermann (1965–1978, deputy chairman 1968–73, chairman 1973–78)
Honrath, Kurt (1980–81)
Huhn, Peter (1981–88)
Jähne, Friedrich (1941–45, 2. deputy chairman 1943–45)
Jentzsch, Wolfgang H. (1991–2000)
Jung, Adolf (1882–1902, 1886–89 chairman)
Jung, Gustav (1879–85, 1884–85 chairman)
Jung, Otto (1903–43, deputy chairman 1915–31, chairman 1931–43)
Kämmerer,* Thilo (2003–)
Kammholz, Günter (1977–92)
Kaske, Karlheinz (1991–98)
Katte,* Hans-Dieter (1998–, deputy chairman 2003–)

Kohlhausen, Martin (1999–2003)
Kopper, Hilmar (1993–97)
Kornblum,* Jakob (1978–83)
Krauss, Georg (1879–1906)
Kreß,* Karl (1953–61)
Lang, Carl (1879–84 chairman)
Lang-Puchhof, Carl von (1884–1916)
Leberkern,* Adalbert (1978–85)
Leitmann,* Josef (1982–88)
Liebler,* Reinhold (1988–93)
Linde, Carl von (1890–1934, chairman 1890–1930)
Linde, Friedrich (1948–61, deputy chairman 1948–61)
Linde, Richard (1949–55)
Lohse, Reinhard (1993–97)
Meinhardt, Hans (1997–2003, chairman 1997–2003)
Meyer, Otto (1932–67, deputy chairman 1942–47, chairman 1947–65, Ehren chairman 1965–67)
Müller, Klaus-Peter (2003–)
Nesselmann, Kurt (1967–72)
Neuendorff,* Herbert (1961–65)
Nowakowitsch,* Stefan (1988–93)
Oechelhäuser, Max (1915–27)
Oetken, Friedrich August (1950–68, deputy chairman 1964–65, chairman 1965–68)
Ombeck, Hugo (1961–63, deputy chairman 1961–63)
Otto, Jung (1903–43, deputy chairman 1915–31, chairman 1931–43)
Pfotenhauer, Bernhard (1943–44)
Pietsch,* Kay (2003–)
Piltz, Klaus (1992–93)
Plieninger, Theodor (1917–29)
Plötz, Georg (1986–88)
Proebst, Georg (1921–49)
Ranke, Karl (1935–48)
Rath,* Walter (1978–80)
Rittlinger,* Peter (1993–98)
Sandleitner,* Herbert (1985–98)
Saß Herbert, (1980–88)
Scherm,* Josef (1973–75)
Schieren, Wolfgang (1980–96, chairman 1980–96)
Schipper, Friedrich (1929)
Schlaus, Wilhelm (1993–96)
Schmidt,* Rainer (1988–2002)
Schmitz, Ronaldo H. (1988–91)

Schnaller,* Fritz (1961–73 1977–81)
Schneider, Manfred (2001–03, chairman 2003–)
Schulte-Noelle, Henning (1996–2002, weiterer chairman 1999–2003)
Sedlmayr, Anton (1915–20)
Sedlmayr, Carl (1879–1915, deputy chairman 1910–13)
Sedlmayr, Heinrich (1921–53)
Seipp,* Walter (1983–99)
Staub,* Jakob (1978–2002, deputy chairman 1983–2002)
Streich,* Martin (1983–93)
Strenger, Hermann Josef (1986–2001, chairman 1996–97)

Strube, Jürgen F. (2000–)
Vogelsang, Hans Günter (1981–91)
Weisweiler, Franz Josef (1984–85)
Werner, Winfried (1983–84/1985–86)
Wilhelms, Helmut (1974–80, deputy chairman 1976–78, chairman 1978–80)
Wißmüller,* Johann (1978–83)
Wucherer, Johannes (1972–80, deputy chairman 1973–75)
Wucherer, Rudolf (1955–65)
Zimmermann,* Ottmar (1998–2000)
Zukauski,* Frank (2002–)

* Representatives of the Employees

Linde AG Executive Board members, 1879-2004

Beichert, Karl (1943–51)
Beickler, Joachim (1994–96)
Belloni, Aldo (2000–)
Brahms, Hero (1996–2004)
Diesch, Peter (2003–)
Eggendorfer, Gunnar (1988–99)
Espenmüller, Hermann (1935–44)
Full, Gerhard (1978–2002, CEO 1997–2002)
Goedl, Rainer (1992–99)
Grafoner, Peter (2000–01)
Heitmann, Henrich (1989–90)
Hippenmeyer, Otto (1928–55)
Krebs, August (1902–04)
Krossa, Hubertus (2000–)
Linde, Carl von (1879–90, single board member 1879–90)
Linde, Friedrich (1909–46, CEO 1924–46)
Linde, Hermann (1960–76, spokesman 1972–76)
Linde, Richard (1928–50)
Lohse, Reinhard (1979–91)
Megerlin, Christian (1954–68)
Meinhardt, Hans (1970–97, spokesman 1976–80, CEO 1980–97)

Müller, Joachim (1976–93)
Mundkowski, Rudolf (1993–98)
Münzner, Richard (1935–49)
Müther, Willy (1959–65)
Nesselmann, Kurt (1965–67)
Ombeck, Hugo (1928–61, CEO 1954–61)
Orth, Heinrich (1973–79)
Peltz, Horst (1926–27)
Plötz, Georg (1970–86)
Pöhlein, Max (1963–73)
Reitzle, Wolfgang (2002–, CEO as of 2003)
Ruckdeschel, Walter (1950–70)
Schipper, Friedrich (1890–1928, CEO 1890–1924)
Schling, Falko (1998–2001)
Schmohl, Hans–Peter (2000–03)
Simon, Johannes (1954–71)
Tandler, Gerold (1990–2001)
Volland, Ernst (192–32)
Wagner, Otto (1955–63)
Wucherer, Johannes (1954–72, CEO 1961–72)
Wucherer, Rudolf (1928–54, CEO 1952–54)

Linde AG CEOs and spokesmen*

Linde, Carl von (1879–90)
Schipper, Friedrich (1890–1924)
Linde, Friedrich (1924–46)
Wucherer, Rudolf (1952–54)
Ombeck, Hugo (1954–61)
Wucherer, Johannes (1961–72)

Linde, Hermann (1972–76)
Meinhardt, Hans (1976–97)
Full, Gerhard (1997–2002)
Reitzle Wolfgang (2003–)

* chronologically

Linde AG: A timeline 1879–2004

Purchases and sales of company units and strategic alliances have a tinted background

1879	A joint-stock company in Wiesbaden was established under the name 'Gesellschaft für Linde's Eismaschinen'. Carl Linde is the only member of the Executive Board, five of the six remaining founders form the Supervisory Board.
1880	Linde stock issued; refrigeration machines put into operation in breweries, slaughterhouses and ice factories; construction of Linde's first own ice factory. Rudolf Diesel joins the French branch Fa. Linde.
1882	Construction of the first office building in Wiesbaden begins; presentation of an artificial ice skating track.
1883	First boom for contracts for ice machines. Delivery of the first German meat refrigeration plant.
1885	After a great number of licensees, Linde's first foreign subsidiary is begun with the British Linde Refrigeration Co. in London to sell refrigeration units and construct ice factories in England and the British colonial empire.
1886	Foundation of the Société anonyme des Frigorifères d'Anvers in Belgium.
1888	Delivery of marine refrigeration units; construction of a research station for refrigeration machines in Munich.
1890	Foundation of the Gesellschaft für Markt- und Kühlhallen in Berlin to organize the market of constructing and operating cold houses and ice factories.
1895	Patent granted to Carl Linde for the 'process for liquefaction of atmospheric air or other gases'; two years later, company receives contract for patent utilization.
1897	Creation of two departments: Refrigeration and Gas Liquefaction; Carl Linde is ennobled. Patent registration for Linde's explosive 'Oxyliquit'. Linde's residence in Prinz Ludwigshöhe south of Munich is built.
1900	Construction of a research centre in Höllriegelskreuth near Munich to develop gas liquefaction, from which the divisions of Plant Construction and Industrial Gases emerged. Linde receives Grand Prix at the World Fair in Paris for an air liquifier.
1902	Manufacture of oxygen of adjustable purity.
1903	Linde co-founds Vereinigte Sauerstoffwerke GmbH, Berlin as a trust for selling oxygen from Linde and two other partners. Further Linde gas works established in Barmen, Berlin, Düsseldorf, Mühlheim, Altona, Nuremberg, Dresden as well as in Paris, Toulouse and Antwerp.
1906	Co-foundation of British Oxygen Co. Ltd., London with British partners. Co-foundation of Internationale Sauerstoff-Gesellschaft AG, Berlin, to build oxygen works and sell gases abroad, including Barcelona, Stockholm, and Vienna, and to build subsidiaries and alliances and sell licences in other countries.
1907	Linde launches Linde Air Products Company, Cleveland, USA with American partners.
1908	The first international refrigeration congress in Paris; Carl von Linde lectures on air-conditioning. Linde acquires some stock in the Güldner-Motoren-Gesellschaft mbH in Aschaffenburg, which was founded in 1904 by Hugo Güldner, Carl von Linde and others.
1909	Acetylene business built up.
1912	Delivery of two large installations for nitrogen and hydrogen manufacturing to BASF for ammonia synthesis. Internationale Sauerstoff-Gesellschaft's assets taken over, excluding liquidation.

1918	The First World War prompts a substantial loss in protection rights, important branch offices and partnerships abroad (e.g. Linde Air Products, British Oxygene and others).
1920	Acquisition of Maschinenfabrik Sürth near Cologne.
1922	Linde purchases Heylandt Gesellschaft für Apparatebau mbH in Berlin.
1924	Delivery of the first coke gas separator to Belgium.
1926	Purchase of refrigeration cabinet factory G.H. Walb & Co. in Mainz-Kostheim.
1927	Large fire accident in Sürth.
1929	The Linde Company celebrates 50 years of business.
	Takeover of the Güldner-Motoren Gesellschaft in Aschaffenburg.
1931	Serial construction of small diesel engines at Güldner begun.
	50 per cent stake in Marx & Raube GmbH, Frankfurt/M.
1934	Carl von Linde dies at the age of 92.
	Construction of Cold Store Linde in Munich begun.
	'Carl von Linde Foundation' at the Technical University, Munich is founded.
1935	All shares of Marx & Traube GmbH are acquired; company name changed later to Matra-Werke GmbH.
1936	Opening of the Linde Ice Sports and Swimming Stadium in Nuremberg
1937	Submerged arc welding introduced, 'Ellira'.
1938	Tractors taken into production at Güldner.
1940	Production of commercial refrigeration cabinets in small series up to 300 units is begun at Mainz-Kostheim factory.
1945	Second World War results in destruction of plant facilities in Höllriegelskreuth near Munich, Cologne – Sürth, Mainz-Kostheim, and Aschaffenburg and in renewed loss of nearly all international subsidiaries, investments, and patent rights.
1947	Co-foundation of Sauerstoff Wilhelmshaven GmbH, its first investment after the Second World War.
1949	Central Administration (later Group Headquarters) is established.
1950	The production of household refrigerators is begun in Mainz-Kostheim.
1951	Foundation of the Wohnungsbau GmbH Linde as a non-profit housing company in Wiesbaden.
1952	Foundation of the company's own health insurance, Betriebskrankenkasse Linde.
1953	Completion of Europe's largest air separator for delivery to the USA; air-cooled diesel engines taken into production at Güldner.
1955	Launch of the first Hydrocar, a platform truck equipped with hydrostatic drive.
1956	Co-foundation of the TEGA-Technische Gase GmbH, Obereggendorf in Austria as the first large foreign investment after the Second World War; reconstruction of foreign business is begun.
1957	Rationalization measures lead to relocation of sales and distribution of ready-to-plug-in appliances (in particular household refrigerators) from Sürth to Kostheim, where they are manufactured.
1958	Güldner launches standard production of hydraulic units and industrial trucks (forklifts); Unterstützungseinrichtung GmbH is combined with four benefit associations to create the relief company Unterstützungseinrichtung GmbH der Gesellschaft für Linde's Eismaschinen.
1959	Refrigerator export business is transferred from Sürth to Kostheim.
1960	Commercial refrigeration relocated from Wiesbaden to Sürth.
	Radial turbo fans are taken into production at Sürth.
	Household freezer is taken into production at Kostheim.
1961	Shares offered and sold to employees for the first time.
1964	The factories in Sürth and Mainz-Kostheim merge to form the Refrigeration Division.

Agreement with Union Carbide Corp. USA, which had acquired Linde Air Products (established 1907), on regional rights to the 'Linde' trademark.

1965   The company's name is changed to 'Linde AG'; first step is taken in international plant engineering; cold store constructed north of Munich for the new Cold Stores Division.

Foundation of the Linde Hausgeräte GmbH in Wiesbaden and the Linde Nürnberg GmbH.

Cold storage warehouse at Linde Nürnberg GmbH established.

1967   Linde's house appliance business is sold to AEG (75 per cent ownership), the household refrigerator factory in Mainz-Kostheim is included.

1969   Termination of tractor and diesel engines production at Güldner Division; focus on growth sectors Materials Handling and Hydraulics.

1971   Construction of Linde's own large oxygen plants.

100-year celebration at Sürth.

Major fire at Mainz-Kostheim. Erecting hall, varnish plants and a multiple-storeyed storage building are destroyed.

Purchase of in-store construction company Variant GmbH, Bad Hersfeld to expand in installation engineering.

1972   The division Sürth is split into divisions of Industrial Refrigeration and Cooling and Installation Systems; the division Munich is split into Plant Construction (Engineering) and Industrial Gases.

Linde purchases the Messer Griesheim GmbH division of Cryogenics to strengthen the Plant Construction division, and transfers Ellira (Welding) to Messer Griesheim.

Linde founds Likos AG together with Messer Griesheim for international sales of technical gases.

1973   Purchase of complete stock in SE Fahrzeugwerke GmbH, Hamburg (STILL).

1973   Matra-Werke GmbH restructured as a trade company; The Matra factory in Kahl/Main goes to Güldner Aschaffenburg division.

1974   Group Headquarters in Wiesbaden moves into a new building.

Aeroton Gases Industriais Ltda in Rio de Janeiro established.

Linde Gas Pty. Ltd in Sydney established.

1975   Linde Chemical Engineering & Manufacturing Ltd in South Africa established.

Purchase of stock in Selas–Kirchner GmbH, Hamburg to enhance cracker and thermal engineering, first 35 per cent, finally complete stock purchase in 1985.

1976   Purchase of refrigerator manufacturer Tyler Refrigeration International GmbH in Schwelm to improve market position in commercial refrigerators.

1977   Relocation of the Linde-Kühlmöbel to Schwelm/Westfalen.

Linde purchases majority of shares in the American Baker Material Handling Corporation, Cleveland.

1979   The Linde AG celebrates its 100th year.

Industrial Refrigeration and Cooling and Installation Systems are integrated to form Division of Refrigeration and Installation Engineering; Refrigeration and Air Conditioning, Installation Systems as well as Piston and Turbo Engines are subsumed under Refrigeration and Installation Engineering.

1981   Production of ready-to-plug-in cold shelves, displays and freezer islands for the ice cream industry is resumed at factory 2 at Mainz-Kostheim (after the relocation of commercial refrigerators to Schwelm/Westfalen in 1977).

1984   Acquisition of Fenwick, one of the largest French forklift truck manufacturers and foundation of Fenwick–Linde S.A.R.L.

1986   Acquisition of shares in Wagner Fördertechnik in Reutlingen; Termination of joint international gas marketing with Messer-Griesheim in Likos AG.

1989   Consolidation of forklift manufacturing in the Linde Group – WGA group (Güldner Aschaffenburg, Fenwick–Linde, Lansing Linde) and STILL.
First publication of a worldwide financial report for the group.
Linde takes over the British forklift manufacturer Lansing.

1990   Construction of a trade/commerce network and facilities for Materials Handling and Refrigeration Technology in the former GDR.
The production of commercial refrigerators and freezers is taken into production and plant expanded at Mainz-Kostheim.
Linde receives a contract to build one of the world's largest air separation plants from BASF AG, Ludwigshafen.
Linde-WGA Group renamed Linde-FH Group; divisions and group companies are assigned to Plant Engineering, Materials Handling, Refrigeration, and Industrial Gases.
Purchase of the GDR plant engineering company KCA and foundation of Linde-KCA-Dresden GmbH only four months after the fall of the Berlin Wall.
Co-operation with the Leuna Werke, which relinquishes its Industrial Gases section to Linde.

1991   Complete takeover of Wagner Fördertechnik, Reutlingen.
Founding of Indumat GmbH & Co. KG, Reutlingen, which takes on manufacture of automated guided vehicles from Wagner.

1992   Linde acquires 51 per cent of Italian FIAT OM Carrelli Elevatori and majority shareholding in Italian manufacturer of refrigerator cabinets CRIOSBANC S.p.A..
Increase of participation (held since 1974) to over 60 per cent in the largest supplier of industrial gases in the Netherlands, nv W.A. Hoek's Machine- en Zuurstoffabriek.

1993   Joint venture with the second largest Chinese forklift manufacturer in Xiamen.
Linde produces a turnkey-ready steamcracker for BASF-Antwerp. This was the largest single contract ever received by Linde (contract volume 1.3 billion DM).

1994   Inauguration of the JULI Motorworks in Brno in the Czech Republic, a joint venture between Linde and Jungheinrich for the production of electric engines.
Linde purchases Italian gas company Caracciolossigeno s.r.l.
Founding of joint venture in Dalian, China, with a Chinese manufacturer of system units for the design and construction of air separators in Asia.

1995   Acquisition of PanGas, Switzerland.
After completing the shift to polyurethane insulation foam (Cyclopentan) production at Beroun and Torreglia, Linde becomes one of the first worldwide manufacturers of environmentally sound refrigeration cabinets and systems.

1996   Operations begin at new forklift production factory in Xiamen, China; selling company established in Yokohama for industrial trucks and sales organization expanded throughout Asia.
Acquisition of the refrigeration company Frigorex AG, Lucerne, from the Swiss Sulzer AG, Winterthur.
Linde purchases the US plant construction company The Pro-Quip Corporation (TPQ), Tulsa, Oklahoma, which is world market leader for small hydrogen plants.

1997   Linde takes over Radford Retail Systems, one of the leading suppliers of refrigerated cabinets, shelving and other retail equipment in Great Britain.
Order from the USA (Kansas) for the world's largest liquid helium plant.
Construction of world's four largest air separation plants for the national oil company Pemex in Mexico (contract volume 150 Mio. US$).

1998   Linde acquires a 75 per cent stake in Seral do Brasil S.A., Osasco/Sao Paulo. Seral belongs to the leading suppliers of refrigerated cabinets in Brazil and is market leader for shelving and checkout systems for supermarkets.

Takeover of majority shares in the Chief Group, one of the leading distributors of refrigeration cabinets in France, UK, Belgium, and the Netherlands. Linde takes over Gephal S.A., Paris, the principal shareholder of the Chief Group, Paris. Chief is one of the leading distributors of refrigeration cabinets in France, the UK, Belgium, and the Netherlands.

1999   From 1 January 1999, the North-American Group companies are able to use the Linde name. Linde AG, Wiesbaden, Germany, repossesses the worldwide rights for the 'Linde' name and trademark, 44 years after the war's end.

Linde takes over the Polish industrial gases business of the American Airgas Inc., Radnor, Pennsylvania. With this takeover, Linde becomes the largest supplier of industrial gases in Poland.

2000   On 9 February the European Commission gives the go-ahead for Linde's takeover of the Swedish gas company AGA. With this merger Linde ranks as one of the largest gas suppliers in the world.

Negotiations for a possible takeover of the German gas company Messer Griesheim are terminated by mutual agreement as it is doubtful whether the Commission would give its approval because of tougher competition regulation.

At the end of June Linde acquires the remaining outstanding shares of the Dutch market leader in industrial gases n.v. Hoek Loos. Up to then, Linde had already owned 65 per cent of the company's shares.

2001   Lotepro and Pro-Quip are merged into Linde Process Plants. Together with the gas companies BOC and Praxair, LPP supplies large hydrogen plants to US refineries. LPP is to cover future BOC-US demand for air separation plants.

Linde takes over the majority stake in Ameise Comércio e Indústria S.Al, the leading manufacturer of warehousing equipment in Brazil and importer of forklifts for the regional market.

The European Commission approves the foundation of Supralift GmbH & Co. KG, Hofheim (at Frankfurt/Main). The company is a joint venture of Jungheinrich AG, Hamburg, and the Linde AG, Wiesbaden aiming to create the largest European Internet platform for used industrial trucks (forklift and warehouse trucks).

The Linde Gas AG is reintegrated into Linde AG in October. The spin-off had become necessary because of an envisioned takeover of Messer Griesheim GmbH in order to give Messer Industriegesellschaft mbH direct participation in the gases businesses.

2002   Linde and The BOC Group, Windlesham, UK, announce their co-operation in air separation and synthetic gas plants. According to the terms of the agreement, Linde is the exclusive supplier of plants of this type to BOC, while BOC takes a 30 per cent stake in the Linde Process Plants, Inc., Tulsa, USA. The new company trades under Linde BOC Process Plants LLC, integrates know-how and resources of BOC's Technology Centre at Murray Hill, New Jersey.

2003   Linde bundles its gas activities in the USA. In the future, Group's US gas business will be controlled from Cleveland, Ohio. This means that Linde is bringing the three existing gas companies AGA Gas, Inc. in Cleveland, Holox in Atlanta, Georgia, and Linde Gas in La Porte, Texas, together under one roof. At the same time, the Group's medical gases business will be brought together under one management.

Joint venture is founded between Linde and the Algerian state energy company Sonatrach for tapping and using a helium natural gas source on Algeria's eastern coast. Linde thus becomes one of the world's largest suppliers of helium.

2004   On March 15, 2004, Linde sells its refrigeration branch to the American Carrier Corporation, a subsidiary of United Technologies Corporation in Hartford/Connecticut.

Turnover, profits, and employee numbers, Linde AG 1879–2002

Year	Currency	Turnover[1]	Annual Net Profit/Group Profits	Employees[2]
1879	RM	29.000,00	n.a.	3
1880	RM	n.a.	n.a.	n.a.
1881	RM	n.a.	100.000,00	n.a.
1882	RM	n.a.	200.000,00	n.a.
1883	RM	n.a.	350.000,00	n.a.
1884	RM	n.a.	1.000.000,00	n.a.
1885	RM	n.a.	500.000,00	n.a.
1886	RM	n.a.	650.000,00	n.a.
1887	RM	n.a.	600.000,00	n.a.
1888	RM	n.a.	700.000,00	n.a.
1889	RM	n.a.	890.000,00	n.a.
1890	RM	6.445.998,42	690.155,00	over 70
1891	RM	6.637.206,85	774.635,45	n.a.
1892	RM	4.700.000,00	421.378,18	n.a.
1893	RM	4.500.000,00	378.472,79	n.a.
1894	RM	5.144.000,00	495.710,99	n.a.
1895	RM	5.280.125,00	415.925,69	n.a.
1896	RM	8.902.000,00	790.962,37	n.a.
1897	RM	9.200.000,00	848.798,72	n.a.
1898	RM	13.780.000,00	1.408.456,00	n.a.
1899	RM	11.426.000,00	1.240.418,19	n.a.
1900	RM	9.200.000,00	838.891,50	over 100
1901	RM	6.423.000,00	n.a.	n.a.
1902	RM	5.970.000,00	384.264,01	n.a.
1903	RM	4.250.000,00	136.657,67	n.a.
1904	RM	4.044.000,00	n.a.	n.a.
1905	RM	5.232.000,00	470.058,08	n.a.
1906	RM	6.530.000,00	667.953,06	n.a.
1907	RM	7.600.000,00	714.419,18	n.a.
1908	RM	6.105.000,00	729.329,37	n.a.
1909	RM	4.100.000,00	595.640,72	n.a.
1910	RM	6.273.903,00	812.344,34	over 500
1911	RM	7.500.170,00	997.562,25	n.a.
1912	RM	10.636.677,63	1.197.991,41	n.a.
1913	RM	10.859.394,34	1.199.516,19	n.a.
1914	RM	5.888.500,00	687.385,79	n.a.
1915	RM	7.836.034,00	1.186.420,94	n.a.
1916	RM	9.017.260,00	1.577.056,67	n.a.
1917	RM	9.584.980,00	1.993.532,08	n.a.
1918	RM	9.600.000,00	1.823.512,80	n.a.
1919	RM	8.833.000,00	1.794.568,23	n.a.
1920	RM	72.267.000,00	5.602.218,95	1.500
1921	RM	98.683.830,00	9.875.705,15	n.a.
1922	RM	n.a.	388.142.051,00	n.a.
1923	RM	n.a.	2.675.492.318.375.570.000,00	n.a.
1924	RM	7.900.000,00	2.289.278,30	n.a.
1925	RM	13.100.000,00	1.973.353,17	n.a.
1926	RM	22.000.000,00	1.873.240,05	n.a.
1927	RM	28.500.000,00	2.405.823,68	n.a.

Year	Currency	Turnover[1]	Annual Net Profit/Group Profits	Employees[2]
1928	RM	40.000.000,00	2.793.144,93	n.a.
1929	RM	45.000.000,00	3.200.505,32	n.a.
1930	RM	45.000.000,00	3.191.643,41	2.873
1931	RM	31.000.000,00	1.296.197,30	n.a.
1932	RM	19.000.000,00	1.041.913,06	n.a.
1933	RM	19.000.000,00	1.215.564,79	n.a.
1934	RM	24.000.000,00	1.703.093,26	n.a.
1935	RM	34.000.000,00	1.912.706,01	n.a.
1936	RM	47.700.000,00	2.141.307,09	n.a.
1937	RM	50.000.000,00	1.490.129,60	n.a.
1938	RM	60.000.000,00	1.643.365,94	n.a.
1939	RM	76.000.000,00	1.655.609,50	n.a.
1940	RM	70.000.000,00	1.664.648,52	5.184
1941	RM	89.000.000,00	1.744.725,05	n.a.
1942	RM	86.000.000,00	1.765.466,52	n.a.
1943	RM	93.000.000,00	1.731.031,78	n.a.
1944	RM	95.000.000,00	354.836,63	n.a.
1945	RM	23.000.000,00	-3.678.275,30	1.669
1946	RM	40.800.000,00	-1.971.786,86	n.a.
1947	RM	40.800.000,00	-39.182,97	n.a.
1948	RM	Six months-23.300.000,00	Six months-170.947,28	4.100
1949	DM	87.000.000,00	1.426.182,95	4.991
1950	DM	103.000.000,00	1.436.850,67	5.728
1951	DM	150.000.000,00	1.800.391,87	6.475
1952	DM	182.000.000,00	2.182.788,15	6.641
1953	DM	197.000.000,00	3.081.460,88	6.942
1954	DM	231.000.000,00	3.220.941,22	7.716
1955	DM	283.000.000,00	3.560.497,02	8.761
1956	DM	322.000.000,00	4.416.685,78	9.309
1957	DM	366.000.000,00	6.224.621,13	9.923
1958	DM	392.000.000,00	7.773.136,02	10.339
1959	DM	433.000.000,00	8.794.879,29	10.849
1960	DM	486.648.495,57	9.215.575,22	12.312
1961	DM	534.788.302,71	11.714.665,41	12.305
1962	DM	560.740.512,92	11.725.910,00	12.584
1963	DM	605.480.842,98	11.716.850,00	12.753
1964	DM	661.695.561,80	11.722.710,00	12.954
1965	DM	688.420.988,01	11.726.650,00	13.080
1966	DM	679.931.848,70	9.781.782,00	12.896
1967	DM	649.863.172,00	9.633.060,00	10.219
1968	DM	677.659.035,00	10.974.366,00	10.411
1969	DM	727.106.515,00	11.292.842,00	10.342
1970	DM	760.866.446,00	12.920.327,00	10.865
1971	DM	981.173.403,00	11.206.272,00	11.050
1972	DM	998.302.820,00	12.120.864,00	10.384
1973	DM	1.229.328.497,00	15.887.369,00	15.605
1974	DM	1.297.719.418,00	19.635.346,00	15.481
1975	DM	1.387.363.772,00	22.691.935,00	14.505
1976	DM	1.553.735.231,00	23.908.195,00	14.757
1977	DM	1.667.494.102,00	23.039.130,00	14.905

Year	Currency	Turnover[1]	Annual Net Profit/Group Profits	Employees[2]
1978	DM	1.825.191.136,00	27.604.834,00	15.338
1979	DM	2.012.290.754,00	25.265.645,00	15.679
1980	DM	2.174.775.866,00	32.477.471,00	15.765
1981	DM	2.484.919.674,00	32.531.606,00	15.534
1982	DM	2.512.716.835,00	32.556.509,00	15.094
1983	DM	2.671.146.246,00	32.685.379,00	15.051
1984	DM	2.602.389.033,00	36.119.156,00	13.907
1985	DM	2.707.472.384,00	45.059.409,00	14.362
1986	DM	3.877.000.000,00	57.125.911,00	19.252
1987	DM	4.133.000.000,00	63.657.000,00	19.646
1988	DM	4.505.454.000,00	147.541.000,00	21.222
1989	DM	5.453.144.000,00	186.319.000,00	25.679
1990	DM	6.068.849.000,00	212.315.000,00	27.676
1991	DM	6.911.521.000,00	252.414.000,00	28.535
1992	DM	7.534.257.000,00	254.514.000,00	30.424
1993	DM	7.172.216.000,00	178.158.000,00	29.636
1994	DM	7.967.646.000,00	245.758.000,00	29.618
1995	DM	8.283.757.000,00	358.514.000,00	30.068
1996	DM	8.800.904.000,00	395.924.000,00	30.746
1997	DM	9.545.920.000,00	447.452.000,00	32.112
1998	DM	10.738.326.000,00	507.007.000,00	33.371
1999	Euro	6.193.923.000,00	260.722.000,00	35.597
2000	Euro	8.450.279.000,00	274.377.000,00	47.126
2001	Euro	8.833.000.000,00	240.000.000,00	46.400
2002	Euro	8.726.000.000,00	241.000.000,00	46.521

Source: Linde Business Reports and informations of Linde AG.
[1,2]*As of 1986, values are for the world group.*

Organization charts

Organization Chart 1937

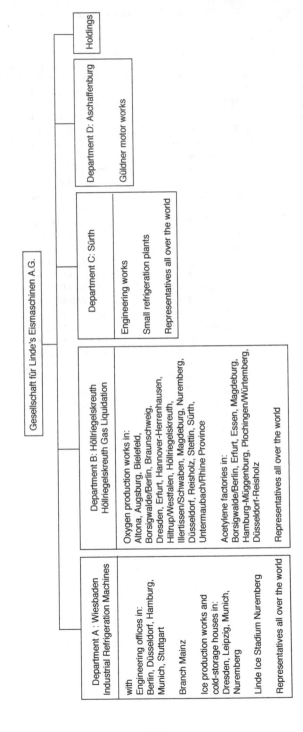

Gesellschaft für Linde's Eismaschinen A.G.

Department A : Wiesbaden
Industrial Refrigeration Machines

with

Engineering offices in:
Berlin, Düsseldorf, Hamburg,
Munich, Stuttgart

Branch Mainz

Ice production works and
cold-storage houses in:
Dresden, Leipzig, Munich,
Nuremberg

Linde Ice Stadium Nuremberg

Representatives all over the world

Department B: Höllriegelskreuth
Höllriegelskreuth Gas Liquidation

Oxygen production works in:
Altona, Augsburg, Bielefeld,
Borsigwalde/Berlin, Braunschweig,
Dresden, Erfurt, Hannover-Herrenhausen,
Hiltrup/Westfalen, Höllriegelskreuth,
Illertissen/Schwaben, Magdeburg, Nuremberg,
Düsseldorf, Reisholz, Stettin, Sürth,
Untermaubach/Rhine Province

Acetylene factories in:
Borsigwalde/Berlin, Erfurt, Essen, Magdeburg,
Hamburg-Müggenburg, Plochingen/Würtemberg,
Düsseldorf-Reisholz

Representatives all over the world

Department C: Sürth

Engineering works

Small refrigeration plants

Representatives all over the world

Department D: Aschaffenburg

Güldner motor works

Holdings

Source: Linde AG archives, Wiesbaden

Organization Chart 1954

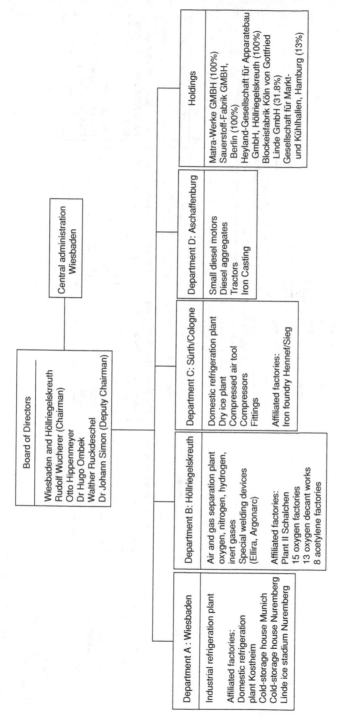

Board of Directors

Wiesbaden and Höllriegelskreuth
Rudolf Wucherer (Chairman)
Otto Hippenmeyer
Dr Hugo Ombek
Walther Ruckdeschel
Dr Johann Simon (Deputy Chairman)

Central administration
Wiesbaden

Department A : Wiesbaden

Industrial refrigeration plant

Affiliated factories:
Domestic refrigeration
plant Kostheim
Cold-storage house Munich
Cold-storage house Nuremberg
Linde ice stadium Nuremberg

Department B: Höllriegelskreuth

Air and gas separation plant
oxygen, nitrogen, hydrogen,
inert gases
Special welding devices
(Ellira, Argonarc)

Affiliated factories:
Plant II Schalchen
15 oxygen factories
13 oxygen decant works
8 acetylene factories

Department C: Sürth/Cologne

Domestic refrigeration plant
Dry ice plant
Compressed air tool
Compressors
Fittings

Affiliated factories:
Iron foundry Hennef/Sieg

Department D: Aschaffenburg

Small diesel motors
Diesel aggregates
Tractors
Iron Casting

Holdings

Matra-Werke GMBH (100%)
Sauerstoff-Fabrik GMBH,
Berlin (100%)
Heyland-Gesellschaft für Apparatebau
GmbH, Höllriegelskreuth (100%)
Blockeisfabrik Köln von Gottfried
Linde GmbH (31.8%)
Gesellschaft für Markt-
und Kühlhallen, Hamburg (13%)

Source: 75 Jahre Linde, 1954, p. 4

341

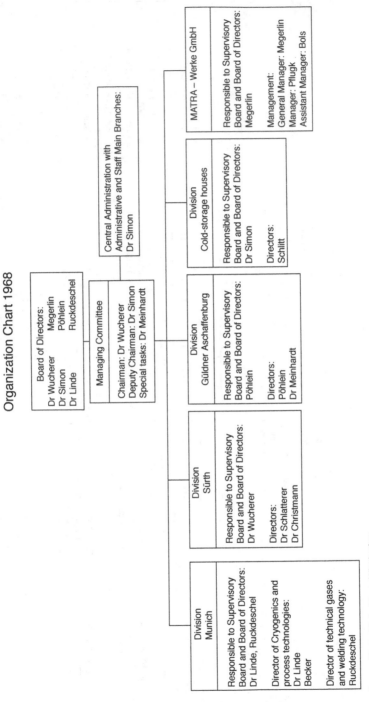

Organization Chart 1968

Board of Directors:
Dr Wucherer   Megerlin
Dr Simon   Pöhlein
Dr Linde   Ruckdeschel

Managing Committee

Chairman: Dr Wucherer
Deputy Chairman: Dr Simon
Special tasks: Dr Meinhardt

Central Administration with
Administrative and Staff Main Branches:
Dr Simon

Division Munich

Responsible to Supervisory
Board and Board of Directors:
Dr Linde, Ruckdeschel

Director of Cryogenics and
process technologies:
Dr Linde
Becker

Director of technical gases
and welding technology:
Ruckdeschel

Division Sürth

Responsible to Supervisory
Board and Board of Directors:
Dr Wucherer

Directors:
Dr Schlatterer
Dr Christmann

Division Güldner Aschaffenburg

Responsible to Supervisory
Board and Board of Directors:
Pöhlein

Directors:
Pöhlein
Dr Meinhardt

Division Cold-storage houses

Responsible to Supervisory
Board and Board of Directors:
Dr Simon

Directors:
Schlitt

MATRA – Werke GmbH

Responsible to Supervisory
Board and Board of Directors:
Megerlin

Management:
General Manager: Megerlin
Manager: Pflugk
Assistant Manager: Bols

Source: Linde AG Archives, Wiesbaden

Organization Chart 1971

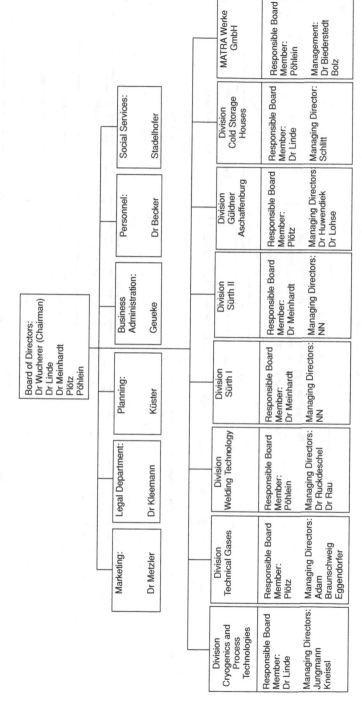

Board of Directors:
Dr Wucherer (Chairman)
Dr Linde
Dr Meinhardt
Plötz
Pöhlein

Marketing:	Legal Department:	Planning:	Business Administration:	Personnel:	Social Services:
Dr Metzler	Dr Kleemann	Küster	Geueke	Dr Becker	Stadelhofer

Division Cryogenics and Process Technologies	Division Technical Gases	Division Welding Technology	Division Sürth I	Division Sürth II	Division Güldner Aschaffenburg	Division Cold Storage Houses	MATRA Werke GmbH
Responsible Board Member: Dr Linde	Responsible Board Member: Plötz	Responsible Board Member: Pöhlein	Responsible Board Member: Dr Meinhardt	Responsible Board Member: Dr Meinhardt	Responsible Board Member: Plötz	Responsible Board Member: Dr Linde	Responsible Board Member: Pöhlein
Managing Directors: Jungmann Kneissl	Managing Directors: Adam Braunschweig Eggendorfer	Managing Directors: Dr Ruckdeschel Dr Rau	Managing Directors: NN	Managing Directors: NN	Managing Directors: Dr Huwendiek Dr Lohse	Managing Director: Schlitt	Management: Dr Biederstedt Bolz

Source: Linde AG Archives, Wiesbaden

Organization Chart 1989

Dr Meinhardt (Chairman)
Central Administration
Executive

Board of Directors:

Full	Müller
Refrigeration, Hydraulics,	Plant Engineering and
Central Technical Functions	Process Technology
Dr Lohse (Labour Director)	Patents
Material Handling	Dr Eggendorfer (Deputy)
Personnel	Technical Gases

Central Administration

Schmitz:	Accounting, Holdings, Taxes, Business Management
	Finance
Dr Nolte:	Marketing, Public Relations, Planning
(Dr Meinhardt)	Personnel
Dr Becker	Legal Department
Haeseler	Inspections, Insurance
Dr Rau	Engineering and Organization
(Full)	

Division Refrigeration and Equipment Technologies

Refrigeration

Managing Directors:
Dietter
Dr Thielmann

Factory Group TVT Munich

Plant Engineering and Process Technology Patents

Managing Directors:
Dorner
Jungmann
Kersten
Wege

Division Güldner Aschaffenburg

Material Handling and Hydraulics

Managing Directors:
Dr Megerlin
Dr Metzler
Mundkowski

Factory Group Technical Gases

Technical Gases

Managing Directors:
Braunschweig
Dr Goedl
Dr Klauser

Still GmbH*

Material Handling

Managing Directors:
Fillbach
Schröder
Trainer

* Most important company in group, not assigned to a factory group
Source: Jahresbericht Linde AG, 1988, p. 6

Organization Chart 1995

Dr Meinhardt (Chairman)
Central Administration
Executive

Board of Directors:
Beickler (Deputy)
Accounting, Taxes, Business Management,
Auditing
Dr Goedl (Deputy)
Material Handling (S)*
Mundkowski (Deputy)
Material Handling (E)*
Central Technical Functions

Dr Eggendorfer
Technical Gases
Full
Plant Engineering, Patents
Tandler
Refrigeration
Finance
Personnel (Labour Director)

Central Administration

(Dr Meinhardt): Marketing, Public Relations, Planning
Department Management
Inspections, Business Management
Schmitz   Accounting, Taxes
Dr Gröhn   Personnel
Dr Klingelhöfer Legal Department
Haeseler   Insurance
Dr Rau   Engineering, Organization,
Fillbach   Information Systems
(Provisional)

Construction

Factory Group
Process Technology
and Construction

Director:
Dr Belloni
Dorner (Chairman)
Habicht
Wege

Material Handling

Factory Group
Material Handling and
Hydraulics

Director:
Dr Klasper
Klein
Dr Megerlin

Sales and Distribution East Asia:
Winzl

Refrigeration

Factory Group
Refrigeration and
Equipment Technologies

Director:
Dr Schuppar
Dr Thielmann

Technical Gases

Factory Group
Technical Gases

Director:
Braunschweig
Dr Metzler

Still GmbH**

Management:
Pfeiffer
Schröder
Trainer

Fiat OM Carrelli
Elevatori SpA**

Management:
Dr Prati

---

nv WA Hoek's Machine-
en Zuurstoffabriek**

Board of Directors:
Jager Bruining
Van Tienhoven

---

* (S) = Sales and Distribution Department, (E) = Engineering
** Important company in group, not assigned to a factory group
Source: Jahresbericht Linde AG 1994, p. 6

Organization Chart 2000

Gerhard Full (Chairman)
Technical Gases
Central Administration, Executive

Board of Directors:

Dr Aldo Belloni	**Falko Schling**
Construction, Patents	Refrigeration (E)*,
Hero Brahms	Central Technical Functions
Accounting, Taxes	**Dr Hans-Peter Schmohl**
Business Management, Inspections	Material Handling (E)*
Hubertus Krossa	**Gerold Tandler**
Marketing, Material Handling (S)*	Refrigeration (S)**
	Finance, Personnel (Director of Works)

Central Administration

(Gerhard Full)	Organization, Information Systems,
	Corporate Communications, Planning
Heinrich Schmitz	Business Management, Inspections
Dr Klaus Gröhn	Accounting, Taxes
Erhard Wehlen	Finances
(Hubertus Krossa)	Marketing
Dr Lutz Klinggelhöfer	Personnel
Hans-Georg Haeseler	Legal Department
Stephan Schaller	Engineering

Construction

Factory Group
Process Technology
and Construction

Directors:
(Dr Aldo Belloni)
Franz Habicht
Ulrich Wege

Material Handling

Factory Group
Material Handling and
Hydraulics

Directors:
Erwin Bruckmoser
Dr Ferdinand Megerlin
(Spokesman)
Ralf Mock

Still GmbH
Management:
Horst Peter Jäger
Norbert Pfeiffer
Jens Reinecke

**Fiat OM Carrelli
Elevatori SpA**
Management:
Udo Kleine

Refrigeration

Factory Group
Refrigeration and
Equipment
Technologies

Directors:
(Hubertus Krossa)
Dr Helmut Schuppar

Technical Gases

Linde Technical
Gases GmbH

Directors:
Dr Gunnar Eggendorfer
Dr Rainer Goedl
Dr Folker Metzler

AGA AB
Chief Executive Officer:
Lennart Selander

**nv WA Hoek's Machine-
en Zuurstoffabriek**
Board of Directors:
Bernard Fortuyn
Fred Van Beuningen

* (E) = Engineering, * (S) = Sales and Distribution Department,
** Important company in group
Source: Das Geschäftsjahr 1999 – Geschäftsbericht der Linde
Aktiengesellschaft, p. 6

Printed sources and literature

Abels, T., Hydrostatisches Gabelstaplergetriebe mit Primär und Sekundärverstellung' *Ölhydraulik und Pneumatik. Unabhängige Zeitschrift für Kraftübertragung, Regelung und Steuerung* 15 (1971) 51–5.

Abelshauser, W. (ed.) *Die BASF – Eine Unternehmensgeschichte* (München 2002).

Adreßbuch der Bierbrauer des gesamten Deutschland und Österreich-Ungarn (Berlin 1874).

AGA Annual Report (1913–98).

AGA Gas GmbH *AGA – 75 Jahre Fortschritt* (Hamburg o. J. [1973]).

AGA Now. Internal Information Nos 1–11 (1997–2000).

Agthe, K. *Unternehmensplanung* (Baden-Baden 1963).

Ahlborn AG (ed.) *Einhundert Jahre Ahlborn* (Hildesheim 1956).

Albach, H. *Beiträge zur Unternehmensplanung* (Wiesbaden 1969).

Albrecht, U./Heinemann-Grüder, A./Wellmann, A. *Die Spezialisten. Deutsche Naturwissenschaftler und Techniker in der Sowjetunion nach 1945* (Berlin 1992).

Albring, W. *Gorodomlia. Deutsche Raketenforscher in Rußland* (Hamburg/Zürich 1991).

Almqvist, E. *Technological Changes in a Company. AGA, the First 80 years* (Lidingö 1992, first published in *Daedalus*, the Technical Museum annual, 1992).

Altenkirch, E. 'Reversible Absorptionsmaschinen und Resorptionsmaschinen' *Zeitschrift für Eis- und Kältetechnik* 6 (1913/14) 29–34.

Ambrosius, G./Kaelble, H. 'Gesellschaftliche und wirtschaftliche Folgen des Booms der 1950er und 1960er-Jahre (Einleitung)', in Kaelble, H. (ed.) *Der Boom 1948–1973. Gesellschaftliche und wirtschaftliche Folgen in der Bundesrepublik Deutschland und in Europa.* (Opladen 1992) 10–35.

Anderson, John W. *Refrigeration* (London 1908).

Andritzky, M. *Oikos. Von der Feuerstelle zur Mikrowelle. Haushalt und Wohnen im Wandel* (Gießen 1992).

Appel, S. K. 'Artificial refrigeration and the architecture of 19th century American breweries' *Industrial Archeology* 16 (1990) 21–39.

Artmann, F. *Die Lehre von den Nahrungsmitteln, ihrer Verfälschung und Conservierung* (Prag 1859).

Barqây, A. *Das Wirtschaftssystem des Nationalsozialismus – Ideologie, Theorie, Politik 1933– 1939* (Frankfurt a.M. 1988).

Barth, E. *Entwicklungslinien der deutschen Maschinenbauindustrie von 1870–1914* (Berlin 1973).

Barth, E. 'Zur Geschichte der Maschinenfabrik Germania, Karl-Marx-Stadt' *Sächsische Heimatblätter* 11 (1965) 273–83.

Bäumler, E. *Ein Jahrhundert Chemie* (Düsseldorf 1963).

Behringer, W. *Löwenbräu – Von den Anfängen des Münchner Brauwesens bis zur Gegenwart* (München 1989).

Belani, A. 'Absorptionskältemaschinen' *Zeitschrift des Vereins Deutscher Ingenieure* 36 (1892) 711–18, 741–6.

Berghoff, H. *Zwischen Kleinstadt und Weltmarkt. Hohner und die Harmonika 1857–1961. Unternehmensgeschichte als Gesellschaftsgeschichte* (Paderborn 1997).

Bergmeier, M. *Umweltgeschichte der Boomjahre 1949–1973 – Das Beispiel Bayern* (Münster 2002).

Bertz, I. *Keine Feier ohne Meyer. Die Geschichte der Firma Hermann Meyer & Co. 1890–1990* (Berlin 1990).

Beutler E. 'Wesen, Aufgaben und Begriffe der Wehrwirtschaft' *Betrieb und Wehr. Monatsblätter für wehrwirtschaftliche und wehrpsychologische Betriebsführung und Arbeitsgestaltung. Beiblatt zu Zeitschrift Deutsche Technik* Folge 18, Juni (1937) 301–5.

Blauhorn, K. *Erdteil zweiter Klasse? – Europas technologische Lücke* (Gütersloh 1970).

Bode, V./Kaiser, G. *Raketenspuren. Peenemünde 1936–1996* (Augsburg 1997).

Bölkow, L. *Energie im nächsten Jahrhundert* (Melle 1987).

Bölkow, L. *Entscheidungen für eine langfristige Energiepolitik* (Ottobrunn 1982).

Borscheid, P. 'Der ökonomische Kern der Unternehmensgeschichte' *Zeitschrift für Unternehmensgeschichte* 46 (2001) 5–11.

Braun, H.-J. *An Uneasy Relationship. Technology Transfer between the Sulzer Co. in Winterthur, Switzerland and the Busch-Sulzer Brothers Diesel Engine Co.* (St Louis 1911–36. Hamburg 1991).

Brüggemeier, F.-J./Rommelspacher, T. *Blauer Himmel über der Ruhr -Geschichte der Umwelt im Ruhrgebiet 1840–1990* (Essen 1992).

Büchmann, R. 'Steigende Weltgeltung der deutschen Industrie im Jahre 1937' *Deutsche Technik, Amtliches Organ des Hauptamtes für Technik der Reichsleitung der NSDAP*, 5 (1937) 588–90.

Büchner, F. *100 Jahre. Geschichte der MAN 1840–1940* (Augsburg/Nürnberg 1940).

Butrica, A. *Out of Thin Air. A History of Air Products and Chemical Inc. 1940–1990* (New York/London 1990).

Bützler 'Die Notwendigkeit der Gefrierhäuser für die Fleischversorgung' *Zeitschrift für Eis- und Kälteindustrie*, 10 (1917) 1–4.

Carson, R. *Der stumme Frühling* (München 1962), *Silent Spring* (Boston 1962).

Chandler, A. D./Hikino T. *Scale and Scope. The Dynamics of Industrial Capitalism.* (Cambridge/ London 1990).

Clausius, R. *Abhandlungen über die mechanische Wärmetheorie.* 2 Bände (Braunschweig 1864–67).

Cooper, G. *Air-conditioning America. Engineers and the Controlled Environment, 1900–1960.* (Baltimore 1998).

Cummings, R. O. *The American Ice Harvests. A Historical Study in Technology, 1800–1918* (Berkeley 1949).

Davies, M. 'William Hampson (1854–1926). A note' *Journal for the History of Science*, 22 (1989) 63–73.

Deutscher Kälteverein (ed.) *Bilder aus der Deutschen Kälteindustrie. Stiftung zum III. internationalen Kältekongreß in Chicago 1913* (München/Berlin 1913).

Dienel, C. *Frauen in Führungspositionen in Europa* (München 1996).

Dienel, H.-L. 'Das Bild kleiner und mittlerer Unternehmen in der bundesdeutschen Forschungs- und Wirtschaftspolitik 1949–1999' Reith, R./Schmidt, D. (ed.) *Kleine Betriebe, angepasste Technologie? Hoffnungen, Erfahrungen und Ernüchterungen aus sozial- und technikhistorischer Sicht* (Münster/New York 2002) 100–23.

Dienel, H.-L. (ed.) *Der Optimismus der Ingenieure. Triumph der Technik in der Krise der Moderne um 1900* (Stuttgart 1998).

Dienel, H.-L. 'Die deutsch-ungarische Zusammenarbeit im Nutzfahrzeugbau 1945–1995 – Eine vergleichende Betrachtung industrieller Kooperationen im Rahmen des RGW und auf der Basis bilateraler Abkommen zwischen Ungarn und beiden deutschen Staaten', Fischer, H. (ed.) *Deutsch-ungarische Beziehungen in Naturwissenschaft und Technik nach dem zweiten Weltkrieg* (München 1999) 481–521.

Dienel, H.-L. 'Eis mit Stiel. Die Eigenarten deutscher und amerikanischer Kältetechnik' Centrum für Industriekultur Nürnberg/Münchner Stadtmuseum (ed.) *Unter Null. Kunsteis Kälte und Kultur* (München 1991) 101–11.

Dienel, H.-L. *Herrschaft über die Natur? Naturvorstellungen deutscher Ingenieure 1871–1914* (Stuttgart 1992).

Dienel, H.-L. 'Praktiker und Theoretiker in der technischen Thermodynamik Die Unternehmeringenieure Rudolf Diesel und Carl von Linde' Schneider, I./Trischler, H./ Wengenroth, U. (ed.) *Oszillationen. Naturwissenschaftler und Ingenieure zwischen Forschung und Markt* (München 2000) 237–67.

Dienel, H.-L. 'Räumliche Bedingungen heterogener Forschungskooperationen' Strübing, J. et al. (ed.) *Kooperation im Niemandsland. Neue Perspektiven auf die Zusammenarbeit in Wissenschaft und Technik* (Opladen 2004) 210–32.

Dienel, H.-L. 'Techniktüftler? Forschung und Technik in der mittelständischen Industrie' Frieß P./Steiner P. (ed.) *Forschung und Technik in Deutschland nach 1945* (München 1995) 170–85.

Diesel, E. *Diesel. Der Mensch, das Werk, das Schicksal* (München 1983, orig. 1937).

Doederlein, G. 'Künstliche Eislaufbahnen' *Zeitschrift für die gesamte Kälteindustrie* 5 (1898) 77–8.

Dräger, H. *Lebenserinnerungen* (Hamburg 1915).

Drägerwerk (ed.) *Der Retter Sauerstoff* (Lübeck 4th edn 1907).

Drägerwerk (ed.) *Laboratorium für technische Physik, Chemie und Physiologie. Fabrik und Konstruktionsanstalt für Sauerstoffapparate* (Lübeck 1910).

Eisenmann, Harry J. *Charles E. Brush Pioneer Innovation in Electrical Technology* Diss. Case Western Reserve University (Cleveland 1976).

Eizenstat, Stuart E. *Unvollkommene Gerechtigkeit. Der Streit um die Opfer von Zwangsarbeit und Enteignung* (München 2003).

Erker, P. *Dampflok, Daimler, DAX – Die deutsche Wirtschaft im 19. und 20. Jahrhundert* (Stuttgart/München 2001).

Erker, P./Pierenkemper, T. (ed.) *Deutsche Unternehmer zwischen Kriegswirtschaft und Wiederaufbau – Studien zur Erfahrungsbildung von Industrie-Eliten* (München 1999).

Esser, C. *Die 'Nürnberger Gesetze' oder die Verwaltung des Rassenwahns 1933–1945* (Paderborn 2002).

Everson, J. G. *Tidewater Ice of the Kennebec River* (Freeport/Maine 1981).

Feldman, G. D. *Die Allianz und die deutsche Versicherungswirtschaft 1933–1945* (München 2001).

Feldman, G. D. *Hugo Stinnes. Biographie eines Industriellen 1870–1924* (München 1998).

Feldman, G. D. *Vom Weltkrieg zur Weltwirtschaftskrise. Studien zur deutschen Wirtschafts- und Sozialgeschichte 1914–1932* (Göttingen 1984).

Feldman, G./James, H. *Deutschland in der Weltwirtschaftskrise 1924–1936* (Stuttgart 1988).

Forstmann, W. 'Die Chemische Fabrik Griesheim in der großen Depression' *Zeitschrift für Unternehmensgeschichte* 26 (1981) 42–60.

Frischer, D. *Le Moïse des Amériques. Vies et œuvres du munificent Baron de Hirsch* (Paris 2002).

Gall, L. (ed.) *Die Deutsche Bank 1870–1995* (München 1995).

Gall, L. *Krupp im 20. Jahrhundert. Vom ersten Weltkrieg bis zur Gründung der Stiftung* (Berlin 2002).

Gall, L./Pohl, M. (ed.) *Unternehmen im Nationalsozialismus* (München 1998).

Ganzenmüller, T. 'Über den gegenwärtigen Stand der Kältetechnik in der Brauerei' *Zeitschrift für die gesamte Kälteindustrie* 16 (1909) 189–94.

Gebrüder Sulzer (ed.) *100 Jahre Gebrüder Sulzer 1834–1934* (Winterthur 1934).

Gesellschaft für Geschichte und Bibliographie des Brauwesens e.V. (ed.) *Carl von Lindes Kältemaschine und ihre Bedeutung für die Entwicklung der modernen Lagerbierbrauerei* (Berlin 1929).

Gesellschaft für Linde's Eismaschinen (ed.) *50 Jahre Kältetechnik 1879–1929. Geschichte der Gesellschaft für Linde's Eismaschinen AG* (Wiesbaden 1929).

Gesellschaft für Linde's Eismaschinen (ed.) *Arbeit bei Linde* (Wiesbaden o.J. [1961]).

Gesellschaft für Linde's Eismaschinen (ed.) *Lowest Temperatures in Industry. Presented to the Members of the III. International Congress of Refrigeration* (München 1913).

Gesellschaft für Linde's Eismaschinen AG (ed.) *75 Jahre Linde* (Wiesbaden 1954).

Gesellschaft für Linde's Eismaschinen AG (ed.) *Das ist Linde* (Wiesbaden 1964).

Gilroy, B. M. *Networking in Multinational Enterprises. The Importance of Strategic Alliances* (Columbia 1993).

Goschler, C./Ther, P. (ed.) *Raub und Restitution. Zur 'Arisierung' und Rückerstattung jüdischen Eigentums in Europa* (Frankfurt 2003).

Göttsche, Georg *Die Kühlmaschinen und ihre Anlagen* (Hamburg 1915).

Gräffkes, van 'Zur Geschichte der Schweißbrenner' *Autogene Metallbearbeitung* 12 (1909) 91–3.

Gries, R./Ilgen, V./Schindelbeck, D. 'Kursorische Überlegungen zu einer Werbegeschichte als Mentalitätsgeschichte' Gries, R./Ilgen, V./Schindelbeck, D. *Ins Gehirn der Masse kriechen! Werbung und Mentalitätsgeschichte* (Darmstadt 1995) 1–28.

Groß, A. *50 Betriebsjahre der Maschinenfabrik Esslingen 1846–1897* (Esslingen 1897).

Gumz, W./Regul, R. *Die Kohle. Entstehung, Eigenschaften, Gewinnung und Verwendung* (Essen 1954).

Haase, R. *Das bißchen Haushalt. Zur Geschichte der Technisierung und Rationalisierung der Hausarbeit.* (Stuttgart 1992).

Hachmeister, L. 'Die Rolle des SD-Personals in der Nachkriegszeit. Zur nationalsozialistischen Durchdringung der Bundesrepublik' *Mittelweg 36* (2002), H. 2, 17–35.

Hachtmann, R. *Industriearbeit im 'Dritten Reich'. Untersuchungen zu den Lohn- und Arbeitsbedingungen in Deutschland 1933–1945* (Göttingen 1989).

Hall, H. 'The ice-industry of the United States' United States Department of the Interior. Census Office (ed.) *Report on Power and Machinery Employed in Manufactures* (Washington 1888, Reprint Northhampton 1974).

Haltmeier, A. *Auskühlung ebener und zylindrischer Wände aus dem Beharrungszustand* (München 1926).

Handbuch der großen GmbH, KG, OHG, Einzelfirmen *Die Großunternehmen im Deutschen Reich*, Bd. 7 (Berlin 1944).

Handbuch der süddeutschen Aktiengesellschaften und der an süddeutschen Börsen kurshabenden Staatspapiere (Berlin/Leipzig/Hamburg 29th edn 1912).

Hanslian, R. *Der chemische Krieg. Gasangriff, Gasabwehr und Raucherzeugung* (Berlin 1925).

Hård, M. *Machines are Frozen Spirit. The Scientification of Refrigeration and Brewing in the 19th Century. A Weberian Interpretation* (Frankfurt/Boulder 1994, = reworked version of *In the Icy Waters of Calculation. The Scientification of Refrigeration Technology and the Rationalization of the Brewing Industry in the 19th Century.* Diss. Göteborg 1988).

Hardach, K. *Wirtschaftsgeschichte Deutschlands im 20. Jahrhundert* (Göttingen 1976).

Haßler, F. *Geschichte der L.A. Riedinger Maschinen und Bronzewarenfabrik AG* (Augsburg 1928).

Haubold GmbH (ed.) *100 Jahre Haubold* (Chemnitz 1938).

Haubold, S. *Entwicklung und Organisation einer Chemnitzer Maschinenfabrik* Diss. (Bonn 1939).

Hausbrand, E. *Die Wirkungsweise der Rektifizier- und Destillierapparate mit Hilfe einfacher mathematischer Betrachtungen dargestellt* (Berlin 1903).

Hausen, H. 'Austauschvorgänge bei der Zerlegung von Gasgemischen' *Zeitschrift für die gesamte Kälteindustrie* 43 (1937) 248–53.

Hausen, H. *Der Thompson/Joule-Effekt und die Zustandsgrößen der Luft bei Drücken bis zu 200 at und Temperaturen zwischen 10 Grand und -175 Grad Celsius* (Berlin 1926).

Hausen, H. 'Die physikalischen Grundlagen der Gasverflüssigung und Rektifikation' Gesellschaft für Linde's Eismaschinen AG (ed.) *Geschichte der GFLE Wiesbaden 1879 – 50 Jahre Kältetechnik* (Wiesbaden 1929) 135–63.

Hausen, H. 'Gedanken und Erkenntnisse Carl von Lindes auf dem Gebiete der Luftverflüssigung und Gaszerlegung' *Zeitschrift für die gesamte Kälteindustrie* 42 (1935) 220–5.

Hausen, H. 'Über die Theorie des Wärmeaustausches in Regeneratoren' *Zeitschrift für angewandte Mathematik und Mechanik* 9 (1929) 173–200.

Hausen H./Linde H. *Tieftemperaturtechnik. Erzeugung sehr tiefer Temperaturen, Gasverflüssigung und Zerlegung von Gasgemischen* (Berlin 2nd edn 1985).

Häutle, C. *75 Jahre Schlacht- und Viehhof München 1878–1953* (München 1953).

Hayes, P. *IG Farben in the NAZI Era* (New York 1987).

Heckhorn, E./Wiebe H. *München und sein Bier. Vom Brauhandwerk zur Bierindustrie* (München 1989).

Heinel, C. *Bau und Betrieb von Kältemaschinenanlagen. Zahlenstoff und Winke für Ingenieure* (München 1908 (= Diss. TH Berlin 1906).

Heinzerling; C. *Austreibung von Naphthalin aus dem Stadtgas, in Zeitschrift für die gesamte Kälteindustrie* 5 (1898) 157–9.

Heiss, R. 'Einrichtungen und Aufgaben des Reichsinstituts für Lebensmittelfrischhaltung' Plank, Rudolf (ed.) *Festschrift zum 10jährigen Bestehen des Kältetechnischen Instituts der Technischen Hochschule Karlsruhe 1926–1936* (Dissen 1936) 55–63.

Heller, D. *Zwischen Unternehmertum, Politik und Überleben – Emil G. Bührle und die Werkzeugmaschinenfabrik Oerlikon, Bührle & Co. 1924–1945* (Frauenfeld 2002).

Hellmann, U. 'Höchst unauffällig. Der Aufstieg des Kühlschranks zur Unabdingbarkeit', Centrum für Industriekultur Nürnberg/Münchner Stadtmuseum (ed.) *Unter Null. Kunsteis Kälte und Kultur* (München 1991).

Hellmann, U. *Künstliche Kälte. Die Geschichte der Kühlung im Haushalt* (Gießen 1990).

Hentschel, V. *Wirtschaft und Wirtschaftspolitik im wilhelminischen Deutschland. Organisierter Kapitalismus und Interventionsstaat?* (Stuttgart 1978).

Hentschel, V. *Wirtschaftsgeschichte der Maschinenfabrik Esslingen AG 1846–1918* (Stuttgart 1977).

Herbert, U. *Fremdarbeiter. Politik und Praxis des 'Ausländereinsatzes' in der Kriegswirtschaft des Dritten Reiches* (Berlin 1985).

Herbert, U. (ed.) *Wandlungsprozesse in Westdeutschland. Belastung, Integration, Liberalisierung 1945–1980* (Göttingen 2002).

Herbst, L. 'Die nationalsozialistische Wirtschaftspolitik im internationalen Vergleich' Benz, W. et al (ed.) *Der Nationalsozialismus. Studien zur Ideologie und Herrschaft* (Frankfurt a. M. 1993) 153–76.

Herrmann, K. *Traktoren in Deutschland 1907 bis heute* (Frankfurt a. M. 1987).

Herzog, M. *Strukturwandel der Linde AG als Lernprozess Überschneidung von (modifizierter) Kontingenztheorie und Theorie des Organisationslernens.* Diplomarbeit (Berlin 2002).

Hesselmann, H. *Das Wirtschaftsbürgertum in Bayern 1890–1914. Ein Beitrag zur Analyse der Wechselbeziehungen zwischen Wirtschaft und Politik am Beispiel des Wirtschaftsbürgertums im Bayern der Prinzregentenzeit* (Wiesbaden 1985).

Heuss, T./Bosch, R. *Leben und Leistung* (Stuttgart 8th edn 2002, at first 1946).

Heylandt, P. 'Die Verwertung der Luft' *Chemikerzeitung* (1937) 10–11.

Hochgesand, C. P. 'Das Linde-Fränkl-Verfahren zur Zerlegung von Gasgemischen' *Mitteilungen aus den Forschungsanstalten des Gutehoffnungshütte-Konzerns* 4 (1935) 14–23.

Hoerbiger Ventilwerke AG (ed.) *70 Jahre Hoerbiger Ventil* (Wien 1966).

Hoffmann, D. *Acht Jahrzehnte Gefrierverfahren. Ein Beitrag zur Geschichte des Schachtabteufens in schwierigen Fällen* (Dortmund 1962).

Hoffmann, D. *Aufbau und Krise der Planwirtschaft. Die Arbeitskräftelenkung in der SBZ/DDR 1945–1963* (München 2002).

Hoffmann, W. G. *Das Wachstum der deutschen Wirtschaft seit der Mitte des 19. Jahrhunderts* (Berlin and others 1965).

Hoffmeyer, E. *Die amerikanische Herausforderung und die Theorie der Forschungsintensität* (Tübingen 1969).

Hohensee, J. *Der erste Ölpreisschock 1973/74. Die politischen und gesellschaftlichen Auswirkungen der arabischen Erdölpolitik auf die Bundesrepublik Deutschland und Westeuropa* (Stuttgart 1996).

Holzner, G. 'Mitteilungen über den verstorbenen Gabriel Sedlmayr' *Zeitschrift für das gesamte Brauwesen* 15 (1892) 5–6.

Hopkins, J. T. *50 Years of Citrus. The Florida Citrus Exchange 1909–1959* (Gainesville/Fl. 1960).

Horn, J. *Zukunftsgestaltung durch Unternehmensplanung* (München 1967).

Hounshell, D. H./Smith, J. K. *Science and Corporate Strategy. Du Pont R&D 1902–1980* (Cambridge 1988).

Hughes, T. P. 'Das technologische Momentum in der Geschichte. Zur Entwicklung des Hydrierverfahrens in Deutschland 1898–1933' Hausen, K./Rürup, R. (ed.) *Moderne Technikgeschichte* (Köln 1975) 358–83.

Hughes, T. P. *Die Erfindung Amerikas* (München 1991).

Ingels, M. *Willis Haviland Carrier. Father of Air Conditioning* (Garden City 1952).

Jäckel, R. *Statistik über die Lage der technischen Privatbeamten in Groß-Berlin* (Berlin 1907).

Jaeger, E. *Aus der Geschichte der Werksgruppe TVT.* Maschr.schr. (München 1979).

Jaeger, E. *Über Mathias Fränkl und die Anfänge des Linde-Fränkl-Verfahrens* Masch. Schr. (Höllriegelskreuth 1966).

Jaeger, G. *Der Auslandsbezug des Betriebsverfassungsgesetzes. Die Anwendbarkeit des Betriebsve rfassungsgesetzes auf ausländische Betriebe und auf die Ausstrahlungen inländischer Betriebe ins Ausland* (Frankfurt a. M. 1983).

Jaeger, H. *Geschichte der Wirtschaftsordnung in Deutschland* (Frankfurt a. M. 1990).

Jakobsmeier, W. *Linde Werksgruppe Verfahrenstechnik und Anlagenbau. Werbung und Public Relation.* Typescript (Höllriegelskreuth o. J.)

James, H. *Die Deutsche Bank und die 'Arisierung'* (München 2001).

Jerchow, F. *Deutschland in der Weltwirtschaft 1944–1947. Alliierte Deutschland- und Reparationspolitik und die Anfänge der westdeutschen Außenwirtschaft* (Düsseldorf 1978).

Jones, J. C. *American Iceboxes* (Humble/Texas 1981).

Jones, J. C. *America's Icemen. An Illustrative History of the United States Natural Ice Industry 1665–1925* (Humble/Texas 1984).

Jungbluth, R. *Die Quandts. Ihr leiser Aufstieg zur mächtigsten Wirtschaftsdynastie Deutschlands* (Frankfurt a. M. 2002).

Jungnickel, H. 'Carl von Linde, ein Pionier der Kältetechnik' *Luft- und Kältetechnik* 28 (1992) 55–9.

Kaelble, H. *Industrielle Interessenpolitik in der wilhelminischen Gesellschaft – Der Centralverband deutscher Industrieller 1895–1914* (Berlin 1967).

Karlsch, R. *Allein bezahlt? Die Reparationsleistungen der SBZ/DDR 1945–1953* (Berlin 1993).

Karlsch, R./Stokes, R.G. *Faktor Öl. Die Mineralölwirtschaft in Deutschland 1859–1974* (München 2003).

Kessler *Aus den Anfängen der Maschinenfabrik Esslingen. Emil Keßler und sein Werk* (Esslingen 1938).

Kieser A. (ed.) *Organisationstheorien* (Stuttgart 1999).

Kiesewetter, H. *Industrielle Revolution in Deutschland. 1815–1914* (Frankfurt a. M. 1995).

Kirchner, U. *Der Hochtemperaturreaktor – Konflikte, Interessen, Entscheidungen* (Frankfurt a. M. 1991).

Klein, H. *Das Geld liegt auf der Straße. Praxis und neue Techniken der Verkaufsförderung* (Düsseldorf 1963).

Kling, Michael 'Borsig. Ein Abriß der Unternehmensgeschichte' Technische Universität Berlin (ed.) *Berlin. Von der Residenzstadt zur Industriemetropole* (Berlin 1981), Band 3, 193–8.

Klöckner-Humboldt Deutz AG (ed.) *100 Jahre Humboldt* (Köln 1956).

Knappich, J. 'Die autogene Schweißung' *Südwestdeutsche Industriezeitung* (1909) 248–52.

Knapsack-Griesheim AG (ed.) *Unter uns.* Werkzeitung. 1 (1951)–4 (1961).

Knie, A. *Diesel. Karriere einer Technik. Genese und Formierungsprozesse im Motorenbau* (Berlin 1991).

Kocka, J. *Die Angestellten in der deutschen Geschichte 1850–1980* (Göttingen 1981).

Kocka, J. *Unternehmer in der deutschen Industrialisierung* (Göttingen 1975).

Kocka, J. *Unternehmer in Deutschland seit 1945* (Essen 2002).

König, W. *Geschichte der Konsumgesellschaft* (Stuttgart 2000).

Krabbe, W. R. *Die deutsche Stadt im 19. und 20. Jahrhundert. Eine Einführung* (Göttingen 1989).

Krause, M./Plank/Ehrenbaum 'Konservierung von Fischen' *Zeitschrift für die gesamte Kälteindustrie* (1916) 64.

Kreuter, H. *Beiträge zur Geschichte der Werksgruppe Verfahrenstechnik und Anlagenbau 1965–1998.* Typescript (München o. J.)

Landes, D. S. *Der entfesselte Prometheus. Technologischer Wandel und industrielle Entwicklung in Westeuropa von 1750 bis zur Gegenwart* (Köln 1973).

Laschin, M. *Der flüssige Sauerstoff. Seine Aufbewahrung, sein Transport und seine Erzeugung* (Halle 1929).

Ledoux, C. E. *Théorie des machines a froid* (Paris 1878).

Lehnert, W. M. *Leitfaden der modernen Kältetechnik. Ihr Anwendungsgebiet, ihre Maschinen und Apparate* (Leipzig 1905).

Lerche, C. *Der Europäische Betriebsrat und der deutsche Wirtschaftsausschuss. Eine vergleichende Analyse der betrieblichen Mitwirkung der Arbeitnehmer vor dem Hintergrund der Globalisierung der Märkte* (Frankfurt a. M. 1997).

Levi, P. *Ist das ein Mensch?* (München 1992).

Linde AG (ed.) *Dr Hans Meinhardt. 40 Jahre im Dienste der Linde AG* (Aschaffenburg 1994).

Linde AG (ed.) *Geschäftsberichte.* (Wiesbaden 1880–2003).

Linde AG (ed.) *Linde 1879–1979* (Wiesbaden 1979).

Linde AG (ed.) Niederlassung Höllriegelskreuth *Gelieferte Gross-Sauerstoffanlagen. Verzeichnis B* (Höllriegelskreuth n.d.).

Linde AG (ed.) *Werkzeitschrift für die Betriebe der Gesellschaft für Linde's Eismaschinen* (1939–52), *Werkzeitung Linde* (1953–70), *Linde heute* (1971–2002), since 2003 *Linde Today*.

Linde, C. 'Verbesserte Eis- und Kühlmaschine' *Bayerisches Industrie- und Gewerbeblatt* 3 (1871) 264–72.

Linde, C. v. *Aus meinem Leben und von meiner Arbeit* (München 1979) (Unveränderter Neudruck der 1916 erschienenen Aufzeichnungen. Mit 36 Bildtafeln und einem Anhang Ausgewählte Briefe C. Lindes zwischen 1861 und 1910 sowie einem Nachwort Hermann und Gerhard Lindes zur 100. Wiederkehr des Gründungstages der Linde AG).

Linde, C. v. *Aus meinem Leben und von meiner Arbeit* (München 1998). (Neudruck der 1916 erschienenen Aufzeichnungen. Mit 36 Bildtafeln und einem Anhang Ausgewählte Briefe C. Lindes zwischen 1861 und 1910, Vorwort von Dr Hermann Linde und Dr Gerhard Linde).

Linde, C. v. *Aus meinem Leben und von meiner Arbeit. Aufzeichnungen für meine Kinder und meine Mitarbeiter* (München o.J. [1916]).

Linde, C. v. 'Das Berliner Werk der Gesellschaft für Markt und Kühlhallen' *Zeitschrift für die gesamte Kälteindustrie* (1903) 105–9.

Linde, C. v. 'Stickoxydulkältemaschinen' *Zeitschrift des Vereins deutscher Ingenieure* 47 (1903) 1075–8.

Linde, C. v. 'Trocknung des Hochofenwindes mittels Kältemaschinen' *Stahl und Eisen* 25 (1905) 3–13.

Linde, C. v. 'Über die Verflüssigung der Gase' *Bayerisches Industrie- und Gewerbeblatt* 26 (1893).

Linde, C. v. 'Über die Wärmeentziehung bei niedrigen Temperaturen durch mechanische Mittel' *Bayerisches Industrie- und Gewerbeblatt* 2 (1870) 205–11, 321–6, 363–7.

Linde, C. v. 'Versuchsstation der Gesellschaft für Lindes Eismaschinen AG in Höllriegelskreuth bei München' *Zeitschrift des Vereins deutscher Ingenieure* 48 (1903) 1362 p.

Linde, C. v. 'Wirtschaftliche Wirkungen der Kältetechnik' *Zeitschrift des Vereins Deutscher Ingenieure* 50 (1906) 1036.

Linde, F. 'Die wirtschaftliche Entwicklung des gesamten Unternehmens', GFLE, *50 Jahre Kältetechnik*, 175–3.

Linde, F. *Messung der Dielektrizitätskonstanten verflüssigter Gase und die Mossotti-Clausiussche Formel* (Leipzig 1895, Diss. München 1895).

Linde, G. 'Schlachthofgesetzgebung und ihre Wirkung auf die Privatkälteindustrie' *International Institute of Refrigeration. Premier Congrès International de Froid.* 3 Bände. (Paris 1908) Bd. 2 845–7.

Linde, H. 'Aus der Geschichte der Tieftemperaturtechnik' *Luft- und Kältetechnik* (1995) 144–50.

Linde, H. *Eine Münchener Familie im Dritten Reich. Briefe von Richard, Helmut und Werner Linde. Ausgewählt und mit Kommentaren versehen von Hermann Linde. 1. Teil Bis zum Beginn des 2. Weltkriegs.* Typescript (München o.J.).

Linde, H. *Im dritten Rang.* Typescript (München 2003).

Linde, R. *Meine Tätigkeit bei der Abteilung B der Gesellschaft Linde* Typescript 1954/1955 (München o. J.).

Lindner, S.H. *Den Faden verloren. Die westdeutsche und französische Textilindustrie auf dem Rückzug 1930/45–1990* (München 2001).

Lintner, C. *Lehrbuch der Bierbrauerei. Nach dem heutigen Stande der Theorie und Praxis* (Braunschweig 1878).

Lorange, P./Roos, J. *Strategic Alliances. Formation, Implementation and Evolution* (Cambridge/ Mass. 1993).

Lorentz, B. *Industrieelite und Wirtschaftspolitik 1928–1950. Heinrich Dräger und das Drägerwerk* (Paderborn 2001).

Lorenz H. 'Die erste Benutzung des Laternenstopfbüchse' *Zeitschrift für die gesamte Kälteindustrie* 4 (1897) 69–71.

Lorenz, H. 'Die erste Benutzung der Laternenstopfbüchse für Kompressoren' *Zeitschrift für die gesamte Kälteindustrie* 4 (1897) 69–71.

Lorenz, H. 'Die Wirkungsweise und Berechnung der Ammoniak-Absorptionsmaschinen' *Zeitschrift für die gesamte Kälteindustrie* 2 (1895) 21–9.

Lorenz, H. 'Neuere Fortschritte auf dem Gebiet der Kälteerzeugung' *Zeitschrift des Verein Deutscher Ingenieure* 39 (1895) 697–9.

Lorenz, H. *Praxis, Lehre und Forschung. Akademische Erinnerungen und Erfahrungen.* Typescript o.J. (*ca.* 1935).

Lorenz, H. 'Vergleichende Theorie und Berechnung der Kompressionskühlmaschinen' *Zeitschrift für komprimierte und flüssige Gase* 2 (1898/99) 215–17.

Lorenz, H. *Vergleichende Theorie und Berechnung der Kompressionskühlmaschinen* (München 1897).

Lütgert, S.A. *Eiskeller, Eiswerke und Kühlhäuser in Schleswig-Holstein und Hamburg Zusatz – ein Beitrag zur Kulturlandschaftsforschung und Industriearchäologie* (Husum 2000).

Manchester, W. *Krupp. Zwölf Generationen* (München 1968, Manchester, W. *The Arms of Krupp*, 1587–1968. Boston 1968).

Marsch, U. *Zwischen Wissenschaft und Wirtschaft. Industrieforschung in Deutschland und Großbritannien 1880–1936* (Paderborn 2000).

Maschinenbauanstalt Humboldt (ed.) *Führer durch die Maschinenbauanstalt Humboldt. 60 Jahre technische Entwicklung 1856–1916* (Köln-Kalk 1916).

Maschinenfabrik AG, vormals Tanner, Laetsch & Co. (ed.) *Eis- und Kühlmaschinen System Linde* (Wien 1903).

Maschinenfabrik Germania (ed.) *Geschichte der Maschinenfabrik Germania, vormals J.S. Schwalbe & Sohn in Chemnitz 1811–1911* (Chemnitz 1911).

Matschoß, C. *Die Entwicklung der Dampfmaschine. Eine Geschichte der ortsfesten Dampfmaschine und der Lokomobile, der Schiffsmaschine und der Lokomotive* (Berlin 1908).

Matschoß, C. 'Vom Ingenieur. Sein Werden und seine Arbeit in Deutschland' *Beiträge zur Geschichte der Technik und Industrie* 20 (1930) 1–21.

Matthews, F. E. *Elementary Mechanical Refrigeration* (New York 1912).

Mausbach, W. *Zwischen Morgenthau und Marshall. Das wirtschaftspolitische Deutschlandkonzept der USA 1944–1947* (Düsseldorf 1996).

Meadows, D. (ed.) *Die Grenzen des Wachstums. Bericht des Club of Rome zur Lage der Menschheit* (Reinbek bei Hamburg 1973).

Meidinger, H. 'Die Aufbewahrung des Eises' *Badische Gewerbezeitung* 4 (1870/1871) 49–64.

Meidinger, H. 'Die Fortschritte in der künstlichen Erzeugung von Kälte und Eis' *Dinglers Polytechnisches Journal* 218 (1875) 49–59, 140–8, 230–43, 471–8.

Meidinger, H. 'Eismaschinen für technischen Betrieb' *Badische Gewerbezeitung* 3 (1869) Beilage, 1–14.

Meinhardt, H. 'Optimierung des Portfolios in diversifizierten Unternehmen' Henzler, H. A. (ed.) *Handbuch strategische Führung* (Wiesbaden 1988) 134–55.

Mesarovic, M./Pestel, E. (ed.) *Menschheit am Wendepunkt – Zweiter Bericht an den Club of Rome zur Weltlage* (Stuttgart 1974).

Messer, A. 'Interessante Neuerungen bei der Herstellung von industriellem Sauerstoff' *Autogene Metallbearbeitung* (1923) 33–5.

Michaelis, L. 'Die autogene Schweißung im Dampfkesselbetriebe' *Zeitschrift für komprimierte und flüssige Gase* 11 (1908) 151–61.

Michels, J. *Peenemünde und seine Erben in Ost und West – Entwicklung und Weg deutscher Geheimwaffen* (Bonn 1997).

Milert, W./Tschirbs, R. *Von den Arbeiterausschüssen zum Betriebsverfassungsgesetz – Geschichte der betrieblichen Interessensvertretung in Deutschland* (Köln 1991).

Moll, E. 'Aktiengesellschaften. III. Statistik der Aktiengesellschaften. A. Die Aktiengesellschaften in Deutschland' *Handwörterbuch der Staatswissenschaften* Bd. 1. Jena (4th edn 1924) 141–60.

Molvaer, J./Stein, K. *Ingenieurin, warum nicht? Berufsbild und Berufsmotivation von zukünftigen Ingenieurinnen und Ingenieuren. Ein interkultureller Vergleich* (Frankfurt 1994).

Mommsen, H. 'Konnten Unternehmer im Nationalsozialismus apolitisch bleiben?' Gall, L./Pohl, M. (ed.) *Unternehmen im Nationalsozialismus* (München 1998) 69–72.

Mommsen, H./Grieger, W. *Das Volkswagenwerk und seine Arbeiter in der NS-Zeit* (Düsseldorf 1996).

Mosolff, H. *Der Aufbau der deutschen Gefrierindustrie* (Hamburg 1941).

Müller, J. *Aus Entwicklung der Werksgruppe TVT München Anlagenbau 1960–1993.* Typescript (Höllriegelskreuth 1993).

Neebe, R. *Überseemärkte und Exportstrategien in der westdeutschen Wirtschaft 1945 bis 1966* (Stuttgart 1991).

Niethammer, L. *Die Mitläuferfabrik. Die Entnazifizierung am Beispiel Bayerns* (Bonn 1982).

Nimax 'Über Kühlanlagen für Fleisch und andere Lebensmittel' *Deutsche Bau-Zeitung* 26 (1892) 602–4, 614–16, 628–30.

Nipperdey, T. *Deutsche Geschichte 1866–1918. Band 1 Arbeitswelt und Bürgergeist* (München 1990).

Nipperdey, T. *Gesellschaft, Kultur, Theorie* (Göttingen 1975).

North, D. C. *Institutions, Institutional Change and Economic Performance* (Cambridge 1990).

Oertzen, C. v. *Teilzeitarbeit und die Lust am Zuverdienen. Geschlechterpolitik und gesellschaftlicher Wandel in Westdeutschland 1948–1969* (Göttingen 1999).

Ostwald, W. *Deutsche Autos 1920–1945* (Stuttgart 1990).

Pachtner, F. *Deutscher Maschinenbau 1837–1937 im Spiegel des Werkes Borsig* (Berlin 1937).

Pavlis F. E. *The Aggressive Exploitation of Worldwide Opportunities. A Study in Corporate Culture Development.* Typescript (o.O. 1980).

Pavlis, F.E. *A Prologue to the History of Air Products.* Typescript (o.O. 1980).

Petz, F. 'Elektrolytische Sauerstoffgewinnung' *Dinglers Polytechnisches Journal* 89 (1908) 414–16.

Petzina, D. *Die deutsche Wirtschaft in der Zwischenkriegszeit* (Wiesbaden 1977).

Pfeiffer, K./Niebergall, W. '50 Jahre Borsig Kältemaschinen' *Kältetechnik* 1 (1949) 74–81.

Pfister, C. (ed.) *Das 1950er Syndrom. Der Weg in die Konsumgesellschaft* (Bern 1995).

Philippe, M. 'Bierbrauen einst und jetzt' Gesellschaft für Geschichte und Bibliographie des Brauwesens (ed.) *Carl von Lindes Kältemaschine und ihre Bedeutung für die Entwicklung der modernen Lagerbierbrauerei* (Berlin 1929) 8–17.

Pierenkemper, T. 'Sechs Thesen zum gegenwärtigen Stand der deutschen Unterne hmensgeschichtsschreibung. Eine Entgegnung auf Manfred Pohl' *Zeitschrift für Unternehmensgeschichte* 45 (2000) 158–66.

Pierenkemper, T. *Unternehmensgeschichte. Eine Einführung in ihre Methoden und Ergebnisse* (Stuttgart 2000).

Pierenkemper, T. 'Was kann eine moderne Unternehmensgeschichtsschreibung leisten? Und was sollte sie tunlichst vermeiden' *Zeitschrift für Unternehmensgeschichte* 44 (1999) 15–31.

Pistor, G. *100 Jahre Griesheim 1856–1956. Ein Beitrag zur Geschichte der chemischen Industrie* (Tegernsee 1958).

Plank, R. 'Bedeutung und Zukunftsaufgaben der Kältetechnik' Gesellschaft für Lindes Eismaschinen (ed.) *50 Jahre Kältetechnik* (Wiesbaden 1929) 184–92.

Plank, R. 'Geschichte der Kälteerzeugung und Kälteanwendung' Plank R. (ed.) *Handbuch der Kältetechnik. Band 1* (Berlin/Göttingen/Heidelberg 1954) 1–161.

Plank R./Kallert E. *Über die Behandlung und Verarbeitung von gefrorenem Schweinefleisch. Im Auftrage der ZEG durchgeführte Untersuchungen* (Berlin 1915).

Plank, R./Kuprianoff, J. *Haushaltskältemaschinen und kleingewerbliche Kühlanlagen* (Berlin 1934).

Plumpe, G. *Die I.G. Farbenindustrie AG. Wirtschaft, Technik und Politik 1904–1945* (Berlin 1990).

Pohl, M. 'Zwischen Weihrauch und Wissenschaft? Zum Standort der modernen Unternehmensgeschichte. Eine Replik auf Toni Pierenkemper' *Zeitschrift für Unternehmensgeschichte* 44 (1999) 150–63.

Pross, C. *Wiedergutmachung. Der Kleinkrieg gegen die Opfer* (Frankfurt 1988).

Pursell, C. W. Jr. 'Science and industry. Modern industrial research' Daniels, G.H. (ed.) *Nineteenth-Century American Science. A Reappraisal* (Evanston 1972) S. 231–48.

Radkau, J. *Technik in Deutschland. Vom 18. Jahrhundert bis zur Gegenwart* (Frankfurt a.M. 1989).

Raschke, J. (ed.) *Die Grünen. Wie sie wurden, was sie sind* (Köln 1993).

Reichel, P. *Vergangenheitsbewältigung in Deutschland. Die Auseinandersetzung mit der NS-Diktatur von 1945 bis heute* (München 2001).

Reif, E. G. *Störungen an Kältemaschinen, insbesondere deren Ursachen und Beseitigung* (Leipzig 1925).

Reinhardt, D. *Von der Reklame zum Marketing. Geschichte der Wirtschaftswerbung in Deutschland* (Berlin 1993).

Reiß, A. *Die Entwicklung des Industriegase-Sektors am Beispiel Sauerstoff. Markt- und Unternehmensstrukturen, ca. 1900–1939* Magisterarbeit (Münster 2002).

Reitzle, W. *Luxus schafft Wohlstand. Die Zukunft der globalen Wirtschaft* (Reinbek bei Hamburg 2001).

Rheinmetall-Borsig AG (ed.) *Deutscher Maschinenbau 1837–1937* (Berlin 1937).

Richter, R./Furubotn, E. *Neue Institutionenökonomik* (Tübingen 1996).

Ritschl, A. *Deutschlands Krise und Konjunktur 1924–1934. Binnenkonjunktur, Auslandsverschuldung und Reparationsproblem zwischen Dawes-Plan und Transfersperre* (Berlin 2002).

Rohbeck, W. 'Die Bedeutung der deutschen Eisindustrie' *Kältetechnischer Anzeiger* (1928) 79–81.

Rosenberg, H. *Große Depression und Bismarckzeit. Wirtschaftsablauf, Gesellschaft und Politik in Mitteleuropa* (Berlin 1967).

Ruckdeschel, W. E. *Mein Lebenslauf.* Typescript (München 1989).

Sack, W. *Güldner-Traktoren und –Motoren* (Brilon 1998).

Sass, F. *Geschichte des deutschen Verbrennungsmotorenbaues* (Berlin 1962).

Schenck, R. 'Die Verwendung von Sauerstoff und sauerstoffreicher Luft bei der Roheisenerzeugung' *Stahl und Eisen* 44 (1924) 521–6.

Schildt, A. *Moderne Zeiten – Freizeit, Massenmedien und 'Zeitgeist' in der Bundesrepublik der 50er-Jahre* (Hamburg 1995).

Schildt, A./Sywottek A. (ed.) *Modernisierung im Wiederaufbau – Die westdeutsche Gesellschaft der 50er Jahre* (Bonn 1993).

Schipper, W. 'Bedeutung der Kälteindustrie' *Kältetechnischer Anzeiger* (1929) 65–7.

Schissler, H. (ed.) *The Miracle Years. A Cultural History of West Germany 1949–1968* (Princeton/Oxford 2001).

Schmitz, C. 'Cooperative managerial capitalism. Recent research in German business history' *German History* 10 (1992) 91–103.

Schnabel, F. 'Der Aufbau des Prüfwesens der Maschinenfabrik Sürth' *Linde Berichte aus Wissenschaft und Technik* 4 (1958) 26–33.

Scholl, P. *Die Technik des Kühlschrankes. Einführung in die Kältetechnik für Käufer und Verkäufer von Kühlschränken, Gas- und Elektrizitätswerken, Architekten und das Nahrungsmittelgewerbe* (Berlin 1932).

Scholz, F. *Erinnerungen* (Höllriegelskreuth Archiv der Linde AG, o.J.) 3–12.

Schomerus, H. *Die Arbeiter der Maschinenfabrik Esslingen. Forschungen zur Lage der Arbeiterschaft im 19. Jahrhundert* (Stuttgart 1977).

Schöttler, R. 'Neuere Kältemaschinen' *Zeitschrift des Vereins deutscher Ingenieure* 28 (1884) 130–5, 697–701, 722–4 und 738–40.

Schultz-Gambard, J. 'Maßnahmen deutscher Wirtschaftsunternehmen zu vermehrter Integration von Frauen in den Managementbereich eine Bestandsaufnahme' *Zeitschrift für Frauenforschung* 11 (1993), H. 4, 17–32.

Schumann, D. *Bayerns Unternehmer in Gesellschaft und Staat 1834–1914. Fallstudien zu Herkunft und Familie, politischen Parteien und staatlichen Auszeichnungen* (Göttingen 1992).

Schütz, E./Gruber, E. *Mythos Reichsautobahnen. Bau und Inszenierung der 'Straßen des Führers' 1933–1941* (Berlin 1996).

Schwarz, A. 'Die Fleischversorgung Deutschlands und Österreichs mit Hilfe von Kühlhäusern' *Zeitschrift für Eis- und Kälteindustrie* 8 (1916) 102–9.

Scurlock, Ralph G. 'A matter of degrees. A brief history of cryogenics' *Cryogenics* 30 (1990) 483–500.

Sedlmaier, H. *Firmenjäger. Wie Raider Unternehmen kaufen, zerschlagen, verschachern* (Frankfurt a. M./New York 2003).

Sedlmayr, F. *Die Geschichte der Spatenbrauerei unter Gabriel Sedlmayr dem Älteren und dem Jüngeren 1807–1874 sowie Beiträge zur bayerischen Brauereigeschichte dieser Zeit.* 2 Bände (Nürnberg 1951).

Seidl, J. *Die Bayerischen Motorenwerke (BMW) 1945–1969. Staatlicher Rahmen und unternehmerisches Handeln* (München 2002).

Serchinger, R. 'Walter Schottky und die Forschung bei Siemens' Schneider, I./Trischler, H./ Wengenroth, U. (ed.) *Oszillationen. Naturwissenschaftler und Ingenieure zwischen Forschung und Markt* (München 2000) 167–211.

Servan-Schreiber, J.-J. *Die amerikanische Herausforderung* (Hamburg 1968).

Siegrist, H. 'Deutsche Großunternehmen vom späten 19. Jahrhundert bis zur Weimarer Republik. Integration, Diversifikation und Organisation bei den hundert größten deutschen Industrieunternehmen (1887-1927) in international vergleichender Perspektive' *Geschichte und Gesellschaft* 6 (1980) 60–102.

Siegrist, H. *Vom Familienbetrieb zum Managerunternehmen. Angestellte und industrielle Organisation am Beispiel der Georg Fischer AG Schaffhausen 1797–1930* (Göttingen 1981).

Smith, J. K. *The Biography of a Business. Air Products Then and Now.* Typescript (Bethlehem o.J.).

Sohns, H. 'Sind Heimstoffe zu teuer?' *Deutsche Technik* 4 (1936) 586–9.

Spiliotis, S.-S. *Verantwortung und Rechtsfrieden. Die Stiftungsinitiative der deutschen Wirtschaf* (Frankfurt 2003).

Spoerer, M. *Zwangsarbeit unter dem Hakenkreuz. Ausländische Zivilarbeiter, Kriegsgefangene und Häftlinge im Deutschen Reich und im besetzten Europa 1939–1945* (Stuttgart/München 2001).

Steiner, A. *Die DDR-Wirtschaftsreform der sechziger Jahre. Konflikt zwischen Effizienz- und Machtkalkül* (Berlin 1999).

Stetefeld, R. 'Die deutsche Kälteindustrie im Auslande' *Zeitschrift des Vereins deutscher Ingenieure* 63 (1919) 932.

Stickel, A./Tröster, M. (ed.) *48,98. Tante Emma–Megastore. Fünfzig Jahre Lebensmittelhandel in Deutschland* (Frankfurt a. M. 1998).

Stiller, W. 'Jagdfieber' *Capital* 3 (1988) 157.

Stokes, R. G. *Divide and Prosper. The Heirs of IG Farben under Allied Authority 1945–1951* (Berkeley 1988).

Stokes, R. G. 'Technology transfer and the emergence of the West German petrochemical industry 1945–1955' Diefendorf, Jeffry M. (ed.) *American Policy and the Reconstruction of West Germany* (Washington 1993) 217–36.

Stokes, R. G. 'Von der I.G. Farbenindustrie bis zur Neugründung der BASF 1925–1952' Abelshauser, W. (ed.) *Die BASF – Eine Unternehmensgeschichte* (München 2002) 221–358.

Struve, E. *Zur Entwicklung des bayerischen Braugewerbes im neunzehnten Jahrhundert. Ein Beitrag zur deutschen Gewerbegeschichte der Neuzeit* (Leipzig 1893).

Stuhr, H.-W. *Fahrzeuge mit hydrostatischem Fahrantrieb. 10 Jahre Güldner-Hydrocar. Ein Beitrag zur Geschichte der Transport und Fahrzeugtechnik von Linde und Güldner* (Aschaffenburg 1963).

Stürmer, M. *Das ruhelose Reich. Deutschland 1866–1918* (Berlin 2nd edn 1994).

Swoboda, K. *Die Anlegung und Benutzung transportabler und stabiler Eiskeller und Eisschränke, Eisreservoirs und amerikanischer Eishäuser* (Weimar 1868).

Tettinger, W. 'Aus der Arbeit unserer Versuchsanstalt in Sürth. Höllriegelskreuth 1957' *Linde-Berichte aus Wissenschaft und Technik* 4 (1957), 19–26.

Teuteberg, H.J. 'Zur Geschichte der Kühlkost und des Tiefgefrierens' *Zeitschrift für Unternehmensgeschichte.* 36 (1991) 139–55.

Thévenot, R. *A History of Refrigeration Throughout the World.* (Paris 1979).

Thomas, D. E. *Diesel. Technology and Society in Industrial Germany* (Tuscaloosa 1987).

Thomas, G. *Geschichte der deutschen Wehr- und Rüstungswirtschaft 1918–1943/45* (Boppard 1966).

Tillmann, H. F. *Der Einfluß der Kältetechnik auf die Entwicklung des Braugewerbes.* Typescript (Bochum 1972).

Tilly, R. 'Großunternehmen. Schlüssel zur Wirtschaft- und Sozialgeschichte der Industrieländer?' *Geschichte und Gesellschaft* 19 (1993) 530–49.

Todt, F. 'Aufgaben der deutschen Technik. Rede zur Sondertagung des Amtes für Technik auf dem Reichsparteitag Nürnberg 1936' *Deutsche Technik, Amtliches Organ des Hauptamtes für Technik der Reichsleitung der NSDAP,* 4 (1936) 478–80.

Treue, W. *Die Geschichte der Ilseder Hütte* (Peine 1960).

Tungsram (ed.) *75 Jahre Tungsram* (Budapest 1971).

Uhl, M. *Stalins V 2 Technologietransfer der deutschen Fernlenkwaffentechnik in die UdSSR und der Aufbau der sowjetischen Raketenindustrie 1945–1959* (Bonn 2001).

Wagner, B. C. *IG Auschwitz. Zwangsarbeit und Vernichtung von Häftlingen des Lagers Monowitz 1941–1945* (München 2000).

Walsh, M. *The Rise of the Midwestern Meatpacking Industry* (Lexington 1982).

Weckerle, H. *Eine wirtschaftswissenschaftliche Studie über die Kältemaschinen-Industrie.* Diss. (Stuttgart 1937).

Wegener, F. *Die Kälteindustrie.* Diss. (Greifswald 1921).

Wehler, H.-U. *Deutsche Gesellschaftsgeschichte.* Bd. 3 Von der 'Deutschen Doppelrevolution' bis zum Beginn des Ersten Weltkrieges 1849–1914 (München 1995).

Weimer, W. *Deutsche Wirtschaftsgeschichte. Von der Währungsreform bis zum Euro* (Hamburg 1998).

Weißstein/Riedel *Kommentar zum Militärregierungsgesetz Nr. 59 für die amerikanische Zone* (Koblenz 1953).

Welsch, F. *Geschichte der chemischen Industrie* (Berlin (Ost) 1981).

Westberg, K. *Var optimist! – AGAs innovativa verksamhet 1904–1959* (Stockholm 2002).

Wilde, M. W. *Industrialization of food processing in the United States, 1860–1960* (Delaware 1988).

Williamson, O. E. *Die ökonomischen Institutionen des Kapitalismus* (Tübingen 1990).

Winkel, H. *Die Wirtschaft im geteilten Deutschland 1945–1970* (Wiesbaden 1974).

Winter, F. H./Neufeld, M.J. 'Heylandt's rocket cars and the V2. A Little Known chapter in the history of rocket technology' *History of Rocketry and Aeronautics; Proceedings of the 26th History Symposium of the International Academy of Aeronautics in Washington D.C. 1992* (Univelt 1997) 41–72.

Wyss, E. 'Das autogene Schneiden und das Verfahren der Sauerstofferzeugung nach G. Claude' *Zeitschrift für komprimierte und flüssige Gase* 11 (1909) 104–11.

Zeitschrift des Vereins deutscher Ingenieure 2 'Die Regenerativkaltluftmaschine' (1858) 287–8.

Zeitschrift für komprimierte und flüssige Gase 24 (1931) 61–8.

Zentes, Joachim (ed.) *Kooperationen, Allianzen und Netzwerke. Grundlagen-Ansätze–Perspektiven* (Wiesbaden 2003).

Illustration credits

Fig. 1.1	Private property of Hermann Linde.
Fig. 1.2	Linde AG Photo Archives.
Fig. 1.3	German Museum, Munich. Special Collection.
Fig. 1.4	*50 Years of Refrigeration 1879–1929 – The History of Linde AG's Ice-Making Machines,* Wiesbaden, 1929, p. 27.
Fig. 1.5	Linde AG Photo Archives.
Fig. 1.6	*50 Years of Refrigeration 1879–1929 – The History of Linde AG's Ice-Making Machines,* Wiesbaden, 1929, p. 16.
Fig. 1.7	Excerpts from the general meeting minutes, Linde Archives, Wiesbaden, IV.
Fig. 1.8	Source: MAN Historical Archives, Richard Buz Estate (235/1a), Ice Department main ledger and sales records.
Fig. 2.1	Linde AG, Wiesbaden.
Fig. 2.2	Linde AG Photo Archives.
Fig. 2.3	Private property of Hermann Linde.
Fig. 2.4	Linde AG Photo Archives.
Fig. 2.5	*50 Years of Refrigeration 1879–1929 – The History of Linde AG's Ice-Making Machines,* Wiesbaden, 1929, p. 20.
Fig. 2.6	Private property of Hermann Linde.
Fig. 3.1	Private property of Hermann Linde.
Fig. 3.2	*50 Years of Refrigeration 1879–1929 – The History of Linde AG's Ice-Making Machines,* Wiesbaden, 1929, p. 40.
Fig. 3.3	Linde AG Photo Archives.
Fig. 4.1	*Linde Today,* 2/1951, p. 5.
Fig. 4.2	Linde AG Photo Archives.
Fig. 4.3	Private property of Fritz Orschler.
Fig. 4.4	Linde AG, Wiesbaden.
Fig. 5.1	Linde AG Photo Archives.

Archives listing

AGA Archives (collection today in Föreningen Stockholms Företagsminnen, Stockholm)
Bavarian Public Records Office, Munich
Berlin State Archives
Case Western Reserve University Archives, Cleveland/Ohio
Cornell University Archives
German Museum of Technology Archives, Berlin
Klöckner-Humboldt-Deutz Archives
Linde AG Archives (with collections in Wiesbaden, Höllriegelskreuth and Aschaffenburg)
MAN Historical Archives
MAN Archives

Munich City Archives
National Museum of American History Archives
Private Archive, John K. Smith, Bethlehem
Private Archive, Hermann Linde, Munich
Secret Central Archives of the former Prussian State, Berlin
Siemens Archives, Munich
Technical University Dresden Archives
Technical University Munich Archives

Index